BASIC PROBABILITY AND APPLICATIONS

Miloslav Nosal
Associate Professor
Department of Mathematics and Statistics
University of Calgary
Calgary, Alberta, Canada

1977
W. B. SAUNDERS COMPANY
Philadelphia, London, Toronto

W. B. Saunders Company: West Washington Square
Philadelphia, PA 19105

1 St. Anne's Road
Eastbourne, East Sussex BN21 3UN, England

1 Goldthorne Avenue
Toronto, Ontario M8Z 5T9, Canada

Library of Congress Cataloging in Publication Data

Nosal, Miloslav.

Basic probability and applications.

Includes index.

1. Probabilities. I. Title.

QA273.N62 519.2 76-8582

ISBN 0-7216-6865-8

Basic Probability and Applications ISBN 0-7216-6865-8

Last digit is the print number: 9 8 7 6 5 4 3 2 1

To my parents
Marie and Vilem Nosal

The principal means for ascertaining truth–induction and analogy–are based on probabilities; so that the entire system of human knowledge is connected with the theory of probability.

Pierre Simon, Marquis de Laplace, 1819

No priest or soothsayer that ever lived could hold his own against Old Probabilities.

Oliver Wendell Holmes, 1875

PREFACE

There are so many introductory texts on probability on the market that the question naturally arises: Why should one try to write yet another one since there are some very good features in several of these texts?

A really satisfactory text should be *complex, universal,* and *elegant* in the proper presentation of *all* aspects of the subject. The main objectives of the present text are:

(a) Providing students with a mathematically sound and logically systematic treatment of the foundations of probability theory.

(b) Presenting probability in a proper balance between theoretical and applied aspects by emphasizing multifaceted and diverse practical examples and problems arising in real life.

(c) Giving the student a feeling and flavor for the subject by presenting interesting and often intriguing facts from the history of probability theory, as well as by providing an exposure to the main philosophical problems related to the subject.

(d) Gearing the presentation to the new needs and challenges arising from the wide utilization of pocket calculators and programmable mini-computers.

(e) Last but not least, showing students that probability can be fun by entertaining them with unusual problems, little jokes, and, in general, by keeping a balance between the rigor of theory and the *genre léger* of a storyteller.

This text has evolved from eight years of teaching introductory courses in probability. Some courses were addressed mainly to honors and major students in statistics and pure and applied mathematics and to teachers of mathematics, but some were for students in engineering, biology, physics, economics, business management, computer science, psychology, and other sciences, who require a solid background in probability and statistics. Students at this level usually do not have any background in calculus and it is not really needed even if we want to develop a mathematically sound and logically consistent treatment of the elements of probability theory. For this reason, the main bulk of the text (Chapters I through XVI) *does not require any knowledge of calculus or higher algebra* and can be followed easily by students with a good background in high school mathematics. For example, Chapter XIV, on Markov chains, does not assume any knowledge of matrix algebra. Necessary notions of vectors, matrices, and their multiplication are naturally introduced using basic probabilistic ideas, and this method of treatment appears to be much more acceptable to students than the rather artificial introduction of these notions in linear algebra. The main emphasis in this part of the book is on *axiomatic treatment of finite probability spaces* and their applications.

There are at least two ways of presenting a mathematically rigorous treatment of finite probability spaces. The first method assigns certain weights (or probabilities) to all outcomes, and the probability of any event is then the sum of weights of all favorable outcomes. The main advantage of this approach is that it does not burden the student with any involved notions and ideas. The second method treats probability axiomatically as a normed measure defined on all subjects of a sample space. It can easily be seen (and it is shown in Chapter V) that both methods are mathematically equivalent. The text is based on the latter approach, but makes a liberal use of the first method. There are several advantages in the treatment of probability as a measure, even at this elementary level. Students are usually quite familiar with sets and operations on sets, and probability treated as a set function further develops their perception of modern mathematics. Many students who are introduced to probability theory during later studies in a more general situation (although even perhaps only in the case of a real line) usually encounter some difficulties. However, an early exposition of the basic notions of probability measure and expected value in the transparent and clear case of finite spaces makes them better motivated, receptive, and enthusiastic towards abstract theory. Finally, for students going into probability and statistics programs, the inevitable Kolmogoroff axiomatic system is much more acceptable if they have been exposed to its elements from early stages.

The axiomatic model of probability theory can be easily extended from finite spaces to *countable spaces* using elementary knowledge of infinite series: the necessary facts are usually covered either in high school mathematics or in the very first university calculus course. Elementary ideas of a limit and the definite integral can be also used to investigate the *normal distribution* as an approximation of the binomial distribution. These ideas are covered in Chapters XVII and XVIII of the text. It is clear that, unlike the first sixteen chapters, these two chapters require some elementary background in calculus. The text does not try to pretend that it covers uncountable (continuous) spaces and distributions. Logically and mathematically consistent treatment of such spaces requires an incomparably higher level of mathematical sophistication (a fact which is very often overlooked) and an inconsistent treatment (so frequent in the literature) is against the stated principles of this text.

Great effort has been made to motivate all notions of probability theory by practical examples and to show the students that even elementary probability can be very useful in solving problems of the real world. The theory of probability historically developed from gambling problems and this motivation can be very good in introducing new notions. It might be claimed that it is immoral to use gambling as a motivation in learning and that such an approach may contribute to a further spread of gambling. Nothing, however, can be farther from the truth. A minister can preach that gambling is immoral, a philosopher may argue that it is unethical, but as long as people believe that they can make a "quick buck," they will gamble. Only a probabilist can convince them that chances in all casino games are hopelessly adverse, and this fact is more convincing than anything else. The book contains a complete or partial statistical analysis of the following games: roulette (both Las Vegas and Monte Carlo types), the great martingale gambling system, craps, keno, poker, bridge, bingo, blackjack, dominoes, and Morra games, as well as tests for the detection of loaded dice and cheating card dealers.

However, gambling examples represent only a minor source of practical problems for motivation. The text contains about 300 solved and 550 unsolved problems and examples from many areas of engineering and science, including many "trick and wit" problems based on collections that are not available in English (French, Russian, German, Polish, Czech, and Greek sources). Chapter XVI is devoted to probabilistic analysis of practical situations arising in genetics, engineering quality control, information science, and linguistics. The conclusion of this chapter is devoted to a stochastic model which enables a student to compose his own song by tossing a die. Tests of significance of statistical hypotheses (both parametric and distribution-free) are briefly introduced in Chapters XI and XVIII.

The teaching of mathematics and statistics has changed dramatically within the last several years as a result of the mass availability of computing hardware. A good pocket calculator is available for less than ten dollars, a calculator with scientific functions (sin, cos, log, e^x, and so on) costs less than twenty dollars, and programmable mini-computers (with calculating means, variances, binomial, normal and other distributions, and so on) cost around one hundred dollars. When a majority of students have calculators, it is easy to perform all detailed numerical calculations directly in the class without any loss of time.

It is also possible to motivate fundamental asymptotic and limit properties (limit distributions in Markov chains, central limit problems, and so on) by numerical investigation of the convergence on a mini-computer. A calculator also enables quick topical interactions with students by collecting data from the class and solving related problems (e.g., the birthday problem—Chapter VII, Example 8—can be calculated for the number of students in the class and then verified by questioning). All these ideas have been incorporated in the text and in the selection of problems.

The text can be used either for a half course (i.e., one term or one quarter course) or for a full course (more than one term or quarter). In the first case, only the fundamental Chapters II, III, and V through XII can be covered (with the possible omission of difficult examples). If time permits, selected material from Chapters IV and XIII through XVI can further illustrate applications of fundamental properties of probability in special situations. For a full course, the whole text can be conveniently covered without any omissions. The instructor's manual, available with the book, gives complete solutions of all exercises and supplies further hints and ideas for the use of the text. This book can be used for a theoretical and mathematically systematic course on probability since all theorems and rigorous proofs are included. However it is more likely that the text will be used in practically designed courses without emphasis on mathematical rigor; for this reason all proofs are printed in a small type and can be omitted (with the hope that interested students will still try to read them and penetrate subtle nuances of the subject).

I am deeply indebted to my colleague and friend Dr. Rolf Schassberger for fruitful discussion of the basic ideas and their proper methodological and pedagogical treatment. I am thankful to Dr. Peter Ehlers for interesting arguments and for help with the English language of the text. I appreciate the continuous positive contributions of my colleagues at the Department of Mathematics and Statistics, University of Calgary, who encouraged me to undertake the task of writing this book and to carry it to the end despite the many hardships during the work. Many thanks are due to Ms. Tae Hayashitani for her patience in deciphering my scribbling and putting the text into the final typed form. I am also especially indebted to Mr. John Snyder, Jr. and all workers of W. B. Saunders Company for their contributions and mainly for their tolerance of my irregular working habits and capricious requests. My final thanks go to my parents, without the altruistic help of whom I could never have undertaken any intellectual endeavor.

MILOSLAV NOSAL

CONTENTS

INTRODUCTION

I

A treuthe be known oonli bi probabilness and likelihode, and not sureli.

Reginald Pecock, 1449

The notion of probability, likelihood and chance or "Lady Luck" has intrigued man for centuries. Shortly after *Homo sapiens* had attained affluence, which allowed him to indulge in free time, he invented gambling. Archeologists, unearthing prehistoric sites of the Stone Age, discovered many dice-shaped bones, called astragali, which were quite likely linked to games of chance and perhaps even to prophecy. Ever since these prehistoric times, men and dice have been impartible friends. Nevertheless, much time elapsed before man's reason was able to cope with the puzzles and riddles offered by a die. *Aristotle,* in his all-encompassing works, defined "a probability" as being "what men know to happen or not to happen, to be or not to be, for the most part thus and thus" and he devoted an entire chapter of his *Metaphysics* to the explanation of "chance."

In their practical calculations, the census takers of imperial Rome came very close to the modern notion of a probability distribution. They were required to keep lists of properties so that winners of power struggles could confiscate and divide the assets of the losers. But there was less interest in the abstract notion of a probability distribution than in the material advantages accruing from the proscriptions. The Dark Ages put an end to these investigations—even though man did not throw the dice away.

Commercial insurance in the Italian cities of the early Renaissance in addition to a highly increased interest in gambling led to new attempts to analyze chance. For example, *Dante* (1265–1321) in his *Divina Commedia* discussed different probabilities of various throws with three dice, placing this subject in the Purgatorio. The first minor mathematical success was achieved by *Niccolò Tartaglia* (1499–1559) and *Geronimo Cardano* (1501–1576). The subject was further developed into a "geometry of the die" by *Blaise Pascal* (1623–1662) and *Pierre de Fermat* (1601–1665). In their correspondence with each other they made precise such notions as probability and expected value, both of which are indispensable tools in gambling. It is interesting to realize that their work was initiated by a problem posed by the famous gambler *Chevalier de Méré.* This gentleman used to bet at equal odds on the event that in four throws of a fair die he would get at least one 6. However he was making so much money that very soon he could not find anybody for a game. Therefore he started betting, again at equal odds, on the event that in 24 throws of a pair of dice he would get at least one double 6. He expected to have the same chance of winning since, according to his reasoning, the first game can be expressed as 4 to 6, the second as 24 to 36, and both fractions are equal. However, to his dismay, he started losing money rapidly and accused arithmetic for being at fault.

Pascal's solutions of this problem and some other gambling questions thus laid the historical foundations of the theory of probability (see Chapter IX, Example 3). The first systematic treatise on probability was published in Basel in 1713 by *Jacques Bernoulli* (1654–1705) under the title *Ars Conjectandi.* In the book Bernoulli mainly developed theories of combinations and permutations, infinite series, finite differences and other mathematical topics. His main contribution was the investigation of the limit behavior of relative frequencies (law of large numbers). Only a small part of the book was devoted to gambling problems. Bernoulli anticipated applications of probabilistic methods in insurance, economics and statistics.

The growing demands of the rapidly developing natural sciences presented another impetus for the development of probability theory. A host of great names emerged in this area and many new and sophisticated analytical tools appeared. *Abraham de Moivre* (1667–1754) in his famous book *The Doctrine of Chances* (1718) investigated applications of the binomial theorem, developed elegant methods for infinite series and applied the notion of expectation to the calculation of risks and annuities. The first draft of his book was communicated to the Royal Society and published in the Transactions "not so much as a matter relating to Play, but as containing some general Speculations not unworthy to be considered by the Lovers of Truth," as de Moivre claimed. Abbè *Thomas Bayes* (1702–1761) systematically developed the method of inverse probabilities (à posteriori probabilities) and thus became founder of a whole branch of statistics.

Perhaps the greatest star of probability appeared in the person of *Pierre Simon, Marquis de Laplace* (1749–1827). This almost omnificent giant of reason contributed greatly to many areas of science and his book *Théorie analytique des probabilités* gave birth to modern probability theory. He widely used differential and integral calculus, differential equations, and introduced such fruitful notions as generating functions, the normal distribution and the central limit problem. ("What more do we need?" many an applied probabilist will ask even today.) He also discussed applications in physics, astronomy, justice and so on. His philosophical ideas on probability and determinism (the Laplace demon and the principle of indifference) had far-reaching influence (see Chapter III). The first tables of the normal distribution and an explanation of how to use them for calculations of insurance premiums, life contingencies, annuities and so on were published by *Augustus de Morgan* (1806–1871) in 1838 in his book *An Essay on Probabilities. Karl Friedrich Gauss* (1777–1855) further developed applications of the normal distribution in the theory of errors and introduced the method of least squares.

The latter half of the nineteenth century witnessed a shift of interest in probability to czarist Russia. Contributions of profound significance were made by *P. L. Tchebychev* (1812–1894), *A. A. Markov* (1856–1922) and *A. M. Liapounov* (1857–1918). The achievements of these men consisted mainly of the extensive use of random variables, the investigation of the laws of large numbers and the introduction of the concept of chain-linked events. Ever since this time, the Russian school has played a leading role in the development of probability. Perhaps the greatest achievement in the subject was established by a Soviet mathematician, *A. N. Kolmogorov,* whose *Grundbegriffe der Wahrscheinlichkeitsrechung* (1933) laid the foundations for the overwhelming advancements of modern probability. His axiomatic system, with its use of measure theory, helped in establishing probability as a discipline which could match any other branch of mathematics in its logical precision. Further advancement is due mainly to the tremendous contributions of *Paul Lévy, William Feller* (1906–1970) and *John L. Doob.*

The success of modern analytical methods established probability as one of the most esthetical abstract mathematical theories, with an unlimited scope of applications in natural sciences, industry, management, agriculture, business, and so on. However, many researchers still feel that the essential question has not been fully answered: How can we use probabilities in the description and evaluation of the process of human creative inductive thinking, in acquiring new knowledge, in epistemology? There are many approaches and attempts at solutions and we should mention at least the great contributions of *Rudolf Carnap* (1891–1970). These problems are very closely related to still another basic question, clearly formulated by *Augustus de Morgan* (1806–1871): ". . . the assertion is *more probable,* and wherever we have the notion of *more* and *less,* we feel the possibility of an answer to the question, 'How much more or less?' and which we should produce if we knew how."

This is the old problem of how to assign numerical probabilities to the events when we know only that some are more likely than others. A solution to this problem must precede full-scale applications of the axiomatic analytical probability methods. Here we are entering the sphere of qualitative or comparative probability, as formulated by *Leonard F. Savage* in his book *The Foundations of Statistics* (1954). Unfortunately, the theory of probability does not yet have a fully satisfactory answer to this question, and we can witness a strong effort by many researchers in this direction. Perhaps the future will bring some new answers.

SETS AND FUNCTIONS

The least bitling of it will so far club and fall in with the laws that bind the whole Set.

Nathaniel Fairfax, 1674

If we want to build our investigation of probability on solid foundations, we have to become familiar with several mathematical notions. However, we will not need a detailed and profound analysis of these concepts because we are interested only in their applications to probability. Therefore in no way can we claim any degree of depth and completeness for our treatment. Perhaps the opposite might be true because we realize that many readers interested in probability unfortunately consider formal treatment of mathematical concepts a hindrance and a burden.

The first important notion is that of a *set*. There are many words in everyday English which express a similar idea of togetherness and collectivity: a pack of wolves, a herd of sheep, a school of fish, a crowd of people, a flock of ducks, a colony of ants, a cluster of grapes, a swarm of bees, a bunch of roses, a bundle of hay, a group of children, a pride of lions and many others. We will not give an exact mathematical definition of a set here; an intuitive approach will be sufficient.

A set is any well defined collection or class of objects. Thus one can speak about the set of all United States presidents, the set of all students registered at the University of Hawaii in the winter session of 1976, the set of all members of NATO, the set of all outcomes of tossing an ordinary die, the set of all prime numbers, and also about the set of all grizzlies in Canada and the set of all camels situated this moment north of the Arctic Circle. We do not attempt to specify what we mean by a "collection" or a "class" even though an attentive reader might object that without an exact explanation of these terms, the notion of a set is not any clearer. If we require that a set is any *well defined* collection, we simply mean that it must be at least hypothetically possible to decide whether any conceivable object does or does not belong to the set under consideration. Any violation of this rule could have very serious negative consequences, as we shall see later. Objects belonging to a set are called elements of the set. Sets will usually be denoted by capital letters with or without subscripts:

$$A, B, C, \ldots, X, Y, Z, A_2, A_5, X_{73}, \ldots, \Omega, \ldots$$

and the elements of sets will be denoted by small letters with or without subscripts:

$$a, b, c, \ldots, x, y, z, a_2, a_5, x_{73}, \ldots, \omega, \ldots.$$

The symbol Ω stands for the Greek capital letter omega, and ω for the small letter omega.

The simplest way to specify a set is to list its elements within braces $\{\ \}$. For example,

$$E = \{1, 3, 5, 7, 9\},$$
$$F = \{\text{Alberta, Manitoba, Saskatchewan}\}.$$

Then F is the set consisting of the three Canadian prairie provinces. Another frequently used method of specifying a set is to state a property that all the elements of the set possess. The standard way of writing such a set is

$$A = \{x : x \text{ has a given property}\}.$$

For example

$$G = \{x : x \text{ is an odd natural number less than } 10\},$$
$$H = \{x : x \text{ is a Canadian prairie province}\},$$
$$J = \{x : x^2 - 2x + 7 = 0\}.$$

The colon is read "such that." For example "H is the set of all x such that x is a Canadian prairie province."

If an object x is a member of a set A, we write

$$x \in A$$

which is read "x belongs to A." If x is not a member of A, we write

$$x \notin A$$

which is read "x does not belong to A." For example, referring to the above defined sets,

$$\text{Alberta} \in H, \text{ Texas} \notin H, \ 7 \in E, \text{ and } 21 \notin E.$$

If we compare the sets E and G defined above (or for that matter the sets F and H), then we notice that they look conspicuously similar even though they are given in different ways.

DEFINITION 1: We say that sets A and B are equal if they consist of the same elements, i.e., if every element which belongs to A also belongs to B and every element which belongs to B also belongs to A. In this case we write

$$A = B.$$

Referring to the above defined sets, obviously

$$E = G \quad \text{and} \quad F = H.$$

In the beginning, we spoke about the set of all camels north of the Arctic Circle. This set will have very few elements, if any at all, and we are inclined to believe that this set is void; but we cannot be sure that some crazy skipper did not take his pet camel on his whaling expedition. Therefore, we have to consider the set of all camels north of the Arctic Circle just as any other set, even if we could establish that there are no such camels.

However, let us consider a more serious example. If

$$K = \{x : x \text{ real and } x^2 + 1 = 0\},$$

then only a more detailed look reveals that there is no real number x which satisfies the equation

$$x^2 + 1 = 0$$

and therefore K has no elements. This set is called the *empty set* and is usually denoted by the symbol ϕ. The empty set ϕ is sometimes also called the null set but we shall reserve this term for another situation.

If every element x of a set A is also an element of a set B then we say that A is a *subset* of B. This relation is called inclusion and is denoted by the symbol $\subset$, that is,

$$A \subset B.$$

Sometimes we write $B \supset A$ instead of $A \subset B$.

For example, if

$$E = \{1, 3, 5\},$$
$$\text{and } F = \{1, 2, 3, 4, 5, 6\},$$
$$\text{then } E \subset F, \text{ or equivalently } F \supset E.$$

If, further,

$$G = \{x : x \text{ is an integer divisible by } 6\}$$
$$H = \{x : x \text{ is an integer divisible by } 3\}$$

then obviously

$$G \subset H, \text{ or equivalently } H \supset G.$$

Suppose that $A \subset B$. Then necessarily every element of A also belongs to B. If, in addition, every element of B also belongs to A, then A and B are equal. If, on the other hand, there is an element of B that does not belong to A, then we say that A is a *proper* subset of B. For example, if

$$E = \{x : x \text{ is an odd outcome of a toss of an ordinary die}\},$$
$$F = \{1, 3, 5\},$$
$$G = \{1, 3, 5, 6\},$$

then obviously $E \subset F$ but E is not a proper subset of F; at the same time $E \subset G$ and F is a proper subset of G.

Given any set, one can imagine a larger set. However, the investigation of excessively large sets, such as the set of all sets, is logically troublesome and we shall therefore assume that *all our sets are subsets of some universal set,* usually denoted by Ω or S. The following example, due to *Bertrand Russell,* shows that it is really necessary to make sure that all sets are well defined subsets of some well defined universal set. Let us assume that a student has the rather peculiar summer job of making a list of all books in the University library which do not contain their own title within their own text. This list will be bound as a book with the title "Master List" and placed on the shelves of the library. Should the student include the title "Master List" in his list or not? If he does, he should not because in that case his book contains its title within its text. If he does not, he should because in that case his book does not contain its title within its text. In any case, the collection of all titles to be included in the Master List is not well defined because we cannot decide about the title "Master List" and the apparently easy task can never be accomplished.

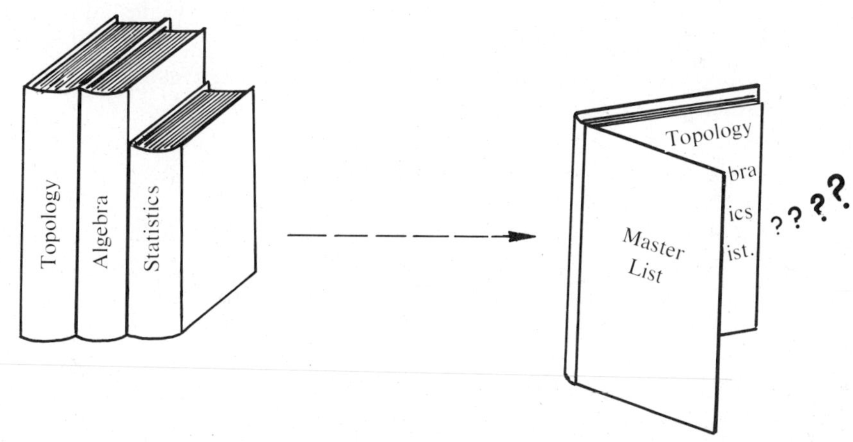

Two sets A and B are equal if and only if $A \subset B$ and, at the same time, $B \subset A$. In other words A equals B if and only if for every $x \in \Omega$, $x \in A$ implies $x \in B$ and at the same time for every $x \in \Omega$, $x \in B$ implies $x \in A$. This is the main idea which we use when we want to prove the equality of two sets and it will occur very often. If it is not true that $A \subset B$ then we simply write

$$A \not\subset B.$$

We can depict relations between sets by using so-called Venn-Euler diagrams.

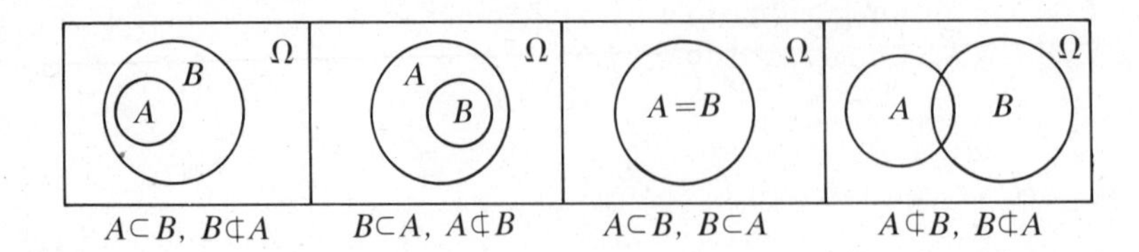

$A \subset B$, $B \not\subset A$ $\quad$ $B \subset A$, $A \not\subset B$ $\quad$ $A \subset B$, $B \subset A$ $\quad$ $A \not\subset B$, $B \not\subset A$

Two sets are said to be *disjoint* if they do not contain any elements in common.

For example,

$$E = \{x : x \text{ is an odd integer}\},$$
$$F = \{x : x \text{ is an even integer}\}$$

are two disjoint sets. Using Venn-Euler diagrams, we can depict two disjoint sets A and B as follows:

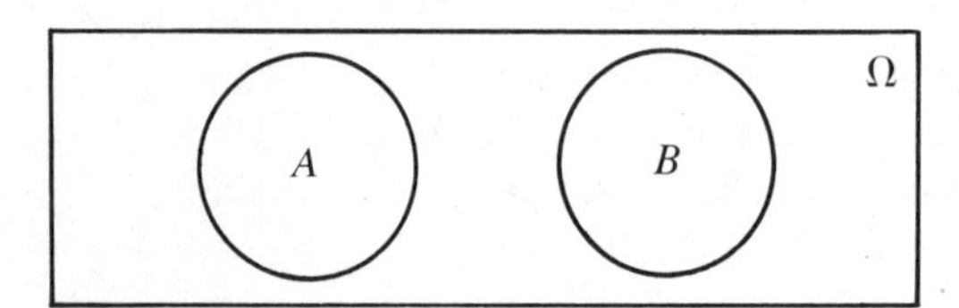

Given sets A and B which are, of course, subsets of Ω, we can form new sets by performing certain operations on these sets.

DEFINITION 2: The union $A \cup B$ of two sets A and B is the set of all elements belonging to either A or B (or to both of them).

Let us investigate the set of all Beethoven symphonies and some of its subsets. (An example from the rock music scene might be more topical, but unfortunately it would certainly be much more prone to deteriorate with time.) Let us denote the symphonies by Roman numerals.

Symphony No. 1 in C major	SI
Symphony No. 2 in D major	SII
Symphony No. 3 in E-flat major, called "Eroica"	SIII
Symphony No. 4 in B-flat major	SIV
Symphony No. 5 in C minor	SV
Symphony No. 6 in F major, called "Pastorale"	SVI
Symphony No. 7 in A major	SVII
Symphony No. 8 in F major	SVIII
Symphony No. 9 in D minor, called "The Ninth with the Hymn to Joy"	SIX

$A = \{x : x \text{ is a Beethoven symphony in a major key}\}$
$B = \{x : x \text{ is a Beethoven symphony with a descriptive name}\}$
$C = \{x : x \text{ is a Beethoven symphony with the number not divisible by three}\}$
$D = \{x : x \text{ is a Beethoven symphony in a minor key}\}$

Then we can see that

$$\Omega = \{\text{SI, SII, SIII, SIV, SV, SVI, SVII, SVIII, SIX}\}$$

$$A = \{\text{SI, SII, SIII, SIV, SVI, SVII, SVIII}\}$$
$$B = \{\text{SIII, SVI, SIX}\}$$
$$C = \{\text{SI, SII, SIV, SV, SVII, SVIII}\}$$
$$D = \{\text{SV, SIX}\}$$

We can see that $A \cup B = \{\text{SI, SII, SIII, SIV, SVI, SVII, SVIII, SIX}\}$
$= \{x : x \text{ is a Beethoven symphony which is either "named" or in a major key}\}$

$$B \cup D = \{\text{SIII, SV, SVI, SIX}\}$$

Graphically, we can express the union of any two sets A, B using a Venn diagram as follows:

The union $A \cup B$ corresponds to the shaded area.

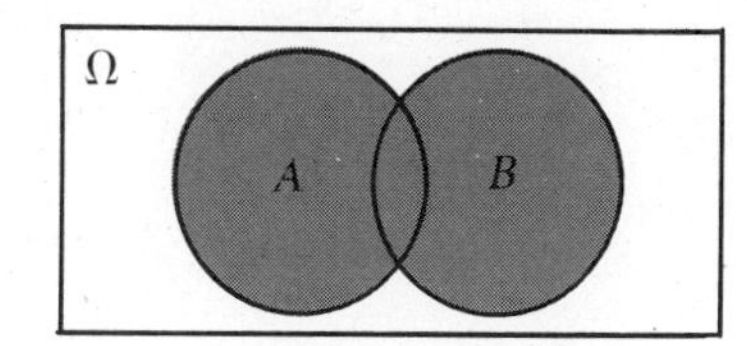

DEFINITION 3: The intersection $A \cap B$ of two sets A and B is the set of all elements belonging to both A and B.

Referring to the above example we have

$A \cap B = \{\text{SIII, SVI}\} = \{x : x \text{ is a "named" Beethoven symphony in a major key}\}$

$B \cap D = \{\text{SIX}\}$

Graphically, we can express the intersection of any two sets A, B as follows:

The intersection $A \cap B$ corresponds to the shaded area.

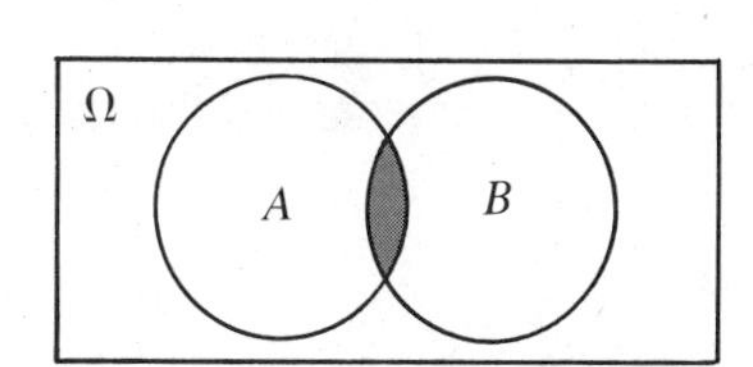

DEFINITION 4: The difference $A - B$ of two sets A and B is the set of those elements of A which do not belong to B.

Referring to the above example,

$A - B = \{\text{SI, SII, SIV, SVII, SVIII}\}$
$= \{x : x \text{ is an "unnamed" Beethoven symphony in a major key}\}$

$B - A = \{\text{SIX}\}$

$= \{x : x \text{ is a "named" Beethoven symphony in a minor key}\}$

Graphically, we can express the difference of any two sets A, B as follows:

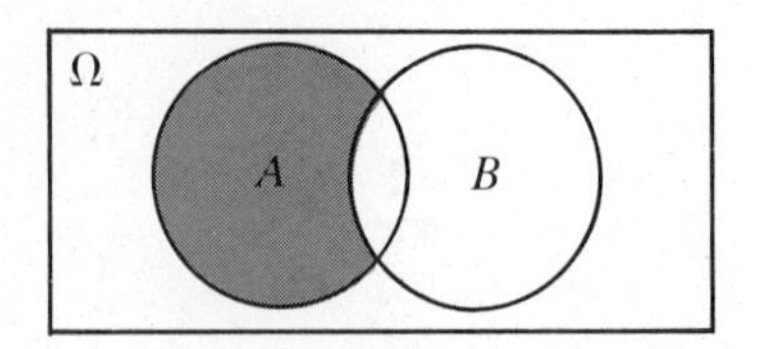

The difference $A - B$ corresponds to the shaded area.

DEFINITION 5: The complement A' of a set A is the set of all elements of Ω which do not belong to A.

$A' = \{\text{SV, SIX}\} = D.$
$B' = C$ (verify this surprising coincidence!)

Let us consider another example.

$$\Omega = \{x : x \text{ is an outcome of a toss of a die}\} = \{1, 2, 3, 4, 5, 6\}$$

$$E = \{x : x \text{ is an even outcome of a toss of a die}\} = \{2, 4, 6\}$$

$$F = \{x : x \text{ is an outcome of a toss of a die less than } 4\} = \{1, 2, 3\}$$

Then apparently $E \cup F = \{1, 2, 3, 4, 6\}$ $\qquad F - E = \{1, 3\}$

$E \cap F = \{2\}$ $\qquad E' = \{1, 3, 5\}$

$E - F = \{4, 6\}$ $\qquad F' = \{4, 5, 6\}$

We said before that any two sets which have no common elements are called disjoint sets. Obviously A and B are disjoint if and only if $A \cap B = \phi$. For example the sets $(E \cap F)$ and $(F - E)$ given above are disjoint; the same is true for the sets $(E - F)$ and E'.

Considering the given definitions, we can immediately write several obvious statements. If A and B are any sets, ϕ is the empty set, and Ω is the universal set, then

$$\phi' = \Omega \qquad \Omega' = \phi$$

$$\phi \cup A = A \qquad \Omega \cup A = \Omega \qquad A \cup A = A$$

$$\phi \cap A = \phi \qquad \Omega \cap A = A \qquad A \cap A = A$$

$$A \subset A \cup B \qquad B \subset A \cup B$$

$$A \cup A' = \Omega \qquad A \cap A' = \phi$$

There are other properties of set operations which are not so obvious and which, perhaps, require a proof. We shall give a detailed proof of the following statement, called the distributive law, in order to show how one can construct such a proof.

→ **THEOREM 1:** For any three sets A, B, C,

$$A \cap (B \cup C) = (A \cap B) \cup (A \cap C).$$

COMMENT. Parentheses ( ) mean, as in algebra, that the operation inside should be performed first.

Proof: In order to prove $A \cap (B \cup C) = (A \cap B) \cup (A \cap C)$, we shall first prove that $A \cap (B \cup C) \subset (A \cap B) \cup (A \cap C)$, then in a second step we shall prove that $A \cap (B \cup C) \supset (A \cap B) \cup (A \cap C)$. (See the above comments about equality of two sets.) Let $x \in A \cap (B \cup C)$, then $x \in A$ and $x \in (B \cup C)$. Therefore $x \in A$ and either $x \in B$ or $x \in C$. This implies that either $x \in (A \cap B)$ or $x \in (A \cap C)$ and therefore, finally, $x \in (A \cap B) \cup (A \cap C)$. This completes the first step; hence $A \cap (B \cup C) \subset (A \cap B) \cup (A \cap C)$. Let us carry out the second step. Let $x \in (A \cap B) \cup (A \cap C)$, then either $x \in (A \cap B)$ or $x \in (A \cap C)$. Obviously, in either case, $x \in A$ and either $x \in B$ or $x \in C$. Hence $x \in A$ and $x \in (B \cup C)$, which implies $x \in A \cap (B \cup C)$. This completes the second step, hence

$$(A \cap B) \cup (A \cap C) = A \cap (B \cup C). \qquad \text{q.e.d.}$$

COMMENT. The letters q.e.d. are abbreviations for the Latin expression "quod erat demonstrandum" which means "which was to be shown." We shall always use these letters to mark the end of a proof.

There are other properties which can be easily proved by the reader, for example:

For any three sets, $A, B, C, A \cup (B \cap C) = (A \cup B) \cap (A \cup C)$.

For any two sets $A, B, A \subset B$ implies $A \cap B = A$ and $A \cup B = B$.

An interested reader might ask why we study these relations. Well, there are even more involved set relations that we shall find exceedingly useful in applications to probability theory. Consider the following situation. Two hunters, say Mr. A and Mr. B, shoot at an elk simultaneously. We are interested in the event that the animal will be hit. Then we are interested in the event that "either A hits or B hits." However, this first event is equivalent to the event saying that "It is not true that neither A nor B hits." Despite the rather involved expression used in the description of the second event, it is often much more convenient and easier to handle (especially in the probability theory). The equivalence of the two events is ascertained by the so-called de Morgan law and we shall have many opportunities to use this trick in the calculation of probabilities.

THEOREM 2: de Morgan Laws: For any two sets A, B,

$$(A \cup B)' = A' \cap B'$$

$$(A \cap B)' = A' \cup B'.$$

COMMENT. In the example about the two hunters, we used the first law, which actually implies $A \cup B = (A' \cap B')'$.

Proof: We shall prove only the first law, leaving the second one to the reader. Let $x \in (A \cup B)'$; then $x \notin (A \cup B)$ and therefore $x \notin A$ and $x \notin B$ (otherwise x would belong to $(A \cup B)$). This implies that $x \in A'$ and $x \in B'$, and finally that $x \in A' \cap B'$. This means that

$$(A \cup B)' \subset A' \cap B'.$$

Let $x \in A' \cap B'$; then $x \in A'$ and $x \in B'$ and therefore $x \notin A$ and $x \notin B$. This implies that $x \notin (A \cup B)$ and therefore that $x \in (A \cup B)'$. This means $A' \cap B' \subset (A \cup B)'$. q.e.d.

The operations of union and intersection can be easily generalized to any class of n sets (where n is a natural number). Let $A_1, A_2, \ldots A_n$ be sets. Then the union of the A_i's, denoted by $\bigcup_{i=1}^{n} A_i$, is the set of all elements belonging to at least one of the sets $A_1, A_2, \ldots A_n$. The intersection of the A_i's, denoted by $\bigcap_{i=1}^{n} A_i$, is the set of all elements belonging to all the sets $A_1, A_2, \ldots A_n$. In other words,

$$\bigcup_{i=1}^{n} A_i = A_1 \cup A_2 \cup A_3 \cup \ldots \cup A_n$$

$$\bigcap_{i=1}^{n} A_i = A_1 \cap A_2 \cap A_3 \cap \ldots \cap A_n.$$

Following are several useful relations:

$$A \cap \left(\bigcup_{i=1}^{n} B_i\right) = \bigcup_{i=1}^{n} (A \cap B_i) \qquad \text{distributive law}$$

$$A \cup \left(\bigcap_{i=1}^{n} B_i\right) = \bigcap_{i=1}^{n} (A \cup B_i) \qquad \text{distributive law}$$

$$\left(\bigcap_{i=1}^{n} A_i\right)' = \bigcup_{i=1}^{n} A_i' \qquad \text{de Morgan law}$$

$$\left(\bigcup_{i=1}^{n} A_i\right)' = \bigcap_{i=1}^{n} A_i' \qquad \text{de Morgan law}$$

Let us recall that two sets A, B are disjoint if and only if $A \cap B = \Phi$.

Let us consider a collection of n sets $A_1, A_2, \ldots A_n$ which are pairwise disjoint, i.e., $A_i \cap A_j = \Phi$ for $i \neq j$. Such sets are called *mutually exclusive* sets. If the sets $A_1, A_2, \ldots A_n$ satisfy the condition $\cup_{i=1}^{n} A_i = \Omega$ then they are said to be *exhaustive* sets. Any class of mutually exclusive and exhaustive sets $A = \{A_1, A_2, \ldots A_n\}$ is called a *partition* of Ω. The notion of a partition is very important in probability theory. Let us toss a coin three times and denote a head by H and a tail by T. Let Ω be the set of all outcomes. Then

$$\Omega = \{HHH, HHT, HTH, THH, TTH, THT, HTT, TTT\}.$$

In this case, the sets

$$A_1 = \{\text{exactly one head}\} = \{TTH, THT, HTT\}$$
$$A_2 = \{\text{exactly two heads}\} = \{HHT, HTH, THH\}$$
$$A_3 = \{\text{exactly three heads}\} = \{HHH\}$$
$$A_4 = \{\text{no heads}\} = \{TTT\}$$

form a partition $A = \{A_1, A_2, A_3, A_4\}$ of the universal set Ω.

Using sets and operations, we can sometimes analyze situations of everyday life. Let us consider coalitions in a voting body (House of Representatives, a parliament, students' meeting). A coalition is formed when certain members group themselves together in order to pass a certain measure. If they have enough votes, then we call the coalition a *winning coalition.* If the members outside of the coalition have enough votes to defeat the measure, we call the original set a *losing coalition.* Finally, if neither the members of the coalition can pass the motion nor the members outside the coalition can defeat the motion, then the coalition is called a *blocking coalition.* Using set theoretical terminology, if Ω is a voting body, then any subset $C \subset \Omega$ can be considered as a coalition. A coalition C is winning, if its members have enough votes to carry the motion. A coalition C is losing, if C' is winning. A coalition C is blocking if neither C nor C' is winning. An example of such a voting body is the Security Council of the United Nations. The Council has 15 members: five permanent (China, France, the U.K., the U.S.S.R., and the U.S., called the Big Five) plus ten nonpermanent. On substantive matters, decisions by the Council are made by an affirmative vote of nine members that must include the Big Five. Thus any set of the members including the Big Five and altogether including at least nine nations is a winning coalition. Any set including at most six small nations is a losing coalition.

Sometimes we have to distinguish between finite and infinite sets. A set is *finite* if there exists (at least hypothetically) a procedure which allows selection of an element that can be called the first element, then a second element, a third element, and so on, that will terminate after a finite number of steps. A set which is not finite is called *infinite.* For example, if we have the sets

$$E = \{x : x \text{ is a real number such that } 2x^2 + 4x - 6 = 0\},$$

$$F = \{x : x \text{ is an odd integer}\},$$

$$G = \{1,3,5,7\},$$

$$H = \{x : x \text{ is a prime number}\},$$

then E, G are finite sets and F, H are infinite sets. The empty set ϕ is considered to be finite.

If we toss a coin once, then the set of all outcomes consists of two points, namely H (for a head) and T (for a tail); that is, we get $\{H, T\}$. If we repeat this experiment twice, then the set of all outcomes is equal to the set of all ordered pairs of symbols H and T, that is, we get $\{HH, HT, TH, TT\}$. Let us investigate the process leading from the first set to the second set.

Imagine that a rich man is going to buy one car for himself and one for his wife. He would like to buy a Ferrari, Lamborghini or Jaguar for himself and a Toyota, Datsun, V.W. or Mazda for his wife. Later, when his friends ask him which cars he has bought, he will always mention his car first and his answer will follow the form: (Ferrari, V.W.), (Lamborghini, Toyota), (Jaguar, V.W.), etc. The set of all possible answers has 12 elements. Each element is an ordered pair (x, y) where x belongs to the set {Ferrari, Lamborghini, Jaguar} and y belongs to the set {Toyota, Datsun, V.W., Mazda}. This operation of forming new sets is called a *Cartesian product.*

DEFINITION 6: Let A and B be sets. Then the set of all ordered pairs (x,y) such that $x \in A$ and $y \in B$ is called the Cartesian product of A and B and is denoted by $A \times B$.

We emphasize the order of the elements (x,y). For example, (Ferrari, V.W.) is a completely different answer from (V.W., Ferrari), the second one indicating that the man is going to buy a V.W. for himself and a Ferrari for his wife, which, of course, is obviously unacceptable and utterly impossible.

If, for example,

$$E = \{2, 4\}$$

$$F = \{1, 2, 3\}$$

then

$$E \times F = \{(2,1), (2,2), (2,3), (4,1), (4,2), (4,3)\}.$$

We can generalize the notion of the Cartesian product to n sets.

DEFINITION 7: Let $A_1, A_2, \ldots A_n$ be sets. Then the Cartesian product of $A_1, A_2, \ldots A_n$ is denoted by $A_1 \times A_2 \ldots \times A_n$ and is the set of all ordered n-tuples $(x_1, x_2, \ldots x_n)$ such that $x_1 \in A_1, x_2 \in A_2, \ldots x_n \in A_n$.

In the case of $n = 2$, both definitions coincide.

DEFINITION 8: Let A and B be sets. Suppose that, to each element x of the set A, there is assigned a unique element y of the set B. We call such an assignment a function f and write

$$f : A \longrightarrow B$$

which reads "f is a function from A into B." If we want to express that y is the element which f assigns to x, we write

$$y = f(x).$$

The set A is called the domain and the set B is called the range of f.

The domain and range can be arbitrary sets. For example, let the sets E, F be given as follows:

$$E = \{x : x \text{ is a student registered in a STAT 201 course}\}$$

$$F = \{A, A^-, B^+, B, B^-, C^+, C, C^-, D, F, Au, W\}.$$

Then the function f might mean an assessment of a student in the STAT 201 course. Each student gets a unique assessment (perhaps even audit—Au or permission to withdraw—W). If $x \in E$ is a student in the STAT 201 class then $f(x)$ is his final letter grade.

We shall frequently investigate functions whose range is the set of all real numbers $R^\#$. Such functions are called real valued functions. Let, for example,

$$E = \{x : x \text{ is a resident of California on January 1, 1977}\}.$$

Then we can investigate a real valued function

$$f : E \longrightarrow R^\#$$

where for every resident $x \in E$, $f(x)$ means his weight in pounds.

The notion of function is so important in probability that we give several more examples.

Let Ω be the set of all outcomes of 3 tosses of a coin. $\Omega = \{HHH, HHT, HTH, THH, TTH, THT, HTT, TTT\}$ and let the function f on Ω mean "the number of tails." Then $f(HHH) = 0$, $f(HHT) = 1$, $f(TTT) = 3$.

Let $E = \{1, 2, 3\}$ and let F be the class of all subsets of E, that is

$$F = \{\{1\}, \{2\}, \{3\}, \{1,2\}, \{1,3\}, \{2,3\}, \{1,2,3\}, \phi\}.$$

Let the function f be defined for every element x of F in such a way that $f(x)$ means "the number of points of x" (we have to realize that elements of F are sets). Then, for example,

$$f(\{2\}) = 1$$

$$f(\{2,3\}) = 2$$

$$f(\phi) = 0$$

$$f(\{1,2,3\}) = 3.$$

Since the domain of this function is a collection of sets, this function is called a *set function.* We shall see later that the concept of probability can be made rigorous using this kind of function.

Very often we investigate functions that have the set of real numbers $R^\#$ for both the range and domain,

$$f : R^\# \longrightarrow R^\#.$$

Examples well known from high school are

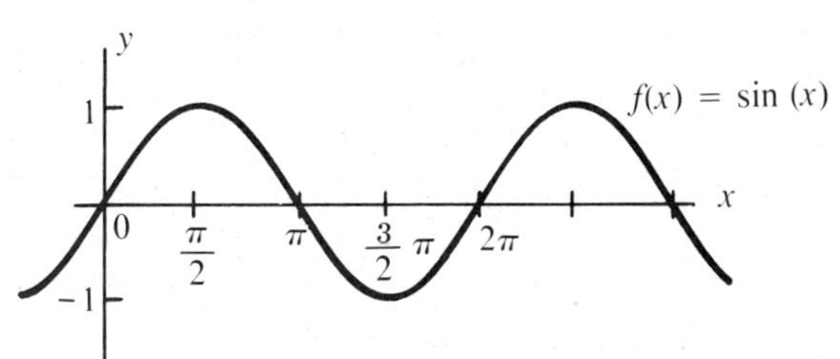

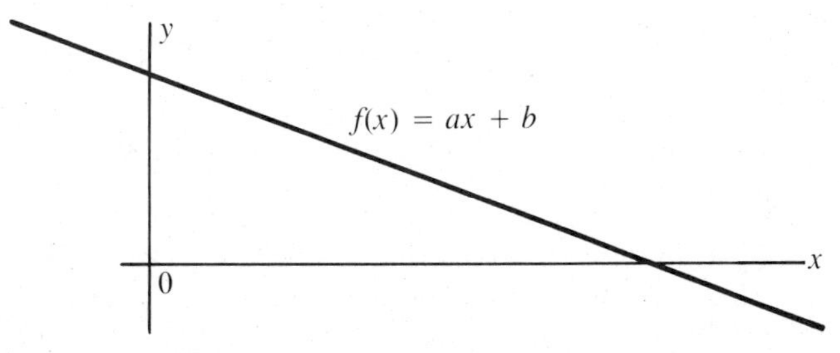

and many others which we are not going to investigate.

Let f and g be two functions with the same domain D and with the same range $R^\#$ (the set of all real numbers).

$$f: D \longrightarrow R^\# \qquad g: D \longrightarrow R^\#$$

Then we can naturally investigate a new function $(f + g)$ which assigns to every $x \in D$ the corresponding value $[f(x) + g(x)]$. For example, if D is the set of all married couples residing in California, then f can mean the income of the older spouse and g the income of the younger spouse (alas, the terms "husband" and "wife" are lately becoming somewhat nebulous and obsolete). Then the function $(f + g)$ means the income per couple.

Similarly, we can investigate other functions, for example, $f \cdot g, f - g, f/g, f^2, (f + g)^2$, etc., which can be defined in the same natural way. All these examples are just special cases of so-called *composite functions,* defined below. This definition should be skipped by the mathematically unsophisticated reader until just before studying Chapter XII.

DEFINITION 9: Let f and g be functions with the same domain A and with the same range B; that is, let

$$f: A \longrightarrow B \quad \text{and} \quad g: A \longrightarrow B.$$

Further, let h be a function with the domain $B \times B$ and with the range C, that is

$$h: B \times B \longrightarrow C.$$

Then the function $k(x)$ with the domain A and with the range C defined as

$$k(x) = h(f(x), g(x))$$

is called a *composite function.*

In our previous example of California couples, we had A equal to the set of all married California couples, $B = R^{\#}$, $C = R^{\#}$, f and g are the incomes of the older and the younger spouses, respectively, and

$$h : R^{\#} \times R^{\#} \longrightarrow R^{\#} \quad \text{such that} \quad h(x,y) = x + y.$$

The notion of a composite function is rather intricate but we cannot avoid it if we want to adhere to any degree of mathematical rigor in our treatment of probability and random variables (Chapters X–XII). Let us give two more examples.

The final overall score in a statistics course is comprised of 40% of the midterm test percentage and 60% of the final test percentage. Let A be the set of all students registered in the course, let $B = R^{\#}$ and $C = R^{\#}$. For every $x \in A$, let $f(x)$ be the midterm test percentage of the student x and $g(x)$ be his final test percentage. Further let

$$h : R^{\#} \times R^{\#} \longrightarrow R^{\#} \quad \text{be such that} \quad h(u,v) = .4u + .6v.$$

If $k(x)$ is the final overall score of the student x, then $k(x) = .4f(x) + .6g(x) = h(f(x), g(x))$, so the final score can be viewed as a composite function.

And one more rather formal example. Let $A = B = C = R^{\#} - \{0\}$.

$$f(x) = 2 \sin x \qquad g(x) = \frac{1}{2} \log |x|$$

$$h(u,v) = \sqrt{\left(\frac{u + 5}{3}\right)(v^2 + 7)}$$

Then the corresponding composite function $k(x)$ is equal to

$$k(x) = h(f(x),g(x)) = \sqrt{\left(\frac{2 \sin x + 5}{3}\right)\left(\frac{1}{4} \log^2 |x| + 7\right)}.$$

We just substituted $u = f(x) = 2 \sin x$, $v = g(x) = \frac{1}{2} \log |x|$.

EXERCISES

*(Answers in Appendix to exercises marked *)*

1. Let Ω be the set of all words listed in the unabridged edition of *The Random House Dictionary* and let the sets A, B, C be given as follows:
   $A = \{\omega : \omega \in \Omega \text{ and } \omega \text{ begins with a vowel}\}$
   $B = \{\omega : \omega \in \Omega \text{ and } \omega \text{ begins with a consonant}\}$
   $C = \{\omega : \omega \in \Omega \text{ and } \omega \text{ has at least two letters}\}$
   $D = \{\text{AND, BY, TRUE, ON}\}$
   Decide which statements are correct:
   (a) $A \cup B = \Omega$
   (b) $C \subset A$
   (c) $A \cap D = \{\text{AND, BY}\}$
   (d) $A \cap B = \phi$
   (e) $D \subset C$
   (f) $C \cup D \subset B$
   (g) $A = B'$
   (h) $C \cap D = D$
   (i) $D \cap B = \{\text{BY, TRUE}\}$
   (j) $C \subset A \cup B$
   (k) $D \subset A$

*2. Express explicitly the following sets (subsets of the set of all real numbers).
   (a) $A = \{\omega : \omega^2 - 2\omega + 1 \leq 0\}$
   (b) $B = \{\omega : \omega = 2n + 1$ for some natural number $n < 6\}$
   (c) $C = \{\omega : \omega^2 - \omega - 6 = 0\}$
   (d) $D = \{\omega : \omega^2 - \omega + 6 = 0\}$
   (e) $E = \{\omega : 2 \sin \omega = 0\}$

3. Determine which of the following statements are correct.
   (a) If $A \subset B$ then $A \cap B = A$.
   (b) If $A \cup B = \Omega$ then $A = B'$.
   (c) If $A = B'$ then A and B are disjoint.
   (d) If $A \cap B = \phi$ and $B \cap C = \phi$ then A and C are disjoint.
   (e) If $A \subset B \subset C$ then $A \cup (B \cap C) = B$.
   (f) If $A \subset B \subset C \subset A$ then $A = B = C = A$.

4. Prove:
   (a) $A \cap (B \cup C \cup D)$
   $= (A \cap B) \cup (A \cap C) \cup (A \cap D)$
   (b) $(A \cap B \cap C)' = A' \cup B' \cup C'$
   (c) $A - (B \cap C) = (A - B) \cup (A - C)$
   (d) $A \cup B = (A - B) \cup (A \cap B) \cup (B - A)$ and the sets $(A - B), (A \cap B), (B - A)$ are disjoint.
   (e) $A \subset B$ if and only if $B' \subset A'$

5. Prove that for any two sets
   (a) $A \subset A \cup B$, $B \subset A \cup B$
   (b) $A \cap B \subset A$, $A \cap B \subset B$
   (c) If $A \subset B$ then $A \cap B = A$ and $A \cup B = B$

6. Prove that for any three sets A, B, C
   $A \cup (B \cap C) = (A \cup B) \cap (A \cup C)$.
   (This is called the second distributive law.)

7. Prove that for any two sets A, B
   $(A \cap B)' = A' \cup B'$. (This is called the second de Morgan law.)

8. Prove that for any n sets $A_1, A_2, \ldots A_n$ and a set B

   (a) $B \cap \left(\bigcup_{i=1}^{n} A_i\right) = \bigcup_{i=1}^{n} (B \cap A_i)$

   (b) $B \cup \left(\bigcap_{i=1}^{n} A_i\right) = \bigcap_{i=1}^{n} (B \cup A_i)$

   (c) $\left(\bigcap_{i=1}^{n} A_i\right)' = \bigcup_{i=1}^{n} A_i'$

   (d) $\left(\bigcup_{i=1}^{n} A_i\right)' = \bigcap_{i=1}^{n} A_i'$

9. Define the symmetric difference $A \Delta B$ as follows:
   $A \Delta B = (A - B) \cup (B - A)$.
   Prove that
   (a) $A \Delta A = \phi$
   (b) If A, B are disjoint then $A \Delta B = A \cup B$
   (c) $A \Delta (B \Delta C) = (A \Delta B) \Delta C$
   (d) $A \cap (B \Delta C) = (A \cap B) \Delta (A \cap C)$
   (e) If $A \Delta B = \phi$ then $A = B$
   (f) $A \Delta \phi = A$

10. Define the Scheffer product $A * B$ as follows:
   $A * B = A' \cap B'$.
   Prove that
   (a) $A * A = A'$
   (b) $A \cap B = (A * A) * (B * B)$
   (c) $A \cup B = (A * B) * (A * B)$
   (d) $A \Delta B = (A * B) * [(A * A) * (B * B)]$
   (e) $A - B = (A * A) * B$

11. Let $A = \{1, 3, 5\}$, $B = \{2, 5, 6\}$ and $C = A \times B$.
   (a) Write C explicitly as a set of ordered pairs.
   (b) Graph C as a set of points of the plane.

*12. Which of the following sets of ordered pairs can be expressed as Cartesian products?
   (a) {(1,2), (1,3), (1,5), (2,5), (2,3), (2,2)}
   (b) {(1,3), (2,4), (3,6)}
   (c) {(1,9), (2,8), (3,7), (2,9), (3,8), (1,7), (3,9), (1,8), (2,7)}

13. Let Ω be the set of all outcomes of a toss of a coin, that is $\Omega = \{H,T\}$ (where H stands for a head and T stands for a tail). Express explicitly
   (a) $\Omega \times \Omega \times \Omega$
   (b) $\Omega \times \Omega \times \Omega \times \Omega \times \Omega$

14. Prove that if A has m elements then there are 2^m subsets of A (use induction).

*15. Let A have m elements and B have n elements. Find the number of nonempty subsets of $A \times B$.

16. Prove $A \times (B \cup C) = (A \times B) \cup (A \times C)$.

17. Is it true that $A \times B = B \times A$ implies $A = B$? (If you agree, give a proof.)

18. Is it true that
   $A \times (B \cap C) = (A \times B) \cap (A \times C)$?
   (Prove or disprove.)

SAMPLE SPACES

Cast any given Sett of Faces with four Cubical Dice.

Richard Bentley, 1692

There are many philosophical views and opinions on the role of probability in the process of human reasoning and analysis of knowledge. Different philosophical views then lead to distinct mathematical models of the theory of probability. We should mention here briefly the most important fundamental ideas and thus simultaneously clarify our point of view in the rest of our treatment of probabilities. However, this area is so vast and perplexing that our short synopsis can in no way claim any degree of completeness.

Optimism about the mechanical materialism of the period of the Enlightenment was perhaps best expressed in the words of *René Descartes:* "Give me extension and motion and I will remake the world. The Universe is a machine in which *everything* happens by figure and motion." This opinion implied that everything in the world is uniquely *determined* by the physical laws, that there is no room for chance if only our information could be complete. How is this claim compatible with the fact that, for example, if you toss a coin you never can predict whether you will get a head or a tail? Perhaps you do not know enough about the laws governing the very complicated flight pattern of the coin: you do not know the momentum of the coin, the original impulse and the exact geometrical configuration of the coin leaving the hand and of the landing pad. If one could know all this information, one should be able to predict exactly the outcome of a toss. As a matter of fact, there is a very sophisticated Variable Probability Coin Tossing Machine by means of which an experimentor can control exactly all these parameters and thus control perfectly each flight of a coin.

Variable Probability Coin Tossing Machine

This basic idea of determinism in nature and the corresponding place of randomness was further analyzed by *Pierre Simon, Marquis de Laplace.* He wrote: "Given for one instant an intelligence which could comprehend all the forces by which nature is animated and the respective situation of the beings who compose it—an intelligence sufficiently vast to submit these data to analysis—it would embrace in the same formula the movements of the greatest bodies of the universe and those of the lightest atom; for it, nothing would be uncertain and the future, as the past, would be present to its eyes. The human mind offers a feeble idea of this intelligence* . . . from which it will always remain infinitely removed. Probability is relative, in part due to this ignorance. The principal means for ascertaining truth–induction and analogy–are based on probabilities; so that the entire system of human knowledge is connected with the theory of probability." This principle of "insufficient reason" or "indifference" therefore blames the imperfection of human intelligence and knowledge for uncertainty in nature. Our ignorance is the sole reason for chance and randomness in observable phenomena. As a matter of fact, this old opinion is enduring even today among some researchers in cybernetics, econometrics, biometrics and other sciences.

We shall not try to criticize or to disprove the above mentioned views of deterministic mechanical materialism–this would go beyond our scope. However, we are going to introduce new ideas and discoveries, some of them not known till the twentieth century, which will lead to a somewhat different outlook on probability and its meaning. When physicists tried to describe the behavior of a gas by investigation of each and every molecule separately (perhaps using differential equations), they could never reach any reasonable result. Only after *J. C. Maxwell* and *L. Boltzmann* introduced the idea of probability distributions to investigate the whole collectives of molecules was it possible to describe the behavior of gases in terms of mass phenomena. Thus the probabilistic methods of statistical mechanics of gases pointed to the fact that probabilistic laws are perhaps imbedded in the innermost structures of nature. Later on the investigation of radioactivity by *Max Planck* emphasized this fact even more. It was discovered that the behavior of elementary particles can never be determined uniquely and that only probability distributions represent an adequate description of their behavior. Further studies in quantum mechanics led Nobel Prize laureate *Werner Heisenberg* to the formulation of the uncertainty principle. According to this principle, it is impossible to specify or determine simultaneously both the position and velocity of a particle with full accuracy. Behavior of a particle can be predicted only statistically and probability is necessary to describe experiments. Thus the new discoveries of physics raised the idea that probabilistic laws are at least as important as any other laws of nature and the universe.

Basic notions of an experiment, an event, randomness, chance and probability became categories of top importance for human reason and analysis of knowledge about the universe. *B. V. Gnedenko,* a leading expert of the Soviet school of probability, expressed this idea of the dialectical materialism in his book on probability: "The basic rule is that *distinct new laws* arise whenever mass phenomena are studied. The theory of probability, in its abstract form, reflects the laws inherent in random events of a mass structure." We usually speak about mass phenomena in the context of experiments involving aggregates of a large number of objects of almost equal status, or in the context of experiments which can be repeated many times (in principle, an unlimited number of times) under virtually the same conditions. *Probability, then, has an objective meaning expressing fundamental intrinsic laws describing mass phenomena and their occurrence in nature.*

However not all uses of probabilistic judgments are of this type. Sometimes probability is used in the sense of a logical corroboration of some uncertain proposition. In other situations probability has a more psychological or intuitive meaning. For example, we can hear often statements such as: "I feel that it is going to rain tomorrow"; "It is very likely that by the year 2001 you can journey to the Moon through a travel agency"; "It's quite likely that I will flunk this course." In all these statements the speaker is expressing his own degree of certainty, inner belief or sureness with respect to some events which might or might not happen. There are two important features in these expressions. First, the degree of certainty cannot be verified by any objective experiment because it is expressing someone else's inner subjective state of mind or reasoning, accessible only through his communication. Second, all these statements speak of

---

*Often called the Laplace demon.

unique situations that cannot be repeated. One cannot repeat "tomorrow" twenty times to see how many times it rained, nor can one repeat the year 2001 fifteen times. Perhaps one can repeat a course ten times but actually every successive repetition is based on different evidence and therefore there are not ten repetitions of the same course but ten different courses. In all these cases, we speak about a subjective degree of certainty or subjective probability. In these situations probabilities of events are not usually expressed by numbers but rather events are ordered by a relation "more likely" or "more probable." There are mathematical techniques for the description of such situations. These methods, investigated by *L. J. Savage,* are referred to as comparative or qualitative probabilities. Having described the ordering of events, one usually tries to express the relation "more likely" in terms of numbers. These are, however, rather subtle and fine procedures with a great deal of mathematical sophistication, which we will not investigate in this book.

It is clear that there are many situations in life that require the concept of likelihood or probability to describe. And there are of course many different opinions, points of view and ideas about the meaning and role of probability in these situations. Mathematics, with its unlimited conceptual wealth, can offer many different techniques or *models of probability* that will suit more or less all these situations and ideas. However, the problem of the correspondence of such models to real life situations remains to be investigated. This idea of an *interpretation* of a model is one of the most complicated of philosophical questions and is beyond the scope of this text. However, one should realize that we are choosing one particular mathematical model of probabilities that corresponds to a certain philosophical point of view, which is adequate for a description of certain types of real life situations. Also, most of the time we will be studying the mathematical rules and properties of our model, leaving the question of interpretations of the model in real life situations to the reader. This will make our task easier (perhaps at the expense of the reader).

Let us begin with several examples of situations that we are going to analyze. "If a perfectly homogeneous, fair coin is freely flipped 200 times, then it is very likely that we obtain about 100 heads." "Let us have two similar, connected containers of helium. If a marked molecule is released in one of the containers, after a sufficiently long period of time the probabilities of it being in both containers are equal." "In poker, it is not very likely that you will be dealt a full house." (A full house in poker is a hand consisting of three of a kind and a pair). Actually, we can deal a hand of cards for poker 100 times and count how many times we get a full house. We can release a marked molecule in one of the containers, and check its location after 3 hours—a procedure that can be repeated perhaps 50 times.

In these situations, we are investigating a physical experiment that can be repeated an unlimited number of times (at least in principle) and the outcomes can be statistically studied. This kind of mass phenomenon is of prime importance in our further investigations. Nevertheless, we have to clarify first what we mean by an "experiment." In general, *an experiment consists of arranging for certain conditions to be present simultaneously or of performing a particular procedure.* (*The Oxford Dictionary* defines an experiment as a tentative procedure; a method, system of things or course of action, adopted in uncertainty whether it will answer the purpose.) A *trial* is an actual performance of an experiment. What interests us most about a certain experiment is its *outcome.* Before we can start the analysis of any particular experiment, we must know what its set of all possible outcomes is. An *outcome is always relative* to an observer. For a man, an outcome of a game of roulette can be winning $25 but for his wife it can be the possibility of flirting with a handsome croupier because the husband decided to use, in full, the favors of Lady Luck and continue his gambling. Or, if investigating the probability that a thumbtack will land on its back (position 1), we can consider as a set of all possible outcomes positions 1 and 2 only. Another observer, however, might insist on position 3 as well.

Position 1 Position 2 Position 3

A sample space Ω *of an experiment is a set which represents uniquely the set of all possible outcomes of the experiment,* i.e., such that there is a one-to-one correspondence between the set of all possible outcomes and Ω. The sample space Ω usually consists of symbols representing the outcomes (for example H for "head of a coin" or 6 for ": : : on a die"). The elements of Ω are sometimes called the sample points. Sample spaces may be of different types and structures. We will be concerned mostly with spaces that have only a finite number of points (playing cards, tossing dice and coins, etc.). However, sometimes a sample space can have infinitely many points. Let us consider an experiment in which a Geiger-Müller counter is exposed to an outburst of radiation. Then the outcome will be a number of radioactive particles scintillating on the screen: perhaps five, ten, perhaps zero, perhaps a million, or a billion, or a billion billion—or who knows how many zillions. The most important fact, however, is that we can count the number of particles—one, fifty, a million, etc.—that is, we can put the outcomes of the experiment into a one-to-one relation with the natural numbers. So there is a one-to-one relation between the set of all outcomes of our experiment and the set of all natural numbers; the set of all outcomes can be represented by the set of all natural numbers. Such a sample space will be called a *countable* space.

There are, of course, experiments which need even bigger sets for their representation—like the set of all real numbers in the interval (0, 1). We will not be interested in such spaces because they are mathematically too complicated. However, countable spaces are encountered in real life very frequently—at any time that it is impractical to impose an upper bound on the number of certain items: the number of molecules of air in a certain volume; the number of tosses of a coin before getting the first head (in principle you can get an arbitrarily long sequence of tails); the number of cars on a turnpike; the number of words uttered by a politician in one hour (it is always finite, of course, but it is very impractical to set any upper limit); the number of practice repetitions by a pianist before mastery of a Chopin Etude, and so on. We are mainly concerned with finite sample spaces but we eventually mention countable spaces also.

Gambler's paraphernalia

There are examples in which the selection of an appropriate set of outcomes, hence the selection of a sample space, plays a very important role.

Let us toss a coin three times. Then we can consider two sample spaces.

$$\Omega_1 = \{HHH, HHT, HTH, THH, THT, HTT, TTH, TTT\}$$

$$\Omega_2 = \{3 \text{ heads}, 2 \text{ heads}, 1 \text{ head}, 0 \text{ head}\}$$

If we are interested in the event that we obtain more than one head, then both Ω_1 and Ω_2 will do, while for consideration of the event that heads and tails alternate, the sample space Ω_2 is not suitable at all. The basic rule says that we must select a sample space large enough to be able to capture and express all the events we have to investigate. However, we do not always have a complete list of interesting events ahead of time; some of them might appear during the investigation. For this reason, it might be best (even though not always most economical) to take a sample space as large as practically possible.

Sometimes we are required to select a sample space even before the experiment has been completely specified and described. It may be to some advantage, or it may perhaps be necessary to keep such a preselected space even if it does not fit the experiment perfectly. Let us consider the following example. A gambler uses special dice for the entertainment of his guests. Besides the ordinary die in the form of a cube with the numbers $\{1,2,3,4,5,6\}$, he also has a tetrahedron with the numbers $\{1,2,3,4\}$ and a dodecahedron with the numbers $\{1,2,3,4,5,6,7,8,9,10,11,12\}$. He is going to select a die and toss it. If we have to specify a sample space before he has selected a die, then the only choice is $\Omega = \{1,2,3,4,5,6,7,8,9,10,11,12\}$. Only this sample space is sufficiently large to cover all the possibilities (even though it is too large if the gambler finally selects the ordinary cube).

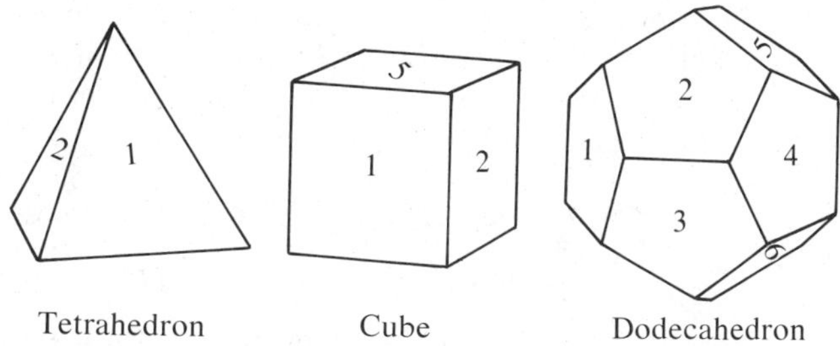

An event is any proposition or statement related to an experiment in such a way that, after any particular realization of the experiment, one can decide whether or not it occurred. (*The Oxford Dictionary* defines an event as any one of possible occurrences, some one of which will happen under stated conditions.) We shall denote events by capital letters

$$A, B, C, \ldots X, Y, Z, A_5, C_{17} \ldots\ldots$$

Having the sample space Ω of an experiment and an event A related to this experiment, one can investigate the set of all outcomes of the experiment for which the event A occurs. For our purposes, *we can identify an event A and the set of all outcomes for which it occurs.* For example, in the case of tossing a die twice, our sample space Ω is

(1,1)	(1,2)	(1,3)	(1,4)	(1,5)	(1,6)
(2,1)	(2,2)	(2,3)	(2,4)	(2,5)	(2,6)
(3,1)	(3,2)	(3,3)	(3,4)	(3,5)	(3,6)
(4,1)	(4,2)	(4,3)	(4,4)	(4,5)	(4,6)
(5,1)	(5,2)	(5,3)	(5,4)	(5,5)	(5,6)
(6,1)	(6,2)	(6,3)	(6,4)	(6,5)	(6,6)

Let us define the following events:

A : the sum is 5

B : the second toss is twice as big as the first one

C : both outcomes are equal

D : the sum is more than 15

Then apparently,

$$A = \{(1,4), (2,3), (3,2), (4,1)\},$$
$$B = \{(1,2), (2,4), (3,6)\},$$
$$C = \{(1,1), (2,2), (3,3), (4,4), (5,5), (6,6)\},$$
$$D = \Phi$$

Or consider a Canadian province which offers the following big game licenses: Elk, Moose, Deer, (Mountain) Goat, (Bighorn) Sheep. The regulations stipulate that "a hunter may purchase any two from the above mentioned licenses but no hunter will be a holder of both Elk and Moose licenses." If we consider a person buying his two big game licenses, then a sample space can be given as follows (the letters stand for the initials):

$$\Omega = \{ED, EG, ES, MD, MG, MS, DG, DS, GS\}$$

Imagine a game warden checking a hunter in the forest, and let us denote the events as follows (assume that the hunter has both big game licenses):

M : the hunter has licenses for two animals of the deer family (E,M,D)*

N : the hunter has no license for a moose

O : the hunter has a license for at least one animal of the deer family

P : the hunter has both licenses for animals of the same family

Then we can see that

$$M = \{ED, MD\}$$
$$N = \{ED, EG, ES, DG, DS, GS\}$$
$$O = \{ED, EG, ES, MD, MG, MS, DG, DS\}$$
$$P = \{ED, MD, GS\}$$

Obviously, *all events are uniquely represented as subsets of the sample space* Ω*; the certain event is* Ω *and the impossible event is* Φ*.*

If we are given events A, B, C, . . . , we can investigate their relations and construct new events using logical connectives like "or," "and," "not," and so on. We say that an event A implies an event B, provided that if A occurs then B also occurs. Having two events A and B, the event "A or B" occurs if and only if A occurs or B occurs (or both) while the event "A and B" occurs if and only if both A and B occur simultaneously. The event "not A" occurs if and only if A does not occur . If the events related to the single toss of a die are given as follows:

E : an outcome less than 3

F : an outcome less than 5

G : an outcome divisible by 2

H : an outcome divisible by 3

*E, M and D belong to the family Cervidae while G and S belong to the family Bovidae.

then E implies F and we can construct new events:

"G and H" : an outcome divisible by 6

"G or F" : anything but five

"not G" : an odd outcome.

We can interpret these events as sets

$$\Omega = \{1, 2, 3, 4, 5, 6\},$$
$$E = \{1, 2\},$$
$$F = \{1, 2, 3, 4\},$$
$$G = \{2, 4, 6\},$$
$$H = \{3, 6\}.$$

Then obviously, $E \subset F$ and

$$\text{"}G \text{ and } H\text{"} = \{6\} = G \cap H$$
$$\text{"}G \text{ or } F\text{"} = \{1, 2, 3, 4, 6\} = G \cup F$$
$$\text{"not } G\text{"} = \{1, 3, 5\} = G'.$$

Or, if we consider the hunter buying his two big game licenses, then M implies O, M implies P and, further, $P \cap O$ means that the hunter has licenses for two animals of the deer family, therefore

$$M \subset O, \quad M \subset P, \quad O \cap P = M.$$

It is quite clear that there is a unique relation between logical connectives and set operations.

Events	*Sets*
$A, B, C, \ldots$	$A, B, C, \ldots$ subsets of Ω
impossible event	ϕ
certain event	Ω
A implies B	$A \subset B$
A or B	$A \cup B$
A and B	$A \cap B$
not A	A'
A but not B	$A - B$

When investigating set operations we mentioned the set of all Beethoven symphonies (see page 6), which can also serve as an example of a sample space. For other examples of a practical character, see the section investigating phenotypes (blood groups), the safety of blood transfusions, and so on, in Chapter XV.

EXERCISES

(Answers in Appendix to exercises marked)*

1. Let a penny, a nickel and a dime be tossed all at a time.
   (a) Find a corresponding sample space.
   (b) Express explicitly the following events:
   A means that at least one head occurred.
   B means that exactly two tails occurred.

C means that both head and tail occurred.
D means that more heads than tails occurred.
E means the penny turns up heads.
F means the nickel and dime turn up tails.
G means that the value of the coins displaying heads is more than five cents.
H means that heads and tails alternate with increasing value of the coins.

2. Two dice are rolled. How many points has a suitable sample space? Express the following events as subsets of the sample space:
(a) The first outcome is bigger than the second one but smaller than 5.
(b) The first minus the second outcome is -1.
(c) The outcomes give a tie.
(d) The maximum is less than 4.
(e) The sum is 7.
(f) At least one 6 is obtained.

3. The final grade in a course can be considered as an outcome of an experiment with possible outcomes A, B, C, D, F and with grades having the following weights: A counts 4 points, B counts 3 points, C counts 2, D counts 1 and F counts zero. A student takes three courses. Express the following events as subsets:
(a) His grade average point is more than 3.
(b) His grade average point is exactly 2.

4. Two hunters, Mr. A and Mr. B, shoot at a bull moose. Each of them fires one shot. Let the event A mean that Mr. A hit the moose and B mean that Mr. B hit. Express the following events:
(a) Both hunters hit the moose.
(b) The moose has been hit.
(c) Mr. A hit the moose but Mr. B did not.
(d) The moose has been hit by one bullet only.
(e) If * is the Scheffer product, what is the meaning of $A * B$? (See Chapter II, Exercise 10.)

5. A light plane flying over the snow-covered Rocky Mountains crashed in a deserted area and only four men and 2 women survived. After several days on the mountain, they decided to make a draw to select an expedition of two people to descend down the valley and search for help. They cut four short and two long sticks and started a draw. Set up the smallest possible sample space describing the selection if:
(a) Everybody draws a stick and those with long ones will go on the expedition (women often having greater physical endurance under conditions of extreme stress).
(b) As in (a) but the first to draw a long stick will get the pilot's warm jacket and sleeping bag, giving him or her a good chance to survive.

What is the relation of sample spaces in (a) and (b)?

6. Referring to the previous example, express the following events (in both cases).
(a) The expedition will consist of 2 men.
(b) The expedition will consist of a man and a woman.
(c) At least one woman will be left on the mountain.

Consider now the following event: the expedition will not consist only of men, but a woman expeditionist will get the jacket and the sleeping bag. Can you express this event using the first sample space? Can you do it with the second space?

7. Just before the elections for Congress, a candidate investigated opinions of the voters in his constituency. He asked a sample of 1000 people the same three questions: whether or not they preferred stronger anti-inflation measures, the bussing of school children and whether or not they were in favor of strict antigun legislation. When the answers were processed by his staff, it was found that
*(a) 100 voters answered yes to all three questions.
*(b) 100 voters answered no to all three questions.
*(c) 300 voters answered yes to the first and either to the second or third question.
*(d) 600 voters answered yes to an odd number of questions.

Denote as A, B and C, respectively, the sets of voters answering yes to the first, second and third question, respectively. Now express the events in (a)–(d) using the sets A, B, C and appropriate set operations (disregard the numbers).

Further, it was observed that
(e) Equal numbers of people answered no to exactly one out of the three questions.
(f) 450 people answered yes to the second question and 350 said yes to the third question.

There are eight possible ways to answer three given questions yes or no. If the candidate decides to follow the opinion of the majority of voters, what answers to these questions should he declare in his election campaign? (Evaluate the numbers of voters giving each possible combination of answers to all three questions.)

8. Consider a town in which there are three piano tuners. On a certain day four residents telephone at random for a tuner. Describe a sample space; how many points are there? Express the events and count the numbers of points:
(a) Exactly one tuner will be telephoned.
(b) All the tuners will be telephoned.

INFORMAL DEFINITIONS OF PROBABILITY

In gatheryng of coniectures that are doubtfull, when probabilitie onely and no assured knowledge boulteth out the truthe of a matter.

T. Wilson, 1551

In the last chapter, we expressed events as subsets of a sample space of an experiment. Our next task is to assign probabilities to these events and to establish the fundamental rules relating the probability of a compound event to the probabilities of its components.

Everybody is familiar with such statements as: "The probability of getting a head when flipping a fair coin is one half" or "The probability of getting a six when tossing a fair die is one sixth." *A probability is therefore a (real) number which is (uniquely) assigned to an event (set).* The first problem is to decide how we are to assign this number. There are several ways to define probabilities, each of which depends on the experiment and sample space under consideration. We shall investigate three of the simplest ways: the classical definition, the geometrical definition and the frequency definition. We shall see that, regardless of their historical importance, all three methods are somewhat inaccurate and nebulous. Despite this, from these definitions, we shall obtain some fundamental *rules* of calculation, which will be perfectly clear and acceptable.

CLASSIC DEFINITION

Let us consider an experiment whose sample space has exactly n equally likely and mutually exclusive outcomes. Let A be an event and let us assume that there are exactly m_A outcomes whose occurrence implies the occurrence of A. Such outcomes are said to be favorable to the event A. In other words, we assume that the sample space has n equally likely points and that A, as a subset of the sample space, has m_A points. The *probability* of the event A is defined by

$$P(A) = \frac{m_A}{n}.$$

If, for example, we toss a die and A means "the outcome is an even number" then there are $n = 6$ outcomes $\{1, 2, 3, 4, 5, 6\}$ of which $m_A = 3$, $\{2, 4, 6\}$, are favorable to A and hence

$$P(A) = \frac{3}{6} = \frac{1}{2}.$$

If we select a card at random from a well shuffled deck of 52 cards and A means "the card selected is an ace," then there are $n = 52$ possible outcomes, of which there are $m_A = 4$ outcomes favorable to A, namely those corresponding to aces of hearts, spades, diamonds and clubs. Hence

$$P(A) = \frac{4}{52} = \frac{1}{13}.$$

Later on we shall investigate hundreds of interesting examples and applications when we know more about counting techniques. However, right now we would like to have a detailed look at the definition itself.

At this time, we can derive and prove from our definition several important statements. First of all, it is obvious that the probability, defined for any subset of a sample space in the way mentioned, is always a *real number.* Let us prove some fundamental statements.

→ **STATEMENT A1:** For any event A (subset of Ω), $P(A) \geq 0$.

Proof: Let Ω be the sample space. Then $A \subset \Omega$. If Ω contains n points (outcomes) of which m_A are favorable to A then apparently

$$m_A \geq 0 \quad \text{and} \quad n > 0$$

so that

$$P(A) = \frac{m_A}{n} \geq 0. \qquad \text{q.e.d.}$$

→ **STATEMENT A2:** For the sample space Ω, $P(\Omega) = 1$.

Proof: If we consider Ω as an event, then every outcome is favorable to Ω and hence

$$P(\Omega) = \frac{n}{n} = 1. \qquad \text{q.e.d.}$$

→ **STATEMENT A3:** Let A, B be disjoint sets, i.e., sets such that $A \cap B = \phi$. Then

$$P(A \cup B) = P(A) + P(B).$$

Proof: If A and B are disjoint, then there is no outcome favorable to both A and B, and every outcome favorable to $(A \cup B)$ is either favorable to A or to B. If m_A, m_B, $m_{A \cup B}$ are the numbers of outcomes favorable to A, B and $A \cup B$, respectively, then

$$m_{A \cup B} = m_A + m_B.$$

Hence

$$P(A \cup B) = \frac{m_{A \cup B}}{n} = \frac{m_A + m_B}{n} = \frac{m_A}{n} + \frac{m_B}{n} = P(A) + P(B). \qquad \text{q.e.d.}$$

Having established the fundamental statements A1, A2 and A3, we can prove some other statements. Notice that in each case we can give two alternative proofs. The first proof uses the classical definition of probability; the second uses the statements A1–A3 only.

→ **STATEMENT S1:** For any set A,

$$P(A) + P(A') = 1.$$

Proof 1: Since each of the n outcomes must be either favorable to A or not favorable to A, we have

$$m_A + m_{A'} = n.$$

Hence

$$P(A) + P(A') = \frac{m_A}{n} + \frac{m_{A'}}{n} = \frac{m_A + m_{A'}}{n} = \frac{n}{n} = 1.$$

Proof 2: We know that A and A' are disjoint sets and that $A \cup A' = \Omega$. Hence by A3, we have

$$P(A \cup A') = P(A) + P(A') = P(\Omega).$$

Now, using A2, we have $$P(\Omega) = 1.$$

Hence $$P(A) + P(A') = 1.$$ q.e.d.

STATEMENT S2: $P(\Phi) = 0.$

Proof 1: Since no outcome can be favorable to the impossible event, $m_\Phi = 0$. Hence

$$P(\Phi) = \frac{0}{n} = 0.$$ q.e.d.

Proof 2: We know that $\Phi' = \Omega$. Hence by S1,

$$P(\Phi) + P(\Phi') = P(\Phi) + P(\Omega) = 1$$

The use of S1 is allowed because the second proof of S1 makes use of only A1–A3. But, by A2,

$$P(\Omega) = 1$$

Hence $$P(\Phi) = 1 - 1 = 0.$$ q.e.d.

We can prove many other statements, but for the time being, we will be content with A1–A3, S1 and S2. It is very important to realize that we can prove S1 and S2, as well as many other statements (see Chapter V), either by using the classical definition of probability, or by using A1–A3 only.

First of all, let us have a closer look at the classical definition itself. This definition assumes that the experiment has only a finite number of outcomes. It cannot be used for experiments with an infinite number of outcomes and finiteness is so important that this definition cannot be generalized to infinite cases. However, we already know (see Chapter III) that there are many situations ("experiments") in real life where the number of possible outcomes is not finite: e.g., accurate measurements of the distance between two atoms of a substance; the number of radioactive particles scintillating in the Geiger–Müller counter; an accurate measurement of the volume of the blood serum in a certain volume of blood; the number of tosses of a coin before the first appearance of a head, and so on. The classical definition is principally incompatible with these situations. The assumption that all outcomes are "equally likely" is even worse. In our attempt to define probability, we have used the concept of *likelihood,* which is not yet defined. The only explanation of equally likely is that the outcomes have the "same probability"—but we do not know yet what probability is.

This is an example of a *circular definition,* where an unknown notion (probability) is defined using another unknown notion (likelihood), which in turn can be defined only by using the first unknown notion (probability). This definition is logically completely unacceptable. There are some situations where the assumption of equally likely is an intrinsic feature of the whole experiment; for example, when we draw a card from a "well shuffled deck" (the assumption that the deck is well shuffled is equivalent to the fact that each card is "equally likely" to be drawn and should be ascertained by appropriate physical means) or when we toss a "fair die" (the assumption of fairness is equivalent to the fact that each face is "equally likely" to appear and should be physically verified). In such situations the classical definition is useful and we shall exploit it to its limits later on. However, such an ingrained and eminent interpretation of "equally likely" is rather special and rare in its occurrence. The real world has a peculiar tendency against symmetry, regularity and harmony and any assumption of a perfectly uniform arrangement must be viewed as an exception. We cannot of course establish our definition of probability on something exceptional, rare and ephemeral.

It is clear that as soon as we try to give a more detailed and profound analysis of the classical definition, we run inadvertently and unavoidably into many adversities and contradictions. The definition itself seems to be quite confused and useless. Nevertheless, the statements A1–A3 and S1, S2 seem still to be quite all right and there are no objections against them. If we should decide to discard our definition, we could develop a reasonable probability theory based on these statements. The only problem would be that we could no longer prove them. At first glance, this approach does not seem to be good; however, the situation is not so bad for if we accept the statements A1–A3 as "the Commandments" (and nobody really doubts them anyway), then the remaining statements S1 and S2 will follow logically from them. Also, we will be able to derive and prove logically many other statements entirely from A1–A3, as we did with S1 and S2. In this way, we can build an extensive theory of probability that is based on a minimal set of obviously and doubtlessly true fundamental statements. This procedure simply means that we try to minimize the extent of our beliefs, and to maximize the extent of our logical reasoning.

GEOMETRICAL DEFINITION

Let us assume that we are given two regions—a large region R and a small region r contained in R.

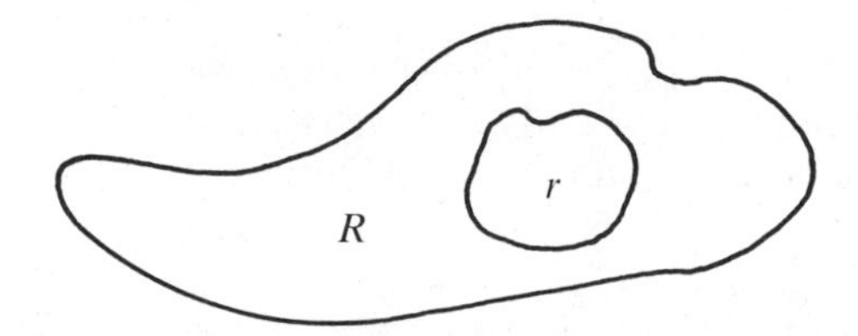

If we toss a point at random at the large region R, then the probability that it hits the small region r is

$$P = \frac{\text{area of } r}{\text{area of } R}$$

For example, if we shoot at random at a circle, what is the probability that we hit closer to the center than to the circumference?

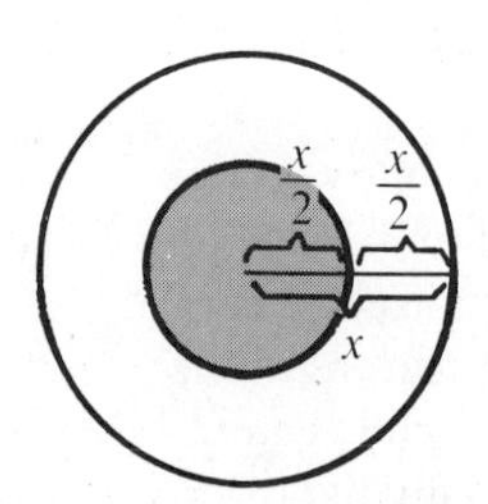

If the radius of the circle is x, then we must hit the shaded area:

$$P = \frac{\pi\left(\frac{x}{2}\right)^2}{\pi x^2} = \frac{1}{4}$$

As another example, suppose two persons A and B agreed to meet at a given place between noon and 1 P.M. The first person to come will wait 20 minutes. If the second one does not show up, the first will leave a message and depart. We assume that the arrivals are at random. What is the probability that they meet? If A arrives x minutes after noon and B arrives y minutes after noon, then the condition for encounter is $|x - y| \leq 20$:

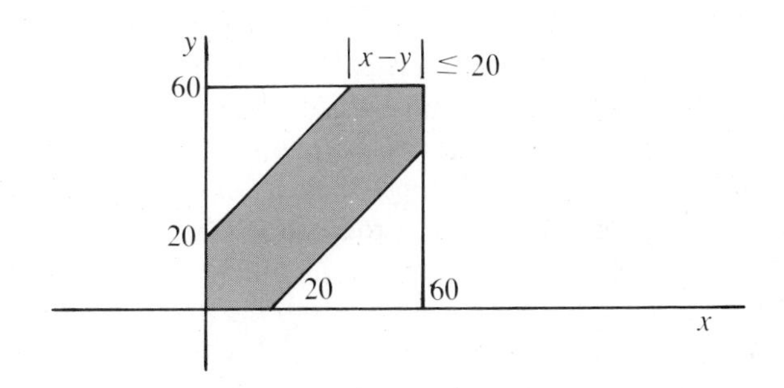

Geometrically, the points (x,y) satisfying the condition $|x - y| \leq 20$ form the shaded region in the accompanying figure. Since the shaded area is $60^2 - 40^2$ and the total area is 60^2, we can see that the probability P of encounter is

$$P = \frac{60^2 - 40^2}{60^2} = \frac{5}{9}.$$

In the case of geometrical definition we can also derive the fundamental properties A1–A3.

STATEMENT A1: For any region r,

$$P(r) \geq 0.$$

Proof: Area of $r \geq 0$, area of $R > 0$,

$$P(r) = \frac{\text{area of } r}{\text{area of } R} \geq 0. \qquad \text{q.e.d.}$$

STATEMENT A2: $P(R) = 1.$

Proof:

$$P(R) = \frac{\text{area of } R}{\text{area of } R} = 1. \qquad \text{q.e.d.}$$

STATEMENT A3: If r_1, r_2 are disjoint regions, then

$$P(r_1 \cup r_2) = P(r_1) + P(r_2).$$

Proof:

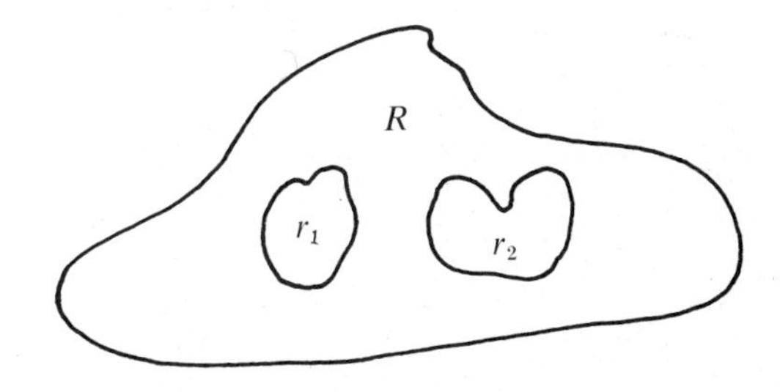

Obviously, area of $(r_1 \cup r_2)$ = area of r_1 + area of r_2.

$$P(r_1 \cup r_2) = \frac{\text{area of } (r_1 + r_2)}{\text{area of } R} = \frac{\text{area of } r_1}{\text{area of } R} + \frac{\text{area of } r_2}{\text{area of } R} = P(r_1) + P(r_2). \qquad \text{q.e.d.}$$

We can easily prove that S1 and S2 follow from the geometrical definition, but this is not necessary since we already know that S1 and S2 follow directly from A1–A3.

As in the case of the classical definition, there are also some serious defects in the geometrical definition of probability. First of all, the assumption that the small object is thrown completely at random is not too sound. For example, in shooting at a target everyone tries to hit the bull's eye (region r) and does not shoot completely at random. Second, this is also a circular and therefore inconsistent definition because the expression at random can only be defined using the notion of probability. Moreover, even if we accept the notion "at random" as "intuitively clear," we shall still run into many discrepancies and contradictions because human common sense intuition is in this respect exceedingly poor . This can be confirmed by the following example, known as *Bertrand's paradox.*

BERTRAND'S PARADOX

Consider a circle with a unit radius. What is the probability P that a random chord of the circle is longer than the side of an equilateral triangle inscribed in the circle (which is equal to $\sqrt{3}$)?

SOLUTION 1. Because of symmetry, we may specify the direction of the chord in advance. If we construct the diameter perpendicular to this direction, then only the chords which intersect this diameter from one quarter to three quarters of its length will have the required property. Therefore $P = \frac{1}{2}$

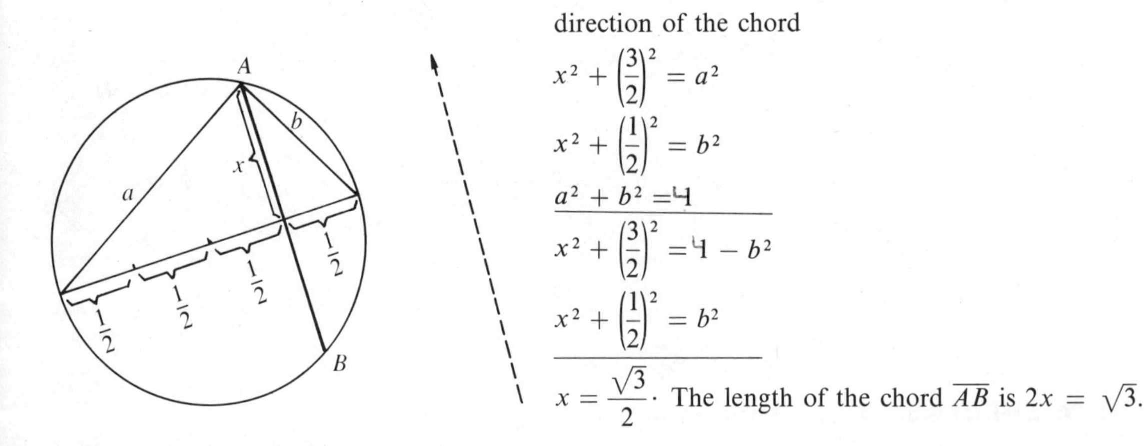

SOLUTION 2. If we fix the end point A of the chord, then we get the following:

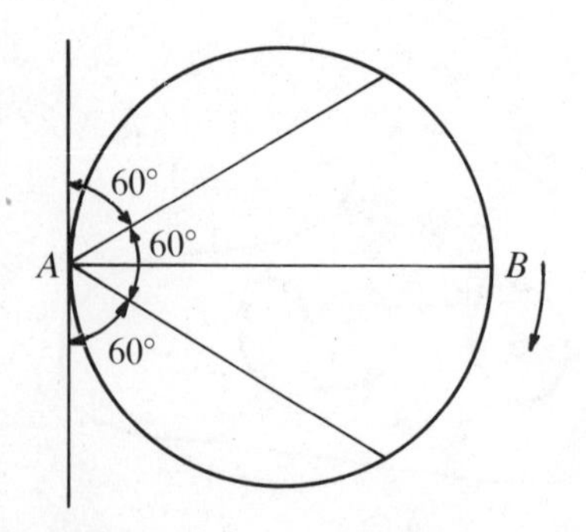

If we let the end point B rotate on the circumference, we get all the chords. Obviously, only the chords passing through the middle 60° angle have the required property, and therefore $P = \frac{1}{3}$.

SOLUTION 3. In order to specify a chord, it is sufficient to give its midpoint M. Obviously only those chords that have the midpoint M within the concentric circle with the radius $\frac{1}{2}$ have the required property, as follows from the calculations given in Solution 1.

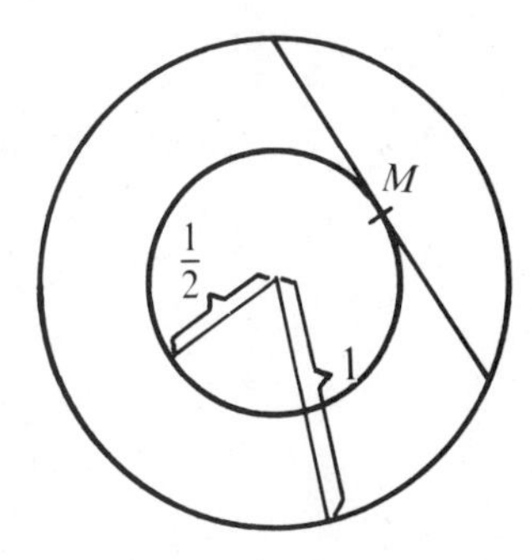

Therefore
$$P = \frac{\pi\left(\frac{1}{2}r\right)^2}{\pi r^2} = \frac{1}{4}.$$

We have obtained three different answers to the same question and all of them look equally trustworthy. The main problem is that we interpreted the notion of a chord chosen at random differently in each case, and none of these interpretations can be claimed to be better than the remaining two. The problem of nebulous and confusing interpretations of the notion "at random" arises very often when we try to apply the geometrical definition, and it is not always so easy to discover discrepancies as in the case of Bertrand's paradox. In many cases it is just about impossible to clarify what is meant by "at random." And even if we were willing to accept the circular and logically inconsistent definition that at random means that "each point has the same probability," we would still be in trouble for there are always infinitely many points in a region, and if they are all to have the same probability, then the probability of each of them must be zero. However, the probability of the whole region consisting of these points is not zero. There are also other objections against the geometrical definition of probability.

In order to calculate a probability, we have to determine the "area of a region." We can do it easily for simple geometrical figures (square, rectangle, circle) but how complicated a region can we admit and how can we measure it? An answer to this question is exceedingly difficult and requires a knowledge of measure theory, which is too "big gun" for our simple problems. The situation stands, therefore, exactly as in the case of the classical definition: accepting the geometrical definition, we can derive the fundamental statements A1–A3 and many others. If we should abandon the geometrical definition because of its inconsistencies, we can still accept A1–A3 as "the Commandments," and then derive additional statements from them.

FREQUENCY DEFINITION

The last commonly used definition of probability, the frequency definition, is based on statistical observations of sequences of trials. Let us assume that we are given an experiment, and that we perform n successive trials under identical conditions. In order to find the probability of an event A, let us assume that this event has occurred in m_A realizations (trials) and let us investigate the *relative frequency.*

$$\frac{m_A}{n}$$

The number that the relative frequency $\frac{m_A}{n}$ approaches as n becomes very large is the probability $P(A)$ of the event A.

For example, while tossing a coin, the following results were obtained:

Number of tosses of a coin	Number of heads	Relative frequency
10	7	.70
50	28	.56
100	54	.54
200	106	.53
1,000	493	.493
10,000	4981	.498

Therefore, we would say that the probability of getting a head is .5 since this appears to be the number that the relative frequencies are approaching. In the case of the frequency definition also, we can prove the fundamental properties A1–A3.

→ **STATEMENT A1:** For any event A

$$P(A) \geq 0.$$

Proof: Obviously, $m_A \geq 0$, $n > 0$ and hence $P(A)$ as a "limit" will also be nonnegative. q.e.d.

→ **STATEMENT A2:** $P(\Omega) = 1.$

Proof: Each time we perform the experiment, the event Ω occurs. Hence the certain event Ω occurs n times in any of n trials. Thus the relative frequency of the certain event is always 1 and hence $P(\Omega) = 1$. q.e.d.

→ **STATEMENT A3:** If A and B are mutually exclusive events, then

$$P(A \cup B) = P(A) + P(B).$$

Proof: If A, B are mutually exclusive events, then

$$m_{A \cup B} = m_A + m_B.$$

Hence

$$\frac{m_{A \cup B}}{n} = \frac{m_A}{n} + \frac{m_B}{n}$$

and also for probabilities

$$P(A \cup B) = P(A) + P(B).$$ q.e.d.

Even though the frequency definition is widely accepted, we can encounter many serious objections against it. First of all, we have to assume that we can perform an unlimited number of trials—otherwise we cannot speak about the "limit" to which $\frac{m_A}{n}$ tends. It is not enough to perform a "very large" number of trials to estimate the probability, for one has to ask immediately: How many trials? Will a hundred do, or perhaps a thousand or a million? Imagine a Martian who has never seen a silver dollar. He likes the coin very much and wants to find the probability of getting a head. He is given one coin and starts to perform the experiment of tossing "many times" according to the frequency definition. He decides to flip it 10,000 times and gets exactly 5000 heads. He feels that he has performed "sufficiently many" trials and concludes, according to the frequency definition, that the probability of getting a head is $\frac{1}{2}$. However he did not notice that he was given a supersophisticated electronic fake coin with a built-in computer brain. This coin was programed to behave like an ordinary coin for the first 10,000 tosses, and then to shift its center of gravity in favor of heads to give a head in nine out of ten tosses. (As a matter of fact, one can achieve this behavior using different settings on the Variable Probability Coin Tossing Machine mentioned on page 17). Now the Martian can toss a million or a billion times and get mostly heads. However, if he stopped after 10,000 tosses, he would never find the truth about his silver dollar. As a matter of fact, even if he decides to toss a million times or more, the coin could have been programed always to cheat on him. The only way to find the truth is to keep on tossing forever and this is very impractical, to say the least. One can never conclude anything about a "limit" of a sequence of observations unless one has the whole sequence available—but this can never happen.

It is of course possible to refrain from such perverted examples and assume that the sequence of observations possesses some intrinsic property of *statistical stability* that allows us to extrapolate its behavior from only a finite number of trials. This point of view is quite common among many experimental scientists in biology, physics, chemistry, the social sciences, and so on. This method is really fruitful when absolutely nothing is known about the statistical nature of the observed phenomena. Then sufficiently extensive statistical data can be used, together with the frequency definition of probability, in order to obtain at least some information with some degree of confidence. However, we have to consider this approach as a last ditch attempt to save whatever can be saved, regardless of its actual logical and cognitive value. For as soon as we start to analyze these attempts logically, we run into many contradictions of a technical and conceptual character.

For example, it is well known that if we perform many tosses of a perfectly fair coin (say a thousand times), there is a certain positive probability that all of them will turn up tails even though this probability is very small. In order to exclude this maverick behavior, we have to declare it practically impossible, even though it can happen. Another problem is with sequences that consistently alternate heads and tails. It can certainly happen, but should we speak in this case about probability when the sequence of trials does not show any sign of randomness? And how do we define sequences that are random enough to speak about probabilities? What about sequences that are "excessively random," and refuse to comply with any requirement of statistical stability? Even this can happen and should not be excluded.

There are many other objections of a more subtle nature; whole books have been written about this subject. The frequency definition has been expanded into a very sophisticated, elaborate and complicated theory that is not very comprehensible, but still cannot claim to be free of defects and shortcomings. For this reason, we will reject the frequency definition of probability, even as we keep in mind some of its merits.

Let us review the situation. In each of the definitions we have discussed, an attempt was made to assign probabilities to certain events. The rules for the calculation of probabilities, namely A1–A3, resulted as a by-product but seemed completely reasonable and natural. On the other hand, the definitions themselves were very objectionable. We now come to the main question of this chapter: What would happen if we abandoned the inconsistent definitions and simply said that *probability is anything that satisfies the fundamental rules* A1–A3? We could not specify probability *numerically,* nor could we prove the rules A1–A3 but then, these rules are so fundamental that nobody would doubt them anyway. On the other hand, we would still be able *to study* probabilities, using the rules A1–A3 and their consequences. Of course, S1 and S2 would still be valid, and, as we shall see in the next chapters, many other rules as well.

This situation occurs very frequently in modern mathematics. Instead of defining the primitive notions, we simply give some fundamental rules governing their behavior, and accept these rules as *axioms.* The entire theory pertaining to the fundamental notions is then derived from the axioms, using logical rules. In such a formal theory, everything must be proved with the exception of the axioms. Historically, the first axiomatic system was developed for geometry by *Euclid* around 300 B.C. Instead of defining points, lines and other primitive notions, he gave axioms which these notions must satisfy. For example, one axiom was that through any two distinct points, there passes a unique line. Then geometry was developed from these axioms. Today, the axiomatic method is used in many areas: in algebra, set theory and theoretical physics, to name just a few. The most famous axiomatic system for probability theory was given by *A. N. Kolmogoroff* in 1933, and our rules A1–A3 are its simplified version.

In the next chapter, the fundamental rules A1–A3 will be taken as axioms, and our main task will be to derive as many conclusions from them as possible. In this way, we will get a complete theory or model of probability based on these axioms and on principles of logical deductive reasoning.

PROBABILITY MEASURES

Man cannot divine what end followeth beginning,
the nearest is a liklyhood.

Sir W. Cornwallis, 1601

In the previous chapter, we gave some plausible heuristic reasons why we consider the fundamental rules A1 to A3 to have such great importance in the theory of probability. From now on, we shall take the point of view that probability must satisfy these rules; we shall accept them as the *axioms* of our theory. Thus these "Commandments" will be laid down as a foundation, and we shall logically derive from them many interesting statements and important theorems. In order to distinguish our approach from the ordinary everyday use of the word *probability* we shall speak about a *probability measure.*

DEFINITION 1: Let Ω be a finite sample space. A real-valued function P defined for all subsets (events) of Ω is called a probability measure if the following axioms are satisfied:

A1 : For every $A \subset \Omega$, $P(A) \geq 0$.

A2 : $P(\Omega) = 1$.

A3 : For any two disjoint sets $A \subset \Omega$, $B \subset \Omega$,

$$P(A \cup B) = P(A) + P(B).$$

The couple (Ω,P) is called a *probability measure space.*

EXAMPLE 1

Let $\Omega = \{H,T\}$ and

$$P(\phi) = 0, \quad P(\{H\}) = P(\{T\}) = \frac{1}{2}, \quad P(\Omega) = 1.$$

Then the probability measure space (Ω,H) represents an experiment consisting of a single toss of a fair coin (where H means a head and T is a tail).

EXAMPLE 2

Imagine that you decide to undertake a certain action only if you obtain a heart in the draw of a single card from a complete deck of ordinary playing cards. Denote the outcome that you get a heart by H and let T mean that you get a card from the three remaining suits. Then a corresponding sample space representing the draw is

$$\Omega = \{H,T\}$$

and the probability measure is

$$P(\{\phi\}) = 0, \quad P(\{H\}) = \frac{1}{4}, \quad P(\{T\}) = \frac{3}{4}, \quad P(\{\Omega\}) = 1.$$

Compare the sample spaces and the measures in these two examples.

EXAMPLE 3

Let $\Omega = \{1,2,3\}$.

Then
$$P_1(\phi) = 0, \quad P_1(\{1\}) = P_1(\{2\}) = P_1(\{3\}) = \frac{1}{3},$$
$$P_1(\{1,2\}) = P_1(\{1,3\}) = P_1(\{2,3\}) = \frac{2}{3}, \quad P_1(\Omega) = 1$$

is a probability measure on Ω because the axioms A_1–A_3 are satisfied (verify it).

Further
$$P_2(\phi) = 0, \quad P_2(\{1\}) = P_2(\{2\}) = \frac{1}{4}, \quad P_2(\{3\}) = \frac{1}{2},$$
$$P_2(\{1,2\}) = \frac{1}{2}, \quad P_2(\{1,3\}) = P_2(\{2,3\}) = \frac{3}{4}, \quad P(\Omega) = 1.$$

Then also P_2 is a probability measure on Ω and $P_1 \neq P_2$.

These examples show that we can have different probability measures on the same sample space. Our axiomatic definition *does not specify the exact numerical values* of a probability measure; it only stipulates its fundamental properties. At this stage we are interested not so much in how the numerical values of probability are assigned, but in what properties they must have and in how we can manipulate them in accordance with the logical reason. For an actual assignment we still must rely often on one of the inconsistent methods mentioned in the previous chapter. From our definition it follows that probability measure is a *nonnegative* (A1) and *normed* (A2) set function defined for all subsets of Ω. The property A3 is sometimes called *additivity* of probability.

→ **THEOREM 1:** For any set $A \subset \Omega$,

$$P(A) + P(A') = 1.$$

Proof: We know that $A \cup A' = \Omega$ and the sets A, A' are disjoint,

$$A \cap A' = \phi.$$

Using the axiom A3: $\quad P(A) + P(A') = P(A \cup A') = P(\Omega).$

From the axiom A2 $\quad P(\Omega) = 1$

and therefore $\quad P(A) + P(A') = 1.$ q.e.d.

COMMENT 1. This is a very useful trick. Instead of calculating $P(A)$, one can calculate $P(A')$ (that is sometimes easier) and use Theorem 1 to get $P(A)$.

COMMENT 2. Take Example 3. If $A = \{1\}$, then $A' = \{2,3\}$. $P_1(A) = \frac{1}{3}$ and $P_1(A')$

$= \frac{2}{3}$; therefore $P_1(A) + P_1(A') = 1$. Verify similarly all the following theorems using Examples 1–3 and a suitable choice of sets.

THEOREM 2: For the empty set Φ, we have

$$P(\Phi) = 0.$$

Proof: We know that $\Phi' = \Omega$.

Hence, by Theorem 1, $P(\Phi) + P(\Phi') = P(\Phi) + P(\Omega) = 1.$

But by A2 we have $P(\Omega) = 1$

hence $P(\Phi) + 1 = 1.$

Finally $P(\Phi) = 0.$ q.e.d.

Theorem 2 states that if $A = \Phi$ then $P(A) = 0$. However, it need not be true that $P(A) = 0$ always implies $A = \Phi$. Consider for example the gambler entertaining his guests with three dice: a cube, a tetrahedron and a dodecahedron (see page 21). The corresponding sample space is $\Omega = \{1,2,3,4,5,6,7,8,9,10,11,12\}$. If the gambler selects the ordinary six-faced die for a toss, then evidently for

$$A = \{7,8,9,10\} \subset \Omega$$

we get

$$P(A) = 0 \quad \text{but} \quad A \neq \Phi.$$

DEFINITION 2: Any set $A \subset \Omega$ for which $P(A) = 0$ is called a null set.

COMMENT. Theorem 2 simply says that the empty set is a null set. In the case of finite probability spaces, the concept of a null set can be completely avoided. However, this interesting notion is so important in advanced probability theory that it might be of some advantage to mention it here. Properties of null sets are investigated in the Exercises at the end of this chapter (see Exercises 16 and 17).

THEOREM 3: For any two sets $A \subset \Omega$, $B \subset \Omega$

$$P(A - B) = P(A) - P(A \cap B).$$

Proof: We can write

$$A = (A - B) \cup (A \cap B)$$

and both sets on the right hand side are disjoint. Using A3, we get

$$P(A) = P[(A - B) \cup (A \cap B)] = P(A - B) + P(A \cap B).$$

And therefore $P(A - B) = P(A) - P(A \cap B).$ q.e.d.

COROLLARY. If $A \supset B$ then $A \cap B = B$, hence

$$P(A - B) = P(A) - P(B).$$

This property of probability measures is called *subtractivity.*

→ **THEOREM 4:** For any two sets $A \subset B \subset \Omega$ we have $P(A) \leq P(B)$.

Proof: If $A \subset B$, then by the corollary to Theorem 3

$$P(B - A) = P(B) - P(A).$$

By A1 $$P(B - A) \geq 0$$

hence $$P(B) - P(A) \geq 0$$

or $$P(A) \leq P(B).$$ q.e.d.

COMMENT. Theorem 4 says that a probability measure is a *monotonic nondecreasing function* of a set variable.

→ **THEOREM 5:** For any set $A \subset \Omega$,

$$0 \leq P(A) \leq 1.$$

Proof: For any set A, $\Phi \subset A \subset \Omega$. Using Theorem 4 twice, we get

$$0 = P(\Phi) \leq P(A) \leq P(\Omega) = 1.$$ q.e.d.

COMMENT. Theorem 5 gives a simple check for calculation of any probability. If one gets a probability less than 0 or greater than 1, it simply indicates an error.

→ **THEOREM 6:** For any class of pairwise disjoint sets $A_1, A_2, \ldots, A_k \subset \Omega$, i.e., $A_i \cap A_j = \Phi$ for $i \neq j$,

we have $$P\left(\bigcup_{i=1}^{k} A_i\right) = \sum_{i=1}^{k} P(A_i).$$

Proof: We can prove the statement by induction on k. For $k = 2$ the statement is equivalent to the axiom A3. Let us assume that the statement is true for any $2 \leq i \leq k$ and let us have $(k + 1)$ pairwise disjoint sets $(A_1, A_2, \ldots, A_k, A_{k+1})$. Then

$$\bigcup_{i=1}^{k+1} A_i = \bigcup_{i=1}^{k} A_i \cup A_{k+1}.$$

Since the sets A_i, $i = 1,2, \ldots, (k + 1)$ are pairwise disjoint, so are the sets A_i, $i = 1,2, \ldots, k$ and also the sets $\bigcup_{i=1}^{k} A_i$ and A_{k+1}. Using A3, we get

$$P\left(\bigcup_{i=1}^{k+1} A_i\right) = P\left(\bigcup_{i=1}^{k} A_i \cup A_{k+1}\right) = P\left(\bigcup_{i=1}^{k} A_i\right) + P(A_{k+1}).$$

By the induction assumption

$$P\left(\bigcup_{i=1}^{k} A_i\right) = \sum_{i=1}^{k} P(A_i)$$

hence $$P\left(\bigcup_{i=1}^{k} A_i\right) = \sum_{i=1}^{k} P(A_i) + P(A_{k+1}) = \sum_{i=1}^{k+1} P(A_i).$$ q.e.d

→ **THEOREM 7:** For any two sets $A \subset \Omega$, $B \subset \Omega$, we have

$$P(A \cup B) = P(A) + P(B) - P(A \cap B).$$

Proof: Obviously $A \cup B = A \cup (B - A)$. The sets A and $(B - A)$ are disjoint and $A \cap B \subset B$.

Using A3, $$P(A \cup B) = P(A) + P(B - A)$$

and using Theorem 3, $$P(B - A) = P(B) - P(A \cap B).$$

$$P(A \cup B) = P(A) + P(B) - P(A \cap B). \qquad \text{q.e.d.}$$

COMMENT 1. Theorem 7 gives a method for calculating the probability of the union of *any* two sets. In the case of disjoint sets, Theorem 7 reduces to A3. In this case $A \cap B = \phi$, hence $P(A \cap B) = 0$ and therefore,

$$P(A \cup B) = P(A) + P(B) - P(A \cap B) = P(A) + P(B).$$

COMMENT 2. We could generalize Theorem 7 for the case of three, four or more sets. In the case of three sets A, B, C, we would get

$$P(A \cup B \cup C) = P(A) + P(B) + P(C) - P(A \cap B) - P(A \cap C) - P(B \cap C) + P(A \cap B \cap C).$$

A proof of this statement is very similar to that of Theorem 7 and is left to the reader, as well as the generalization of this formula for the case of k arbitrary sets (see Exercise 14).

Our axiomatic system looks really very general, abstract, sophisticated and perhaps even a little bit startling or discouraging. However, we should not forget that the sample space Ω is *finite.* This fact actually makes everything very simple. Let Ω have n points

$$\Omega = \{\omega_1, \omega_2, \ldots, \omega_n\}.$$

The probability P must be defined for all the single point sets $\{\omega_i\} \subset \Omega$. Let us denote these probabilities by the numbers p_i, i.e.,

$$P(\{\omega_i\}) = p_i, \qquad i = 1,2, \ldots, n.$$

Once we know all the numbers $p_1, p_2, \ldots, p_n$, we can calculate very easily the probability of any set $A \subset \Omega$, using Theorem 6.

Let $$A = \{\omega_i,\omega_j,\omega_k, \ldots, \omega_m\} \subset \Omega$$

where $\omega_i,\omega_j,\omega_k, \ldots, \omega_m$ are distinct elements of Ω. Then obviously

$$A = \{\omega_i\} \cup \{\omega_j\} \cup \{\omega_k\} \cup \ldots \cup \{\omega_m\}$$

that is, A is the union of all the single point sets formed from the elements of A. These single point sets are pairwise disjoint, hence

$$\begin{aligned} P(A) &= P(\{\omega_i\} \cup \{\omega_j\} \cup \{\omega_k\} \cup \ldots \cup \{\omega_m\}) \\ &= P(\{\omega_i\}) + P(\{\omega_j\}) + P(\{\omega_k\}) + \ldots + P(\{\omega_m\}) \\ &= p_i + p_j + p_k + \ldots + p_m. \end{aligned}$$

If, for example, $A = \{\omega_1, \omega_{23}, \omega_{203}\}$, then

$$P(A) = P(\{\omega_1\}) + P(\{\omega_{23}\}) + P(\{\omega_{203}\}) = p_1 + p_{23} + p_{203}.$$

Therefore the probability of the set A is the sum of those numbers p_i which correspond to the points of A. It is obvious that the p_i's satisfy the following conditions:

(a) $$p_i \geq 0 \qquad i = 1,2, \ldots ,n$$

(b) $$\sum_{i=1}^{n} p_i = 1$$

It is also obvious that if we have a sample space $\Omega = \{\omega_1, \omega_2, \ldots ,\omega_n\}$, then any set of real numbers p_i $i = 1,2, \ldots ,n$ satisfying (a) and (b) will define a probability measure on the subsets of Ω. It is enough to put

$$P(\{\omega_i\}) = p_i \qquad i = 1,2, \ldots ,n.$$

Let us have a sample space of the outcomes of a single toss of a die, i.e., $\Omega = \{1,2,3,4,5,6\}$. Then we can define a probability measure $P^{(1)}$ by setting

$$p_1 = p_2 = p_3 = p_4 = p_5 = p_6 = \frac{1}{6}\cdot$$

If $A = \{x : x \text{ is an even outcome}\}$, then $P^{(1)}(A) = P^{(1)}(\{2,4,6\}) = p_2 + p_4 + p_6 = \frac{1}{6} + \frac{1}{6} + \frac{1}{6}$ $= \frac{1}{2}\cdot$ However we can define another probability measure $P^{(2)}$ by setting

$$p_1 = p_3 = \frac{1}{12}$$

$$p_2 = p_4 = \frac{2}{12}$$

$$p_5 = p_6 = \frac{3}{12}$$

Then $P^{(2)}(A) = P^{(2)}(\{2,4,6\}) = p_2 + p_4 + p_6 = \frac{2}{12} + \frac{2}{12} + \frac{3}{12} = \frac{7}{12} \neq \frac{1}{2} = P^{(1)}(A)$. The probability $P^{(1)}$ would correspond to a fair die while the probability $P^{(2)}$ would correspond to a loaded die biased toward large numbers.

The previous example of $P^{(1)}$ is very important. In the general case it may happen that

$$p_1 = p_2 = p_3 = \ldots = p_n = \frac{1}{n}\cdot$$

Such cases deserve special attention.

DEFINITION 3: Let P be a probability measure on the subsets of a sample space $\Omega = \{\omega_1, \omega_2, \ldots ,\omega_n\}$. If for every two points $\omega_i, \omega_j \in \Omega$

$$P(\{\omega_i\}) = P(\{\omega_j\}) \qquad i,j = 1,2, \ldots ,n$$

then P is called the equally likely probability measure.

THEOREM 8: Let P be the equally likely probability measure on the subsets of $\Omega = \{\omega_1, \omega_2, \ldots, \omega_n\}$. Then for every $i = 1, 2, \ldots, n$

$$p_i = P(\{\omega_i\}) = \frac{1}{n}$$

and for every $A \subset \Omega$ if A has m points, then

$$P(A) = \frac{m}{n}.$$

Proof: We know that

$$0 \leqq p_i \leqq 1 \qquad i = 1, 2, \ldots, n$$

$$\sum_{i=1}^{n} p_i = 1, \quad p_i = p_j \quad i, j = 1,2, \ldots, n$$

The only way to satisfy these conditions is that

$$p_i = \frac{1}{n} \qquad i = 1,2, \ldots, n.$$

Assume that A had m points. The probability of A is equal to the sum of probabilities of all the points in A, but each of them has probability $\frac{1}{n}$. Hence

$$P(A) = \frac{m}{n}. \qquad \text{q.e.d.}$$

Let us mention here just one more point. We started with a very sophisticated axiomatic definition, but very soon we showed that the axiomatic system in the case of a finite sample space $\Omega = \{\omega_1, \omega_2, \ldots, \omega_n\}$ is actually very simple and shrinks to knowledge of n numbers, $p_1, p_2, \ldots, p_n$ satisfying (a) and (b). However, the main advantage of our axiomatic system is that it can be very easily generalized to any *abstract probability space* by a small change in axiom A3 and a different assumption about the class of all events. These changes represent analytical difficulties rather than logical ones, and the whole structure of properties of probability measures as expressed in Theorems 1 through 7 remains unchanged. Knowledge of our simple axiomatic system actually enables an interested student to penetrate immediately into the more general system of abstract probability spaces.

EXAMPLE 4

A number is selected at random from the first 100 positive integers. What is the probability that it is divisible by either 9 or 12?

SOLUTION. A sample space has 100 points $1,2, \ldots, 100$. The expression "selected at random" means that each number has the same probability $\frac{1}{100}$ to be chosen. If A means that the selected number is divisible by 9 and B means that it is divisible by 12, then

$$A = \{9, 18, 27, 36, 45, 54, 63, 72, 81, 90, 99\}$$

$$B = \{12, 24, 36, 48, 60, 72, 84, 96\}$$

$$A \cap B = \{36, 72\}$$

Hence $P(A) = \frac{11}{100}$, $P(B) = \frac{8}{100}$, $P(A \cap B) = \frac{2}{100}$.

We are interested in the probability $P(A \cup B)$ that the chosen number is divisible either by 9 or by 12. Using Theorem 7, we get

$$P(A \cup B) = P(A) + P(B) - P(A \cap B) = \frac{11}{100} + \frac{8}{100} - \frac{2}{100} = \frac{17}{100}.$$

EXAMPLE 5

In a class of 10 men and 20 women, half the men and half the women have brown eyes. Find the probability that a person chosen at random is a man or has brown eyes. (By the expression "at random," we mean that any person has the same chance to be selected.)

SOLUTION. Let A denote the event "chosen person is a man" and let B denote the event "chosen person has brown eyes." Then we seek $P(A \cup B)$, which we can find as

$$P(A \cup B) = P(A) + P(B) - P(A \cap B).$$

There are 30 people, of whom 10 are men, hence $P(A) = \frac{10}{30} = \frac{1}{3}$. There are 15 people with brown eyes, hence $P(B) = \frac{15}{30} = \frac{1}{2}$. There are 5 men with brown eyes, hence $P(A \cap B) = \frac{5}{30} = \frac{1}{6}$. Thus

$$P(A \cup B) = \frac{1}{3} + \frac{1}{2} - \frac{1}{6} = \frac{2}{3}.$$

EXAMPLE 6

A gambler "loads" a six-faced die in such a way that the probability of turning up a number of points is directly proportional to the number of points. Find the probability of
(a) getting a six with this die.
(b) getting an odd number of points.

SOLUTION. Let p_i be the probability of getting i points.

If $p_1 = k$ then $$p_2 = 2k, \quad p_3 = 3k, \quad . \ . \ . \ , \quad p_6 = 6k.$$

Therefore $$\sum_{i=1}^{6} p_i = k(1 + 2 + 3 + 4 + 5 + 6) = 21k = 1.$$

We have $$k = \frac{1}{21} \text{ and}$$

(a) the probability of getting a six is $p_6 = 6k = \frac{6}{21}$.
(b) the probability of getting an odd number of points is $p_1 + p_3 + p_5 = k(1 + 3 + 5) = \frac{9}{21}$.

This is an example of a probability measure that is not an equally likely measure.

EXAMPLE 7

A customer in a corner grocery store has two one-dollar bills in his pocket; one bill is counterfeit. He pays for his purchase by randomly selecting one bill, which is thrown into the grocer's money box. We know that the box previously contained three genuine one-dollar bills. The next customer pays with a five dollar bill and has to get two one-dollar bills for change. What is the probability that he will get the counterfeit bill? (Assume that the grocer picks the bills at random.)

SOLUTION. In order to solve this problem with our limited means, we have to set up the sample space very carefully. Denote the good bill in the pocket of the first customer by G and the counterfeit bad bill by B. Denote further the three genuine one-dollar bills in the grocer's box before the purchases by D_1, D_2, D_3, and denote the bill obtained from the first customer by D. Then an outcome of the purchases can be expressed as an ordered triple. For example, one possible outcome is

$$(B, D_1, D)$$

where the first B means that the first customer pays by a counterfeit bill and the second customer gets the counterfeit bill and D_1 for his change. Therefore

$$\Omega = \{(x,y,z) : x \in \{B,G\}\ y,z \in \{D_1,D_2,D_3,D\}, y \neq z\}$$

where x is the bill paid by the first customer and y,z is the change for the second. If we list all the possibilities, we can see that Ω has 24 points that have to be considered equally likely because all the choices are taken at random. The second customer gets the counterfeit bill in the following cases:

$$\{(B,D,D_1),\ (B,D,D_2),\ (B,D,D_3),\ (B,D_1,D),\ (B,D_2,D),\ (B,D_3,D)\}.$$

Therefore the required probability is $\dfrac{6}{24} = \dfrac{1}{4}\cdot$

EXAMPLE 8

What is the probability that a five card hand for poker, dealt from a well shuffled deck, will contain an ace?

SOLUTION. The sample space in this case will consist of all possible quintuples of cards. It was hard to list the sample space in the previous example but in this instance it is virtually impossible. As a matter of fact, we do not need such a list. Since we are assuming that the deck has been well shuffled, it is reasonable to infer that all the hands are therefore equally likely. Thus all we need is the number of points in the sample space and the number of hands containing an ace. There are certain techniques for counting these points and we shall devote the next two chapters to them. We shall also postpone all further examples until then.

EXERCISES

(Answers in Appendix to exercises marked)*

1. Determine which of the following are probability measures on the events of

$$\Omega = \{\omega_1, \omega_2, \omega_3\}. \qquad \text{(Assume } a > 0,\ b > 0,\ c > 0.\text{)}$$

	ϕ	$\{\omega_1\}$	$\{\omega_2\}$	$\{\omega_3\}$	$\{\omega_1,\omega_2\}$	$\{\omega_1,\omega_3\}$	$\{\omega_2,\omega_3\}$	Ω
P_1	0	$\frac{1}{3}$	$\frac{1}{3}$	$\frac{1}{3}$	$\frac{2}{3}$	$\frac{2}{3}$	$\frac{2}{3}$	1
P_2	0	0	$\frac{3}{5}$	$\frac{2}{5}$	$\frac{3}{5}$	$\frac{2}{5}$	1	1
P_3	$\frac{1}{4}$	$\frac{1}{4}$	$\frac{1}{4}$	$\frac{1}{2}$	$\frac{1}{2}$	$\frac{3}{4}$	$\frac{3}{4}$	1
P_4	0	$\frac{1}{3}$	$-\frac{1}{3}$	$\frac{1}{3}$	$\frac{2}{3}$	$\frac{2}{3}$	$\frac{2}{3}$	1
P_5	0	$\frac{1}{2}$	$\frac{1}{3}$	$\frac{2}{12}$	$\frac{5}{6}$	$\frac{4}{6}$	$\frac{4}{6}$	1
P_6	0	$\frac{a}{a+b+c}$	$\frac{b}{a+b+c}$	$\frac{c}{a+b+c}$	$\frac{a+b}{a+b+c}$	$\frac{a+c}{a+b+c}$	$\frac{b+c}{a+b+c}$	1
P_7	0	$\frac{a+b}{2(a+b+c)}$	$\frac{b+c}{2(a+b+c)}$	$\frac{a+c}{2(a+b+c)}$	$\frac{a+2b+c}{2(a+b+c)}$	$\frac{2a+b+c}{2(a+b+c)}$	$\frac{a+b+2c}{2(a+b+c)}$	1

2. Let P be a probability measure on the events of

$$\Omega = \{\omega_1, \omega_2, \omega_3, \omega_4\}.$$

*(a) Find the probability of $\{\omega_1\}$ and $\{\omega_3\}$ if

$$P\{\omega_1,\omega_2,\omega_3\} = .6,$$

$$P\{\omega_1,\omega_3,\omega_4\} = .8,$$

$$P\{\omega_2\omega_3\} = .5.$$

*(b) Find $P\{\omega_1\}$, $P\{\omega_2\}$, $P\{\omega_3\}$, $P\{\omega_4\}$ if

$$P\{\omega_1,\omega_2,\omega_4\} = \frac{6}{7},$$

$$P\{\omega_1,\omega_3\} = \frac{4}{7},$$

$$P\{\omega_1, \omega_4\} = \frac{4}{7}.$$

(c) Find $P\{\omega_1\}, P\{\omega_2\}, P\{\omega_3\}, P\{\omega_4\}$ if

$$P\{\omega_1,\omega_2\} = \frac{5}{6},$$

$$P\{\omega_2,\omega_3\} = \frac{1}{2},$$

$$P\{\omega_2, \omega_4\} = \frac{1}{3}.$$

3. Find $P(A' \cap B)$, $P(A \cap B')$, $P(A \cup B)$, $P(A' \cap B')$ if

*(a) $P(A) = \frac{1}{2}$,

$$P(B) = \frac{1}{3},$$

$$P(A \cap B) = \frac{1}{6}.$$

(b) $P(A) = \frac{3}{7}$,

$$P(B) = \frac{3}{7},$$

$$P(A \cap B) = 0.$$

4. (a) Find $P(A \cap B)$ if
$P(A) = .4, \quad P(A \cup B) = .7, \quad P(B') = .4.$

*(b) Find $P(A \cap C)$ if
$P(A' \cap B' \cap C') = .1,$
$P(A \,\Delta\, B \,\Delta\, C) = .6,$
$P(B \cap (A \cup C)) = .3,$
$P(A \cap B \cap C) = .1.$
(For the definition of Δ see Chapter II, Example 9.)

5. Let (Ω,P) be a probability measure space, $A \subset \Omega$, $B \subset \Omega$ such that $P(A) = \frac{5}{6}$ and $P(B) = \frac{2}{3}$. Show that

(a) $P(A \cup B) \geq \frac{5}{6}$.

(b) $\frac{1}{2} \leq P(A \cap B) \leq \frac{2}{3}$.

(c) $\frac{1}{6} \leq P(A \cap B') \leq \frac{1}{3}$.

6. Let $P(A) = \frac{2}{3}$, $P(B) = \frac{1}{4}$. Show that

(a) $P(A \cup B) \geq \frac{2}{3}$.

(b) $P(A \cap B) \leq \frac{1}{4}$.

(c) $\frac{5}{12} \leq P(A \cap B') \leq \frac{2}{3}$.

*7. An urn contains two red balls, one black ball and three balls painted a combination of black and red. Two balls are drawn simultaneously at random from the urn. Denote the events as follows:

A: At least one ball contains black.
B: Both balls contain red.

Set up a sample space, express A, B and $A \cap B$ and find their probabilities.

8. Tim enjoys his probability classes—perhaps because his classmates in this course are mainly beautiful girls. As a matter of fact, out of 200 students in the class, 90% are girls and Tim noticed that two thirds of them are quite tall. He is also puzzled by the fact that while only every third tall girl wears a miniskirt, fully 50% of the short girls wear miniskirts. Today Tim decided that regardless of all his prejudices he will select a girl at random and ask her for a date. Denote the events as follows:

A: The selected girl is tall.

B: The selected girl is short.

C: The selected girl wears a miniskirt.

Describe in words the following events and find their probabilities:
(a) $A \cap C'$ (b) $B \cup C$ (c) $C - B$
(d) $(A - C) \cup (B \cap C')$

9. Assume that the coin change after a purchase is equally likely to be 0, 1, 2, . . . , 99 cents. The change is made first with as many quarters as possible, then with as many dimes as possible, then with as many nickels as possible, then pennies. Find the probability that among the change a customer will get
   *(a) Exactly two quarters.
   *(b) Exactly two quarters and exactly two dimes.
   *(c) Exactly two dimes.
   *(d) Exactly three pennies and exactly one nickel.
   (e) At least one quarter.
   (f) At least two pennies.
   (g) No dimes.

10. Three men and two women are in a chess tournament. Those of the same sex have equal probabilities of winning, but each woman is twice as likely to win as any man. Find the probability that a woman wins the tournament.

*11. Three horses, A, B and C, are in a race. A is twice as likely to win as B and B is thrice as likely to win as C. What are their respective probabilities of winning?

12. A card is selected from a well shuffled deck of bridge cards. Find the probability that it is
   (a) an ace.
   (b) a heart.
   (c) a card belonging to the two black suits.
   (d) an even numbered card.
   (e) a prime numbered spade card.
   (f) either a spade or an ace.

13. Prove that $P(A \,\Delta\, B) = P(A) + P(B) - 2P(A \cap B)$. (For the definition of Δ, see Chapter II, Example 9.)

14. Prove
   (a) $P(A \cup B \cup C) = P(A) + P(B) + P(C) - P(A \cap B) - P(A \cap C) - P(B \cap C) + P(A \cap B \cap C)$.
   (b) $P(A \cup B \cup C \cup D) = P(A) + P(B) + P(C) + P(D) - P(A \cap B) - P(A \cap C) - P(A \cap D) - P(B \cap C) - P(B \cap D) - P(C \cap D) + P(A \cap B \cap C) + P(A \cap C \cap D) + P(A \cap B \cap D) + P(B \cap C \cap D) - P(A \cap B \cap C \cap D)$.
   (c) In general $P\left(\bigcup_{i=1}^{n} A_i\right) = \sum_{i=1}^{n} P(A_i)$

$$- \sum_{i \neq j} P(A_i \cap A_j)$$

$$+ \sum_{i \neq j \neq k} P(A_i \cap A_j \cap A_k)$$

$$- \dots (-1)^{n+1} P\left(\bigcup_{i=1}^{n} A_i\right).$$

15. Prove that $P\left(\bigcup_{i=1}^{k} A_i\right) \leq \sum_{i=1}^{k} P(A_i)$. (Hint: Use mathematical induction.)

16. Let (Ω, P) be a probability measure space. A set $A \subset \Omega$ is called P-null if $P(A) = 0$. Show that if both A and B are P-null, so are the sets $A \cap B$, $A \cup B$, and $A \cap B'$. Show that if A is P-null and B is any set, then $P(A \cap B) = 0$ and $P(A \cup B) = P(B)$.

17. Let (Ω, P) be a probability measure space. A set $A \subset \Omega$ is called P-sure if $P(A) = 1$. Show that if both A and B are P-sure, so are the sets $A \cap B$, $A \cup B$, and $A \cup B'$. Show that if A is P-sure and B is any set, then $P(A \cup B) = 1$ and $P(A \cap B) = P(B)$.

18. Let us assume that the simultaneous occurrence of events A_1 and A_2 implies the occurrence of event A. Prove that

$$P(A) \geq P(A_1) + P(A_2) - 1.$$

COUNTING VI

A fundamental technique, a basic tool and an important vehicle for the calculation of probabilities is *counting*. The methods of counting do not belong to probability itself; they are only auxiliary techniques for solving some problems arising in combinatorial probability. In a rich variety of applications, an event of interest, say A, may be decomposed into components, each of which has the same probability. The calculation of $P(A)$ then leads to counting the components. (See Chapter V, Theorem 8 and Example 8.)

Counting techniques belong to the branch of mathematics sometimes called combinatorial analysis. Traces of interest in combinatorial problems of configurations can be found among the oldest available texts. For example, the sacred book *Yi-King* (about 2200 B.C.), used in China by Taoists, describes the configuration Lo-Shu (Grand Plan). This design, which according to legend was found on the back of a divine tortoise in the River Lo, is known today as the magic square, Saturn. It is

$$\begin{matrix} 4 & 9 & 2 \\ 3 & 5 & 7 \\ 8 & 1 & 6 \end{matrix}$$

We can see that the row, column and diagonal sums of this remarkable pattern are all equal to 15. (We shall discuss similar problems later in this chapter.)

We mentioned earlier that permutations and combinations were discovered by Pascal and Fermat in the seventeenth century. More precisely, they were discovered for the Western world, which was ignorant of combinations until then. However, some thoughts were given to the problems of combinations in China before 1100 B.C., and *Rabbi Ben Ezra* (c. 1140 A.D.) seems to have known the corresponding formula. The binomial coefficients were studied by the Indian arithmetician *Bhaskra* in the twelfth century and the famous "Pascal's triangle" (see page 54) had been taught by a Persian philosopher, *Nasir-Ad-Din*, in 1265.

We shall begin the discussion of counting techniques with a basic rule, which reflects common sense in everyday observations.

MULTIPLICATION PRINCIPLE: If some procedure can be performed in n_1 different ways, and if, following this procedure, a second procedure can be performed in n_2 different ways and so on for a third procedure, etc., then the total number of ways the procedures can be performed is the product

$$n_1 \times n_2 \times n_3 \times \ldots .$$

COMMENT. All other methods of this chapter will be based on this principle.

EXAMPLE 1

Let us toss a die (the procedure 1 with $n_1 = 6$ different ways) and flip a coin (the procedure 2 with $n_2 = 2$ different ways). Then the total number of ways we can toss a die and flip a coin is

$$n_1 \times n_2 = 6 \times 2 = 12.$$

EXAMPLE 2

A license plate contains two distinct letters followed by three digits with the first digit different from 0. How many different license plates can be made?

SOLUTION. Assuming that our alphabet has 26 letters and that we use 10 digits, the total number of plates is

$$26 \times 25 \times 9 \times 10 \times 10 = 585{,}000.$$

EXAMPLE 3

There is a lamp in each corner of a room. In how many ways can the room be illuminated?

SOLUTION. Assuming an ordinary rectangular room and lamps which can be either on or off, we have the total number of ways equal to

$$2^4 - 1 = 15.$$

(Each of the four lamps has two states, but we have to disregard the case in which all the lamps are off.)

DEFINITION 1: *Factorial Notation:* Let n be a natural number. Then the product of all numbers from 1 to n is denoted by $n!$, i.e.,

$$n! = 1 \times 2 \times 3 \times \ldots \times (n - 1) \times n.$$

Further we define $0! = 1$.

Examples:

$$3! = 1 \times 2 \times 3 = 6$$

$$6! = 1 \times 2 \times 3 \times 4 \times 5 \times 6 = 720$$

$$9 \times 8 \times 7 = \frac{9!}{6!}$$

$$\frac{n!}{(n-1)!} = n$$

Very often we have to find the number of ways in which to arrange certain objects. It is important to realize whether or not the *order* of arranged objects plays any role. For example, if we are getting cards for the game of blackjack, then the order in which they arrive is of top importance—two sequences of the same cards in a different order might represent two different games. However, if we are getting cards for bridge or poker, then the order in which they are dealt does not play any role at all. Or, think of the army of imperial Rome. The legions were sometimes punished by decimation, i.e., every tenth soldier was killed, in order to boost morale and obedience. This harsh measure was carried out in two different ways: The soldiers were marshaled into an echelon, and either every tenth in the line was executed or 10% of them were selected by lot and killed. The order in the first case meant the difference between life and death while in the other case it was irrelevant. When we consider selection of a committee of three which will appoint its chairman (order of selection unimportant), then the situation is very different from the case where we stipulate that the first person selected for the committee will be its chairman (order important). If the order is important we usually speak about *permutations;* otherwise we speak about *combinations.*

DEFINITION 2: *Permutations without Repetitions:* Let us have a set of n (distinct) elements. An arrangement of any $r \leq n$ of these elements in a given order, without repetitions, is called a permutation (without repetitions) of n objects taken r at a time.

EXAMPLE 4

Take $n = 4$ and let $\Omega = \{a,b,c,d\}$. Let us list permutations of these objects, taken two at a time:

$$ab,\ ba,\ ac,\ ca,\ ad,\ da,\ bc,\ cb,\ bd,\ db,\ cd,\ dc.$$

There are 12 permutations of 4 objects taken 2 at a time. Let us list some of the permutations of these objects taken four at a time:

$$abcd,\ bacd,\ dabc,\ \ldots.$$

← **THEOREM 1:** If we denote the number of different permutations (without repetitions) of n objects taken r at a time by ${}_nP_r$ then

$${}_nP_r = \frac{n!}{(n-r)!}.$$

Proof: Having n objects, a first object can be selected in n different ways, the second object can be selected in $(n-1)$ ways, and so on, and the r-th object can be selected in $(n-r+1)$ different ways. Using the multiplication rule we get

$${}_nP_r = n \times (n-1) \times (n-2) \times \ldots (n-r+1) = \frac{n!}{(n-r)!}. \qquad \text{q.e.d.}$$

EXAMPLE 5

How many four letter words can we make from the letters in the word "gamble"?

SOLUTION. We have $n =$ six objects and we are selecting four at a time, hence

$${}_6P_4 = \frac{6!}{(6-4)!} = \frac{6!}{2!} = 6 \times 5 \times 4 \times 3 = 360 \text{ words.}$$

(Most of them, of course, are meaningless.)

EXAMPLE 6

In how many ways can three Americans, four Frenchmen, four Danes and two Canadians be seated in a row so that those of the same nationality sit together?

SOLUTION. The four nationalities can be arranged in 4! ways.

The three Americans can be arranged in 3! ways.
The four Frenchmen can be arranged in 4! ways.
The four Danes can be arranged in 4! ways.
The two Canadians can be arranged in 2! ways.

Altogether there are $4! \times 3! \times 4! \times 4! \times 2! = 165{,}888$ different ways.

EXAMPLE 7

In how many ways can the people in the previous example be seated around a circular table? (Assume that the position of the table with respect to the rest of the room does not play any role.)

SOLUTION. The four nationalities can be seated in 3! different ways; the arrangements within the groups are the same. Altogether there are $3! \times 3! \times 4! \times 4! \times 2! = 41{,}472$ ways.

COMMENT. It is good to realize that n people can be seated in $n!$ different ways along a straight table, but they can be seated only in $(n - 1)!$ different ways around a circular table.

DEFINITION 3: *Permutations with Repetitions:* Let us have n distinct types of objects. An arrangement of any r of these objects in a given order with possible repetitions of objects of the same type is called a permutation with repetitions of n (types of) objects taken r at a time.

COMMENT. This time r can be greater than n because we allow repetitions.

EXAMPLE 8

Toss a coin three times and consider an appropriate sample space,

$$\Omega = \{HHH, HHT, HTH, THH, HTT, THT, TTH, TTT\}.$$

Then Ω is given as the set of all distinguishable permutations with repetitions of two types of objects (H and T) taken three at a time.

→ **THEOREM 2:** There are n^r distinguishable permutations with repetitions of n objects taken r at a time.

Proof: The first object can be taken in n different ways, the second also in n distinct ways, and so on. According to the fundamental counting principle, the total number of ways is

$$\underbrace{n \times n \times n \times \ldots \times n}_{r \text{ times}} = n^r \qquad \text{q.e.d.}$$

EXAMPLE 9

A young man is returning home rather late, worried that his beautiful visitor might be waiting already in his penthouse on the eleventh floor of the apartment building. There are six people in the elevator—no two of them obviously belonging together. What is the probability that all the passengers will leave at different floors, thus delaying the young man as much as possible? (He knows that nobody in the elevator is going to visit him.)

SOLUTION. There are nine different floors on which passengers can leave the elevator, because the elevator is on the first floor and the penthouse is on the eleventh floor. Any permutation with possible repetitions of these, taken six at a time, will represent a way in which the passengers could leave. Therefore there are 9^6 ways. If the passengers are to leave at distinct floors, then we have to take permutations of floors without repetitions—and there are $\frac{9!}{(9-6)!}$ ways. Therefore the required probability is

$$p = \frac{9!}{3!\,9^6} = .1138.$$

Notice that this seemingly unlikely event does happen in reality quite often, to the great exasperation of the people on the elevator. The numerical probability agrees with this experience: We can expect that it will happen once in nine cases.

EXAMPLE 10

Ordered Samples. An urn contains n numbered balls. Somebody chooses r balls one after the other, and records their numbers in the order obtained. In how many ways can it happen if he
(a) replaces the ball after each draw,
(b) does not replace the balls?

SOLUTION. (a) These are permutations with repetitions and there are n^r distinguishable ways to do it.

(b) These are permutations without repetitions and there are $\frac{n!}{(n-r)!}$ ways to do it.

In the previous case of permutations with repetitions, we assumed that each type has an unlimited supply of objects. This is not always true, and we have to calculate the number of permutations with repetitions differently.

THEOREM 3: Let us have n objects, of which n_1 are of the first type, n_2 of the second type, and so on, and n_r of the r-th type; $n_1 + n_2 + \ldots + n_r = n$. Then the total number of permutations of these objects is

$$\frac{n!}{n_1!\,n_2!\,\ldots\,n_r!}.$$

Proof: If we could distinguish all the objects we would have $n!$ permutations. However, if for any given permutation we will change the placement of the n_i objects of the i-th type among themselves, then the overall arrangement will remain the same. The objects of the i-th type can change their order $n_i!$ times and therefore the total number of distinguishable permutations is

$$\frac{n!}{n_1!\,n_2!\,\ldots\,n_r!}.$$

q.e.d.

EXAMPLE 11

How many different words can be formed from the letters of the word "likelihood"?

SOLUTION There are 10 letters, two of them l's, two i's and two o's. Therefore the total number of distinct words is

$$\frac{10!}{2!\,2!\,2!\,1!\,1!\,1!\,1!} = 453{,}600.$$

Most, of course, are nonsense words.

EXAMPLE 12

How many different signals (hanging on a horizontal line) can be formed with three identical red flags, two identical blue flags, and four identical yellow flags?

SOLUTION. The number of distinguishable messages is

$$\frac{9!}{3!\,2!\,4!} = 1260.$$

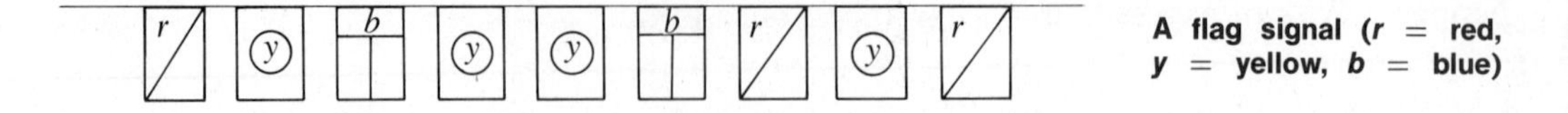

A flag signal (r = red, y = yellow, b = blue)

DEFINITION 4: *Combinations without Repetitions:* Let us have a set of n (distinct) elements. A combination (without repetitions) of these n objects taken r at a time is any subset consisting of r elements (regardless of order).

EXAMPLE 13

When we spoke about permutations, we emphasized the order of the elements; the order is completely irrelevant in the case of combinations. Hence different permutations, *abc*, *acb*, *bac*, *bca*, *cab*, represent the same combination, *abc*. Let us have four elements $\{a,b,c,d\}$:

Combinations	Permutations
abc	*abc, acb, bac, bca, cba, cab*
abd	*abd, adb, bda, bad, dba, dab*
acd	*acd, adc, cad, cda, dac, dca*
bcd	*bcd, bdc, cdb, cbd, dbc, dcb*

This example helps to find the total number of distinct combinations. Notice that from each combination consisting of three letters we can make $3! = 6$ different permutations. We have four combinations and therefore $4 \times 6 = 24$ permutations.

→ **THEOREM 4:** If we denote the total number of distinct combinations (without repetitions) of n objects taken r at a time by ${}_nC_r$ then

$${}_nC_r = \frac{n!}{r!(n-r)!}.$$

Proof: From each combination consisting of r elements we can make $r!$ different permutations by appropriate orderings. Hence

$${}_nP_r = r!\,{}_nC_r,$$

$${}_nC_r = \frac{{}_nP_r}{r!} = \frac{n!}{r!(n-r)!}. \qquad \text{q.e.d.}$$

COMMENT. The numbers ${}_nC_r$ are called combinatorial numbers or binomial coefficients; the symbol ${}_nC_r$ we read "n choose r" and sometimes we use an alternative symbol $\binom{n}{r}$, i.e.,

$$\binom{n}{r} = {}_nC_r.$$

If $r > n$, then we put $\binom{n}{r} = 0$.

EXAMPLE 14

In how many ways can a committee of three men and two women be chosen from seven men and five women?

SOLUTION. We can choose three men out of seven in $\binom{7}{3}$ ways and two women out of five in $\binom{5}{2}$, hence the total number of committees is $\binom{7}{3}\binom{5}{2} = 350$. We assume that the order in which people are selected does not have any influence.

EXAMPLE 15

A young university lecturer contemplates: "I anticipate teaching a statistics course once a year for the next 35 years—God and the Board of Governors permitting. I would not like to become bored by my jokes and I would not like to repeat myself too much. So I will tell three jokes a year, but I do not want to tell the same three jokes in any two different years. What is the minimum number of jokes that will do?"

SOLUTION. If n is the minimum number of jokes, then

$$\binom{n}{3} \geq 35 \quad \text{or} \quad \frac{n(n-1)(n-2)}{2 \times 3} \geq 35.$$

Hence

$$n(n-1)(n-2) \geq 210.$$

We can try to substitute 1,2,3 . . . , etc. for n. Obviously, $n = 7$ is the least number satisfying the condition, and therefore the lecturer needs at least seven jokes.

THEOREM 5: For any nonnegative integers, $0 \leq k \leq n$.

$$\binom{n}{k} = \binom{n}{n-k}.$$

Proof: $\binom{n}{k} = \frac{n!}{k!(n-k)!} = \frac{n!}{(n-k)![n-(n-k)]!} = \binom{n}{n-k}.$ q.e.d.

COMMENT. This theorem is intuitively quite obvious. If we have n objects and choose k of them, then there are exactly $(n - k)$ left. The number of ways in which k objects can be chosen is the same as the number of ways in which $(n - k)$ objects can be left.

THEOREM 6: For any nonnegative integers, $0 \leq k \leq n$.

$$\binom{n}{k-1} + \binom{n}{k} = \binom{n+1}{k}.$$

Proof:

$$\binom{n}{k-1} + \binom{n}{k} = \frac{n!}{(k-1)!(n-k+1)!} + \frac{n!}{k!(n-k)!}$$

$$= \frac{n!k}{k!((n+1)-k)!} + \frac{n!(n+1-k)}{k!((n+1)-k)!} = \frac{(n+1)!}{k!((n+1)-k)!} = \binom{n+1}{k}.$$

q.e.d.

Theorems 5 and 6 will allow us to calculate combinatorial numbers easily. If we arrange these numbers in the form of a triangle we get the following scheme:

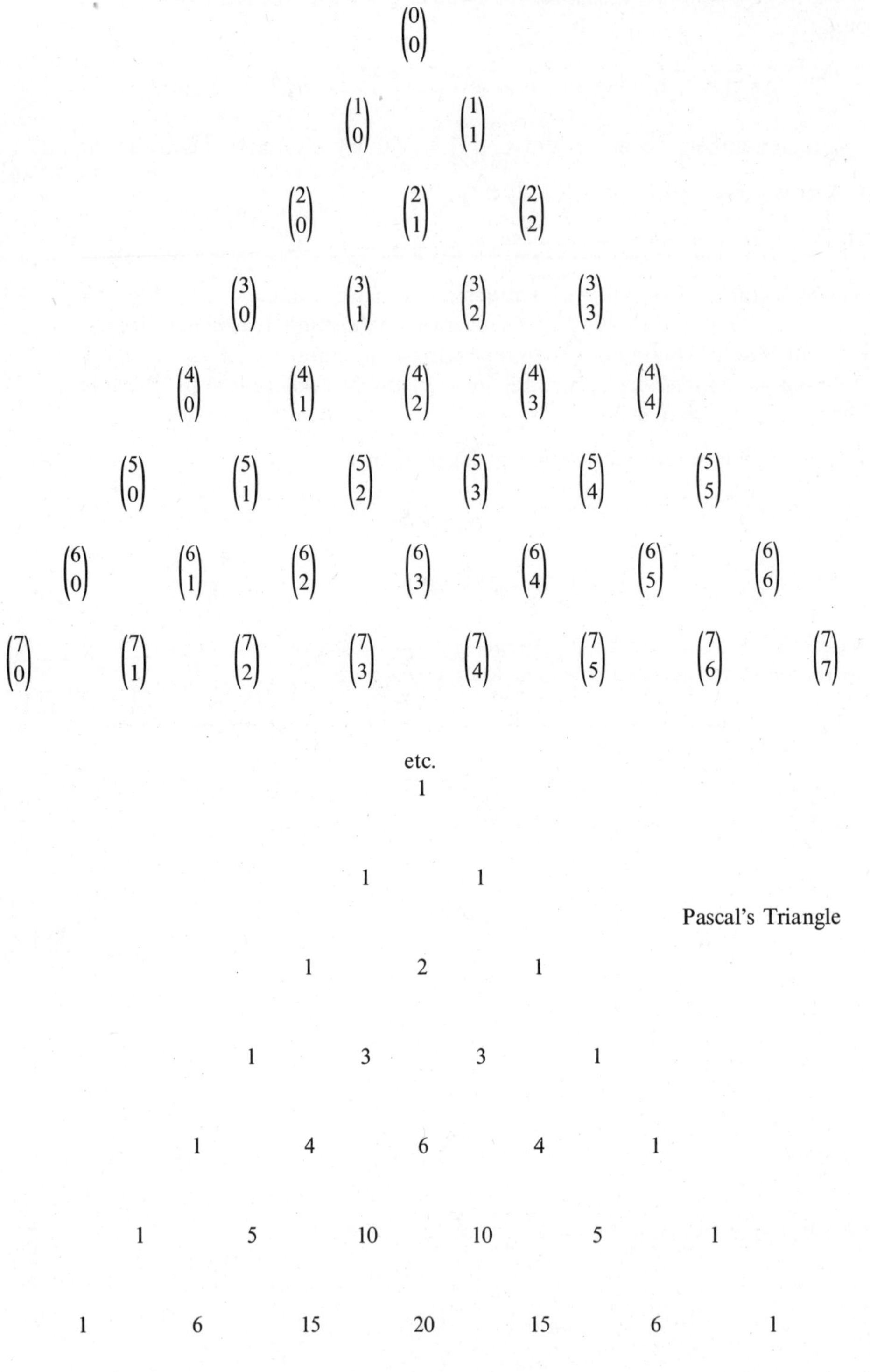

etc.

This scheme is called *Pascal's triangle* and it has the following properties:
(a) The first and the last number in each row is 1.
(b) Every other number in the array can be obtained by adding the two numbers appearing directly above it (this property corresponds to Theorem 6).

Pascal's triangle is used mainly for calculation of the whole sequence of combinatorial numbers $\binom{n}{r}$, where n is fixed and r varies—for example, if we have to find

$$\binom{16}{4}, \binom{16}{5}, \binom{16}{6}, \binom{16}{7}, \binom{16}{8}, \binom{16}{9}, \binom{16}{10}.$$

Blaise Pascal, seventeenth century French philosopher and mathematician. (Courtesy of The Mansell Collection, London.)

From high school we know formulas for the square and the cube of a binomial:

$$(a + b)^2 = a^2 + 2ab + b^2;$$

$$(a + b)^3 = a^3 + 3a^2b + 3ab^2 + b^3.$$

The following formula gives an expansion for the n-th power of a binomial, comprising both above mentioned formulas.

THEOREM 7: *Binomial Theorem:* For any two real numbers a, b and for any natural number n,

$$(a + b)^n = \sum_{k=0}^{n} \binom{n}{k} a^{n-k}b^k.$$

Proof 1. We can prove this theorem by induction.

If $n = 1$ then $(a + b)^1 = a + b = \binom{1}{0} a^{1-0}b^0 + \binom{1}{1} a^{1-1}b^1.$

Assume that the theorem is true for a certain n, that is

$$(a+b)^n = \sum_{k=0}^{n} \binom{n}{k} a^{n-k}b^k$$

and we would like to prove that it is also true for $(n+1)$. Then

$$(a+b)^{n+1} = (a+b)(a+b)^n = (a+b)\sum_{k=0}^{n} \binom{n}{k} a^{n-k}b^k.$$

The term in the product that contains b^k is obtained from

$$\begin{aligned} b\left[\binom{n}{k-1} a^{n-k+1}b^{k-1}\right] + a\left[\binom{n}{k} a^{n-k}b^k\right] &= \binom{n}{k-1} a^{n-k+1}b^k + \binom{n}{k} a^{n-k+1}b^k \\ &= \left[\binom{n}{k-1} + \binom{n}{k}\right] a^{n-k+1}b^k \\ &= \binom{n+1}{k} a^{n-k+1}b^k \\ &= \binom{n+1}{k} a^{(n+1)-k}b^k. \end{aligned}$$

Consequently

$$(a+b)^{n+1} = \sum_{k=0}^{n+1} \binom{n+1}{k} a^{(n+1)-k}b^k. \qquad \text{q.e.d.}$$

Proof 2: This is an alternative proof, using counting ideas.

$$(a+b)^n = \underbrace{(a+b)(a+b)\ \ldots\ (a+b)}_{n \text{ times}}.$$

The result will be a sum consisting of 2^n terms. Each term is a product of n factors a or b. A product containing $(n-k)$ a's and k b's can be written as $a^{n-k}b^k$. But there are exactly $\binom{n}{k}$ such products because we have to choose k b's from n brackets while the remaining $(n-k)$ brackets will yield a's. q.e.d.

COMMENT. We can easily verify that for $n = 2$ and $n = 3$, we get the familiar formulas mentioned above.

EXAMPLE 16

Calculate $(2-3x)^4$.

SOLUTION. Using the binomial theorem for $n = 4$, $a = 2$, $b = -3x$ we get

$$\begin{aligned} (2-3x)^4 &= \binom{4}{0}2^4(-3x)^0 + \binom{4}{1}2^3(-3x)^1 + \binom{4}{2}2^2(-3x)^2 + \binom{4}{3}2^1(-3x)^3 + \binom{4}{4}2^0(-3x)^4 \\ &= 16 - 96x + 216x^2 - 216x^3 + 81x^4. \end{aligned}$$

EXAMPLE 17

Find $\sum_{i=0}^{n} \binom{n}{i} = \binom{n}{0} + \binom{n}{1} + \binom{n}{2} + \ldots + \binom{n}{n}$ and show that it is equal to the number of all subsets of a set with n elements.

SOLUTION. Using the binomial theorem for $a = 1$, $b = 1$ we get

$$(1 + 1)^n = \sum_{i=0}^{n} \binom{n}{i} 1^i 1^{n-i} = \sum_{i=0}^{n} \binom{n}{i} = 2^n.$$

If Ω has n elements, then there are exactly $\binom{n}{k}$ subsets of Ω with exactly k elements. Therefore the total number of subsets of Ω is equal to $\sum_{i=0}^{n} \binom{n}{i} = 2^i$.

EXAMPLE 18

Show that if $|nx|$ is very small, then $(1 + x)^n$ is approximately equal to $1 + nx$:

$$(1 + x)^n \approx 1 + nx.$$

Use this fact to evaluate approximately $(.996)^{16}$.

SOLUTION. Using the binomial theorem, we get for $a = 1$, $b = x$

$$(1 + x)^n = 1 + nx + \frac{n(n-1)}{2!}x^2 + \frac{n(n-1)(n-3)}{3!}x^3 + \text{etc.}$$

Now if $|nx|$ is very small, then the terms

$$\left|\frac{n(n-1)x^2}{2!}\right| \leq \frac{(nx)^2}{2!} \quad \text{and} \quad \left|\frac{n(n-1)(n-2)x^3}{3!}\right| \leq \frac{|nx|^3}{3!}$$

etc., are so small that they can be practically neglected. Therefore

$$(1 + x)^n \approx 1 + nx.$$

$(.996)^{16} = (1 - .004)^{16}$ should be approximately $1 + 16(-.004) = .936$. An accurate evaluation gives $(.996)^{16} = .938$, and therefore our approximation is fairly good.

DEFINITION 5: *Combinations with Repetitions:* Let us have n distinct types of objects. An arrangement of any r of these objects (regardless of order) with possible repetitions of objects of the same type is called a combination with repetitions of n (types of) objects taken r at a time.

COMMENT. r can be greater than n because we admit repetitions of objects, as shown in the following example.

EXAMPLE 19

Let us toss a coin three times and consider the numbers of heads and tails (regardless of order in which they appear). Then we can have the following outcomes:

$$3H0T, \quad 2H1T, \quad 1H2T, \quad 0H3T.$$

→ **THEOREM 8:** The number of combinations with repetitions of n types of objects taken r at a time is equal to

$$\binom{n+r-1}{n-1} = \binom{n+r-1}{r}.$$

Proof: This proof is a bit more demanding and requires a special trick. Imagine that we pick up n boxes and arrange them in a row. Further, let a combination of n types of objects taken r at a time be given. Let us put into the first box as many balls (identical) as there are objects of the first type appearing in the combination. Then let us count the number of objects of the second type and put this many (identical) balls into the second box, and so on. It is clear that any combination defines a unique distribution of r (identical) balls among n boxes, and vice versa. There is a one-to-one correspondence between all combinations and all distributions of balls. Therefore, it is enough to count the number of ways in which we can distribute r (identical) balls among n ordered boxes.

Let us represent each box by its right-hand wall (starting from the left). It is clear that we can omit the wall of the last box without any danger of confusion. If we have $n = 6$ boxes and $r = 8$ balls, then we can get, for example, the following distribution of balls:

$$\cdots\ |\quad\quad |\ \cdot\cdot\ |\quad\quad |\ \cdot\ |\ \cdot\cdot$$

Box 1 Box 2 Box 3 Box 4 Box 5 Box 6

We have, therefore, $(n - 1)$ walls and r balls, which together make $(n + r - 1)$ objects. If we could distinguish them all we would have $(n + r - 1)!$ distinct ordered arrangements. If, however, we permute the order of the balls (and it can be done in $r!$ ways) or the order of the walls (it can be done in $(n - 1)!$ ways) then we cannot discover that the permutation has been performed. The total number of distinguishable arrangements is therefore equal to

$$\frac{(n+r-1)!}{(n-1)!r!} = \binom{n+r-1}{r} = \binom{n+r-1}{n-1}.$$

We get the same answer if we realize that we are actually counting all possible permutations with repetitions of two types of objects: r balls and $(n - 1)$ walls of boxes. (See Theorem 3.) q.e.d.

EXAMPLE 20

There are three piano tuners in a city. On a certain day four people telephone at random for a tuner. In how many ways can the tuners be assigned to jobs? (Disregard the order of the calls.)

SOLUTION. There are three tuners and four calls; we are counting combinations of three types taken four at a time, therefore there are

$$\binom{3+4-1}{4} = \binom{6}{4} = \binom{6}{2} = 15 \text{ ways.}$$

List all the combinations for verification (see Example 21).

EXAMPLE 21

In how many ways can you write a positive integer r as a sum of n nonnegative integers in a given order?

$$r = \alpha_1 + \alpha_2 + \ldots + \alpha_n, \quad r > 0$$

$$\alpha_i \geq 0.$$

SOLUTION. From the proof of Theorem 8, it is clear that if we have α_i balls in the i-th box then we get a one-to-one correspondence between all distributions of r balls into n ordered boxes and all above mentioned partitions. Therefore the number of partitions is

$$\binom{n+r-1}{r} = \binom{n+r-1}{n-1}.$$

Let $r = 4$, $n = 3$. Then we have $\binom{4+3-1}{3-1} = 15$ partitions as follows:

$$4 = 4 + 0 + 0 = 0 + 4 + 0 = 0 + 0 + 4 = 3 + 1 + 0 = 3 + 0 + 1 = 1 + 3 + 0$$
$$= 0 + 3 + 1 = 1 + 0 + 3 = 0 + 1 + 3 = 2 + 1 + 1 = 1 + 2 + 1 = 1 + 1 + 2$$
$$= 2 + 2 + 0 = 2 + 0 + 2 = 0 + 2 + 2.$$

These partitions correspond uniquely to the ways in which three piano tuners can be phoned by four residents in Example 20.

THEOREM 9: The number of combinations with repetitions of n types of objects taken r at a time in such a way that each type is taken at least once is equal to

$$\binom{r-1}{n-1}.$$

COMMENT. Obviously we have to assume $r \geq n$.

Proof: As in Example 21, the total number of combinations of this type is equal to the number of ways in which the number r can be written as a sum of n positive numbers α_i,

$$r = \alpha_1 + \alpha_2 + \ldots + \alpha_n, \quad r > 0$$
$$\alpha_i > 0.$$

But it is equal to the number of ways that we can write

$$(r - n) = (\alpha_1 - 1) + (\alpha_2 - 1) + \ldots + (\alpha_n - 1), \quad r > 0$$
$$\alpha_i > 0.$$

If we write $\beta_i = (\alpha_i - 1)$, then the number of required combinations will be equal to the number of ways in which we can write $(r - n)$ as a sum of n nonnegative β_i's:

$$(r - n) = \beta_1 + \beta_2 + \ldots + \beta_n, \quad r > 0$$
$$\beta_i \geq 0.$$

But according to Example 21, it is equal to

$$\binom{n + (r - n) - 1}{n - 1} = \binom{r-1}{n-1}.$$

q.e.d.

EXAMPLE 22

On a certain day, four residents of a city call at random three available piano tuners (see Example 20). What is the probability that each of the tuners will get a job that day?

SOLUTION. Imagine the calls going through the main switchboard. The first resident can call any one of the three tuners; the same holds for the second resident, and so on. According to Theorem 2, there are $3^4 = 81$ ways in which the calls can come. Consider now the numbers of ways to assign the calls so that each tuner gets a job. $n = 3$ tuners should be chosen by $r = 4$ residents, each of them at least once. There are

$$\binom{r-1}{n-1} = \binom{3}{2} = 3 \text{ ways to do it (Theorem 9).}$$

(Namely 1, 1, 2, 3; 1, 2, 2, 3; 1, 2, 3, 3, where the number refers to the tuner and repetitions mean how many times he is called.) Once we know the tuner who gets two calls (the first, the second or the third) we have to assign these numbers to the four residents. By Theorem 3 there are $\frac{4!}{2!1!1!}$ ways to do it. So the required probability is

$$\frac{\binom{3}{2}\frac{4!}{2!1!1!}}{3^4} = \frac{4}{9}.$$

Let us consider now another type of problem regarding counting.

EXAMPLE 23

Eight students, $A, B, C, D, E, F, G,$ and H, are accepted by the dormitories, and they are to be assigned to the three rooms A1, A2 and A3. Rooms A1 and A2 are triple rooms and room A3 is a double. In how many ways can the assignment be made?

SOLUTION. Let us decide that we assign the students in the order of their arrival, filling first room A1, then A2 and finally A3.

Order of arrival	*CDA*	*BFE*	*GH*
Assignment to the rooms	A1	A2	A3

But even if they arrive in a different order, they still can be assigned to the same rooms.

Order of arrival	*DAC*	*FEB*	*HG*
Assignment to the rooms	A1	A2	A3

They can arrive in 8! different ways. The order of the first three students is insignificant (there are 3! ways), the order of the second three students is insignificant (there are 3! ways) and also the order of the last two students is not important (there are 2! ways). The total number of ways is therefore

$$\frac{8!}{3!3!2!} = 560.$$

This example leads us to the following:

DEFINITION 6: *Ordered Partitions:* Any arrangement of n distinct objects into k cells $[A_1, A_2, \ldots, A_k]$, such that the first cell A_1 contains n_1 elements, the second cell A_2 contains n_2 elements ... and the *k-th* cell A_k contains n_k elements (where $n_1 + n_2 + \ldots + n_k = n$), is called an ordered partition.

→ **THEOREM 10:** The number of ordered partitions is equal to

$$\binom{n}{n_1 n_2 \ldots n_k} = \frac{n!}{n_1! n_2! \ldots n_k!}.$$

Proof: The cell A_1 can be filled in $\binom{n}{n_1}$ ways.

The cell A_2 can be filled in $\binom{n - n_1}{n_2}$ ways.

The cell A_3 can be filled in $\binom{n - n_1 - n_2}{n_3}$ ways.

. . . .

The cell A_k can be filled in $\binom{n - n_1 - n_2 - n_3 - \cdots - n_{k-1}}{n_k}$ ways.

The total number of ways is

$$\binom{n}{n_1}\binom{n - n_1}{n_2}\binom{n - n_1 - n_2}{n_3} \cdots \binom{n - n_1 - n_2 - \cdots - n_{k-1}}{n_k}$$

$$= \frac{n!}{n_1!(n - n_1)!} \frac{(n - n_1)!}{n_2!(n - n_1 - n_2)!} \frac{(n - n_1 - n_2)!}{n_3!(n - n_1 - n_2 - n_3)!} \cdots \frac{(n - n_1 - n_2 - \cdots - n_{k-1})!}{n_k!(n - n - n - \cdots - n_{k-1} - n_k)}$$

but $$(n - n_1 - n_2 - \cdots - n_{k-1} - n_k)! = 0! = 1.$$

Hence if we cancel the common terms in successive fractions, we get

$$\frac{n!}{n_1!n_2! \cdots n_k!}.$$

EXAMPLE 24

In the game of bridge, the hands *N, E, S, W* determine a partition of 52 cards into four cells, each containing 13 cards. The total number of ways to deal the cards is

$$\frac{52!}{13!13!13!13!} = 5.3645 \times 10^{28} \cong 54 \text{ billion billion billion}$$

COMMENT. The numbers $\binom{n}{n_1 n_2 \cdots n_k}$, where $n_1 + n_2 + \cdots + n_k$, are called multinomial coefficients. Using these coefficients, we can prove the following:

THEOREM 11: *Multinomial Theorem:* Let $a_1, a_2, \ldots, a_k$ be real numbers, n be a natural number, then

$$(a_1 + a_2 + \cdots + a_k)^n = \sum \binom{n}{n_1 n_2 \cdots n_k} a_1^{n_1} a_2^{n_2} \cdots a_k^{n_k}$$

where the sum is taken over all nonnegative integers $n_1, n_2, \ldots, n_k$, such that $n_1 + n_2 + \cdots + n_k = n$. A proof can be based on the same ideas as the second proof of the binomial theorem.

THEOREM 12: For any three natural numbers m, n, r, such that $r \leq m + n$, we have

$$\binom{m + n}{r} = \sum_{i=0}^{r} \binom{m}{i}\binom{n}{r - i}.$$

Proof: Imagine a box containing m white and n black balls. There are $\binom{m+n}{r}$ ways to choose r balls. Out of these there are $\binom{m}{i}\binom{n}{r-i}$ ways to get exactly i white and $(r - i)$ black balls for $i = 0,1,2, \ldots, r$. Therefore

$$\binom{m+n}{r} = \sum_{i=0}^{r} \binom{m}{i}\binom{n}{r-i}.$$

q.e.d.

EXERCISES

Exercises for this chapter are located at the end of the next chapter.

SELECTED EXAMPLES OF COMBINATORIAL PROBABILITIES

Probability is the very Guide of Life.

Bishop Joseph Butler, 1736

This chapter contains a series of examples and problems that are solved with the help of the counting techniques and combinatorial methods of the preceding chapter. Some of them are of a rather complex character, others utilize simultaneously several methods of Chapter VI, while still others require special techniques and tricks for their solution. This is one reason why they are presented separately.

EXAMPLE 1

From an ordinary deck of 52 cards, 3 cards are drawn at random. Find the probability P_1 that exactly 1 ace is drawn.

SOLUTION. There are $\binom{52}{3}$ ways to select 3 cards out of 52. There are $\binom{4}{1}$ ways to select 1 ace out of 4 aces and $\binom{48}{2}$ ways to select the 2 remaining non-ace cards out of 48 non-ace cards. Hence

$$P_1 = \frac{\binom{4}{1}\binom{48}{2}}{\binom{52}{3}} = .2042.$$

EXAMPLE 2

Under the same assumptions as in Example 1, find the probability that at least one ace is drawn.

SOLUTION. Let P_1, P_2 and P_3 be the probability of exactly one, exactly two and exactly three aces respectively. We know the probability P_1 from the previous example. Similarly

$$P_2 = \frac{\binom{4}{2}\binom{48}{1}}{\binom{52}{3}} = .0130 \qquad P_3 = \frac{\binom{4}{3}\binom{48}{0}}{\binom{52}{3}} = .0002.$$

Finally, the probability of at least one ace is equal to the probability of exactly one or exactly two or exactly three aces and these events are mutually exclusive.

Hence $$P = P_1 + P_2 + P_3 = .2042 + .0130 + .0002 = .2174.$$

A faster solution is as follows:

$$P(\text{at least one ace}) = 1 - P(\text{no ace drawn}) = 1 - \frac{\binom{48}{3}}{\binom{52}{3}} = .2174.$$

EXAMPLE 3

A set of eight cards contains one joker. There are two players A and B. A chooses five cards at random, and B takes the remaining three cards. What is the probability P that A has the joker?

SOLUTION.

$$P = \frac{\binom{1}{1}\binom{7}{4}}{\binom{8}{5}} = \frac{5}{8}$$

because there are $\binom{8}{5}$ ways for A to get his cards; however he has to get the joker (in $\binom{1}{1}$ ways) and exactly 4 other cards (in $\binom{7}{4}$ ways).

EXAMPLE 4

Find the probability that a *poker* hand will contain:

(a) one pair (two cards of equal face value plus three distinct cards).
(b) two pairs (two pairs of equal face value each plus a different card).
(c) a triple (three cards of equal face value plus two different cards).
(d) a full house (a pair and a triple).
(e) four of a kind (four cards of equal face value).
(f) a straight (including a straight flush) (five cards in a sequence regardless of suit; ace can be either the highest or lowest card).
(g) a flush (including a straight flush) (all cards of the same suit).
(h) a straight flush (five cards in sequence in the same suit).
(i) a royal flush (straight flush of the ten, jack, queen, king, and ace).
(j) nothing.

Royal flush

SOLUTION. There are $N = \binom{52}{5} = 2{,}598{,}960$ distinct hands for poker.

(a) The face value of a pair can be chosen in 13 ways, the suits for the pair in $\binom{4}{2}$ ways, 3 remaining distinct face values can be chosen in $\binom{12}{3}$ ways, and the corresponding suits can be chosen in 4^3 ways. Therefore the probability is

$$13 \times \binom{4}{2} \times \binom{12}{3} \times 4^3/N \quad = .422569$$

(b) Similarly as in (a), $\binom{13}{2} \times \binom{4}{2}^2 \times 11 \times 4/N \quad = .047539$

(c) $13 \times 4 \times \binom{12}{2} \times 4^2/N \quad = .021128$

(d) $13 \times 4 \times 12 \times \binom{4}{2}/N \quad = .001441$

(e) $13 \times 12 \times 4/N \quad = .000240$

(f) $10 \times 4^5/N \quad = .003940$

(g) $4 \times \binom{13}{5}/N \quad = .001981$

(h) $10 \times 4/N \quad = .000015$

(i) $4/N \quad = .000002$

(j) $\left(\binom{13}{5} - 10\right) \times (4^5 - 4)/N \quad = .501177$

We have to get five distinct values, but we have to exclude a straight. Then we have to specify suits for all five cards, but we have to exclude a flush.

EXAMPLE 5

Imagine that cards for bridge have been dealt. If you and your partner together have n cards of one suit (trumps), then there are $(13 - n)$ cards of that suit left, divided between opponents' hands. Find the probability that one of your opponents will have exactly k cards of that suit and the other will have the rest. These probabilities are of top importance for planning the play of a hand of bridge.

SOLUTION. There are 26 cards left for your opponents, out of which there are $(13 - n)$ of the given suit. One opponent (either one) has to get k out of $(13 - n)$ cards of the given suit and $(13 - k)$ out of the remaining $(13 + n)$ cards. The required probability is therefore

$$\frac{2 \times \binom{13-n}{k} \times \binom{13+n}{13-k}}{\binom{26}{13}}.$$

The number 2 in the numerator is expressing the fact that either one of the two opponents has exactly k trumps. The following is the table of these probabilities.

If you and your partner together have in one suit:	The cards of that suit in opponents' hands will be divided:	
6 cards	4–3	62%
	5–2	31%
	6–1	7%
	7–0	Less than $\frac{1}{2}$%
7 cards	4–2	48%
	3–3	36%
	5–1	15%
	6–0	1%
8 cards	3–2	68%
	4–1	28%
	5–0	4%
9 cards	3–1	50%
	2–2	40%
	4–0	10%
10 cards	2–1	78%
	3–0	22%
11 cards	1–1	52%
	2–0	48%

Note: This table is most important if you play bridge, since on every deal you will ever encounter as declarer, you must have between you and your partner at least two seven-card suits, or an eight-card or longer suit.

EXAMPLE 6

A complete domino set consists of 28 black wooden pieces, each of which is divided by a line into two equal parts. Each half contains a certain number of white dots between zero and six (inclusive) in such a way that all the possible combinations of pairs of numbers from zero to six appear exactly once.

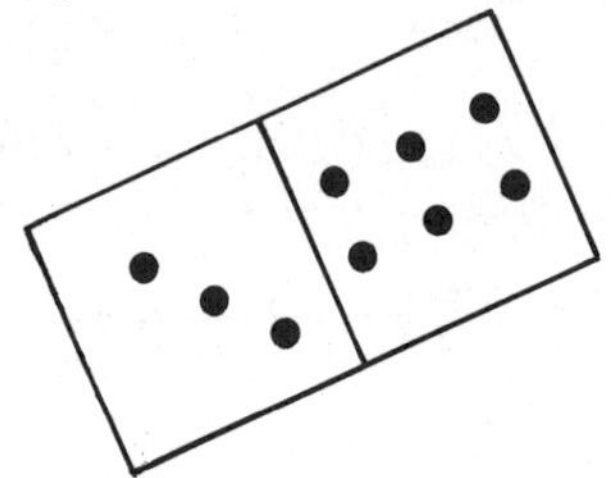

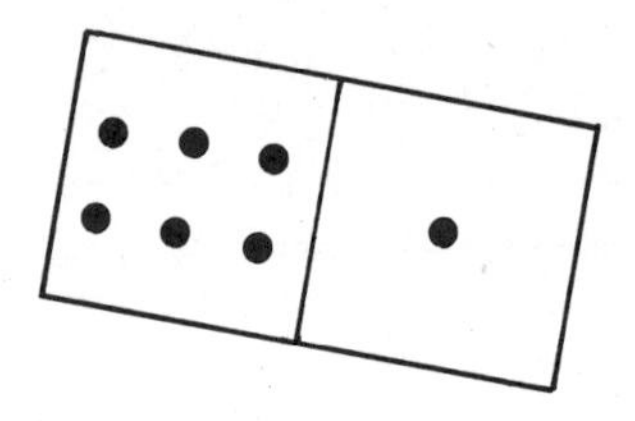

Five pieces are drawn from this set. Find the probability p that at least one of them will have a six marked on it, that is, six dots on at least one of the halves.

SOLUTION. Let us find the probability q of the complementary event. There are 28 pieces in a complete domino set out of which 7 contain a six. Hence

$$q = \frac{\binom{7}{0}\binom{21}{5}}{\binom{28}{5}},$$

$$p = 1 - q = .793.$$

EXAMPLE 7

Find the probability that the last two digits of the cube of a four digit random integer will be two ones. (We assume that any four digit integer has the same chance to be chosen.)

SOLUTION. Let N be a four digit random integer. Then we can write

$$N = a + 10b + 100c + 1000d \quad \text{where } a,b,c,d \text{ are integers,} \quad 0 \leq a,b,c,d \leq 9.$$

Then $N^3 = a^3 + 30a^2b + \ldots$ (this part of the number is not important for the problem). The last two digits, which should be 1, 1, are influenced by a and b only. There are 100 possibilities for the choice of a and b: (0,0), (0,1), (0,2) . . . (9,9). But $a^3 = 1$ and hence a must be one. At the same time, $\frac{N^3 - 1}{10}$ must have the last digit equal to one. Therefore, $\frac{30b + 1 - 1}{10} = 3b$ must end with a 1 and hence b must be 7. The only solution is, therefore, $a = 1, b = 7$ and thus $P = .01$.

EXAMPLE 8

What is the probability P_r that r people attending a party will all have different birthdays? (Assume that a year has 365 days, that the birth rate is constant throughout the year and $r \leqq 365$.)

SOLUTION. The first person can be born on any one of 365 days, the second person on any one of 364 days, and so on. If there is no condition regarding birthdays, then r people can be born in 365^r ways. Hence

$$P_r = \frac{365!}{(365 - r)!365^r}.$$

It can be shown that $P_{22} = .524$, $P_{23} = .493$ and P_r is decreasing. For $r \geq 23$ people attending the party, $P_r < \frac{1}{2}$, that is, if there are more than 22 people at the party it is more likely that at least 2 have the same birthday than that all the birthdays are different. Following is the table of probabilities P_r for some values of r.

Number of people at a party, r	Probability, P_r, that all have different birthdays
5	.973
10	.883
15	.747
20	.589
21	.556
22	.524
23	.493
24	.462
25	.431
30	.294
40	.109
50	.030

EXAMPLE 9

A thief learns that the Marshal Flivvers Company makes only 10 different keys for its cars. He steals five MF cars at random and gets the keys (he does not know whether all the keys are different). The next day he finds your MF rig. What is the probability that he has the key that will open it?

SOLUTION 1 (DIFFICULT). The thief steals keys at random and hence each key can be stolen with the same probability. Let us denote P_i as the probability that he has i distinct keys. If there is no condition about how the keys were stolen, then there are 10^5 possibilities for the set of five keys. The thief can obtain 5 identical keys in 10 different ways (there are 10 different keys which can be stolen), hence

$$P_1 = \frac{10}{10^5}.$$

If he gets two different kinds, then these two types can be specified in $\binom{10}{2}$ ways. However in this case he can get

one key of the first kind and four keys of the second kind in $\binom{5}{1}$ ways,

or two keys of the first kind and three keys of the second kind in $\binom{5}{2}$ ways,

or three keys of the first kind and two keys of the second kind in $\binom{5}{3}$ ways,

or four keys of the first kind and one key of the second kind in $\binom{5}{4}$ ways.

Therefore
$$P_2 = \frac{\binom{10}{2}\left\{\binom{5}{1} + \binom{5}{2} + \binom{5}{3} + \binom{5}{4}\right\}}{10^5} = \frac{1350}{10^5}.$$

If the thief gets exactly three different keys, then these three kinds can be specified in $\binom{10}{3}$ ways. The following table displays how it can happen:

Kind 1 Keys	Kind 2 Keys	Kind 3 Keys	
3	1	1	Each possibility can occur in $\frac{5!}{3!}$ ways.
1	3	1	
1	1	3	
2	2	1	Each possibility can occur in $\frac{5!}{2!2!}$ ways.
2	1	2	
1	1	2	

Hence
$$P_3 = \frac{\binom{10}{3}\left[3\frac{5!}{3!} + 3\frac{5!}{2!2!}\right]}{10^5} = \frac{18000}{10^5}.$$

In the same way, we can calculate P_4.

$$P_4 = \frac{\binom{10}{4} 4 \frac{5!}{2!}}{10^5} = \frac{50400}{10^5}.$$

Finally

$$P_5 = \frac{10 \times 9 \times 8 \times 7 \times 6}{10^5} = \frac{30240}{10^5}.$$

If the thief has n distinct keys, then the probability that he opens your car is $\frac{n}{10}$. Therefore the probability P that he gets your car is

$$P = \sum_{n=1}^{5} P_n \frac{n}{10} = .40951.$$

SOLUTION 2. There is another and much faster way of solving this problem, based on calculation of complements. If the thief steals five keys without restriction, he can do it in 10^5 ways. If he does not get your key, he can get his keys in 9^5 ways. The probability of not getting your key is $\frac{9^5}{10^5}$; the asked probability is

$$P = 1 - \frac{9^5}{10^5} = .40951.$$

This example clearly shows that the same problem can be solved sometimes in several different ways, with considerably different degrees of difficulty. It is up to the student to get a bright idea, and pick the easiest way.

EXAMPLE 10

Banach Matchbox Problem.* A certain mathematician carries two matchboxes with him. Each time he wants to use a match, he selects a box at random. Find the probability that when he discovers that one box is empty, the other contains exactly r matches ($r = 0,1,2, \ldots, n$; n being the number of matches initially contained in each box). Assume that the mathematician never bothers to check how many matches are left in a box after he has taken a match, hence he discovers an empty box only when he selects it but there is no match left.

SOLUTION. The mathematician can have one box in his left pocket L and one box in his right pocket R. If somebody watches him, he can record a sequence of the letters L,R corresponding to the pockets from which the box was selected. If the mathematician used five matches, then possible sequences are, for example, *LLRLR* or *RLLRL*, and so on. The mathematician had $2n$ matches in the beginning and r matches at the end. He made $(2n - r)$ decisions when he got his match, and the last decision when he discovered an empty box. Every possible sequence, therefore, will have $(2n - r + 1)$ symbols, L,R and there are 2^{2n-r+1} such sequences. However, there must be n letters R and $(n - r)$ letters L or n letters L and $(n - r)$ letters R among the first $(2n - r)$ letters. This can be accomplished in $2\binom{2n-r}{n}$ ways.

Hence

$$P = 2\binom{2n-r}{n} 2^{-(2n-r+1)} = \binom{2n-r}{n} 2^{-2n-r}.$$

---

*Banach, Stefan, 1892–1945, Polish mathematician.

EXAMPLE 11

Statistical Mechanics. Imagine a quantity of gas in a bounded region such as a cube or cylinder. The molecules (particles) of gas are moving chaotically in the space and at each moment they are distributed according to some rules at certain locations. If we want to investigate the *equilibrium** of the gas, we have to study these distributions from the point of view of probability. There are several models which are equally sound mathematically, but which have different practical consequences. We shall formulate a general problem of distribution of particles first, and then we shall give three different solutions, which correspond to different assumptions about the behavior of particles.

Imagine n particles, each of which may be found with the same probability in one of N cells ($N > n$). Find the probability that
(a) each of n specified cells contains one particle (p_1).
(b) n arbitrary cells each contain one particle (p_2).
Note: It is not clear whether we can distinguish the particles and how they can fill the cells. Different answers to these questions, therefore, will give us different solutions.

SOLUTION 1. *Maxwell-Boltzmann Statistics.* Let us assume that all the particles are distinguishable, that each cell can contain any number of particles and that all the conceivable distributions of particles among cells are equally likely.

Each particle can be found in any one of N cells, therefore there are N^n distinguishable distributions. Obviously, n distinguishable particles can be distributed among n given cells in $n!$ different ways, therefore

$$p_1 = \frac{n!}{N^n}.$$

Furthermore, we can select n arbitrary cells out of N cells in $\binom{N}{n}$ different ways, therefore

$$p_2 = \binom{N}{n} p_1 = \frac{\binom{N}{n} n!}{N^n} = \frac{N!}{N^n (N-n)!}.$$

SOLUTION 2. *Bose-Einstein Statistics.* In this case, we assume that the particles cannot be distinguished, that is, if two particles change their locations, then we get an identical distribution as before the change. All that matters is how many particles are in a given cell, but not which ones. Further, we assume that all such distributions are equally likely.

It is clear that several distinct distributions in the Boltzmann sense represent an identical distribution in the Bose-Einstein sense. Let us imagine $N = 3$ cells and $n = 2$ particles, which we shall denote a and b.

Distribution Number

	1	2	3	4	5	6	7	8	9
Cell 1	*ab*			*a*	*b*	*a*	*b*		
Cell 2		*ab*		*b*	*a*			*a*	*b*
Cell 3			*ab*			*b*	*a*	*b*	*a*

*According to *The Random House Dictionary,* equilibrium is a state of rest or balance due to the equal action of opposite forces.

The distributions 4 and 5 are distinct in the M-B sense, but they are identical in the B-E sense because they differ by the type of particles but not by the number of particles in the cells. Similarly the pairs 6–7 and 8–9 are identical in the B-E sense. Therefore, there are nine equally likely distributions for M-B statistics, but only six equally likely distributions for B-E statistics.

We have N distinct cells, and we have to choose n of them with possible repetitions in order to allocate the particles. The total number of distinct ways to do it is equal to the number of combinations with repetitions of N types of objects taken n at a time; that is,

$$\binom{N+n-1}{n}.$$

Only one of these distributions is favorable to our first problem, hence

$$p_1 = \frac{1}{\binom{N+n-1}{n}} = \frac{n!(N-1)!}{(N+n-1)!}.$$

Using the same argument as in the previous case, we get

$$p_2 = \binom{N}{n} p_1 = \frac{N!(N-1)!}{(N-n)!(N+n-1)!}.$$

SOLUTION 3. *Fermi-Dirac Statistics.* In this case we assume that all the particles are indistinguishable, and that each cell can contain at most one particle. In order to specify a distribution, we have to select n cells out of N that will contain a particle. This can be done in $\binom{N}{n}$ distinct ways, and only one is favorable to our first problem. Hence

$$p_1 = \frac{1}{\binom{N}{n}} = \frac{n!(N-n)!}{N!}.$$

Similarly,

$$p_2 = \binom{N}{n} p_1 = 1.$$

However, it is obvious even without calculations that in this case p_2 must be equal to one.

An attentive reader may now ask: "Which of these three models corresponds best to the real physical world?" As a matter of fact, each model corresponds to a different situation. When investigating classical systems without the use of quantum mechanics—for example, molecules of gas—then the Maxwell-Boltzmann model is best. When investigating systems in quantum mechanics, one has to decide whether particles spin or not. If the particles spin—for example, electron, proton, neutron, positron, neutrino, and so on—then they have to be placed in different orbits or cells (at most one particle per cell), hence they follow the Fermi-Dirac model. If the particles do not spin—for example, π-meson, ρ-meson, graviton, photon, and so on—then they obey the Bose-Einstein model.

EXAMPLE 12 (DIFFICULT)

$2n$ people are queued up at a theatre box office; n of them have only one five-dollar bill and the remaining n only one ten-dollar bill. There is no cash in the box office when it opens and each patron in turn is going to purchase a single five-dollar ticket. What is the probability that no customer will have to wait for change? (That is, each customer will get appropriate change immediately from the box office and nobody will have to wait until some customers behind him have been served in order to get the change. We assume, of course, that all arrangements of people in the queue are equally likely.)

SOLUTION. All the possible arrangements of customers in the line are equally likely. We make use of the following geometrical approach. Consider the x,y plane, and suppose that the patrons are arranged along the x-axis at the points with abscissas $1,2, \ldots, 2n$ in the same order they occupy in the line. The box office is located at the origin. To each person having a ten-dollar bill we assign an ordinate $+1$, and to each person having a five-dollar bill, an ordinate -1. The ordinates defined in this way at the points with integral values of the abscissa are then summed from left to right, and at each point the partial sum is plotted. It is easy to see that this sum is zero at the point with the abscissa $2n$ (we have n terms equal to $+1$ and n equal to -1). We connect adjacent points by means of line segments, and also connect the leftmost of these points to the origin in the same way. We shall call this line a trajectory.

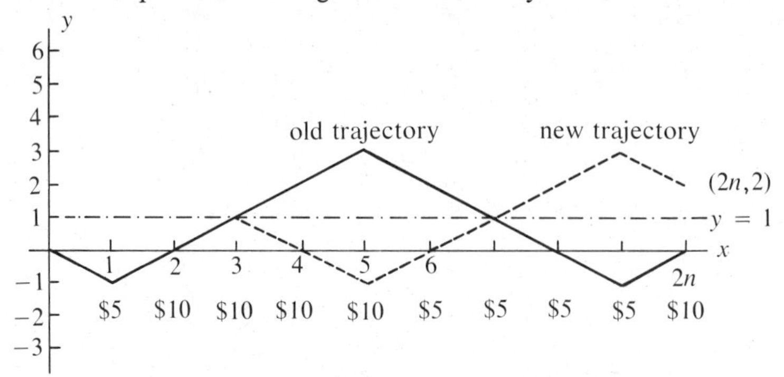

The total number of distinct possible trajectories is equal to $\binom{2n}{n}$ because we have n ascents distributed among $2n$ steps. The trajectories favorable to the event that nobody has to wait will be those and only those trajectories that do not go above the x-axis, i.e., they do not touch or intersect the line $y = 1$. Let us compute the total number of trajectories which touch or intersect the line $y = 1$ at least once. For each such trajectory, we construct a new trajectory that coincides with the old one up to the first point of contact with the line $y = 1$, and from that point it is the mirror image of the old trajectory with respect to the line $y = 1$. The new trajectories corresponding to queues with waiting go from the point (0,0) to the point $(2n,2)$. There is obviously a one-to-one correspondence between queues with waiting, the old trajectories touching the line $y = 1$ and the new trajectories from (0,0) to $(2n,2)$. There are two more ascents than descents in such new trajectories: altogether there are $2n$ steps and therefore there are $(n + 1)$ ascents and $(n - 1)$ descents. The total number of new trajectories corresponding to queues with waiting is hence $\binom{2n}{n+1}$. Finally, the total number of queues without waiting is $\binom{2n}{n} - \binom{2n}{n+1}$, hence

$$P(\text{no waiting}) = \frac{\binom{2n}{n} - \binom{2n}{n+1}}{\binom{2n}{n}} = \frac{1}{n+1}.$$

EXAMPLE 13 (DIFFICULT)

A girl wrote n letters to her n boyfriends and sealed them in envelopes. The next day she wrote addresses at random and sent the letters. Find the probability P_n that at least one boy got the letter written for him.

SOLUTION. Denote A_k the event that the k-th envelope is correctly addressed. Then we are asking for $P\left(\bigcup_{i=1}^{n} A_i\right)$. Let P_n denote this probability. As in Exercise 14 (c) in Chapter V, we can see that

$$P_n = P\left(\bigcup_{i=1}^{n} A_i\right) = \sum_{i=1}^{n} P(A_i) - \sum_{i\neq j} P(A_i \cap A_j) + \sum_{i\neq j\neq k} P(A_i \cap A_j \cap A_k) \ldots$$
$$+ (-1)^{n+1} P\left(\bigcap_{i=1}^{n} A_i\right).$$

Now $P(A_i) = \frac{(n-1)!}{n!}$ because the envelopes can be addressed in $n!$ ways, but if we insist that the envelope i has to be correctly addressed, then there are $(n-1)!$ ways in which one can do it. Similarly, $P(A_i \cap A_j) = \frac{(n-2)!}{n!}$ if two envelopes are to be addressed correctly;

$$P(A_i \cap A_j \cap A_k) = \frac{(n-3)!}{n!}, \text{ three envelopes addressed correctly;}$$

$$\vdots$$

$$P\left(\bigcap_{i=1}^{n} A_i\right) = \frac{1}{n!}, \text{ all envelopes addressed correctly.}$$

Now the first sum contains exactly $\binom{n}{1}$ identical terms $\frac{(n-1)!}{n!}$, the second sum contains $\binom{n}{2}$ identical terms $\frac{(n-2)!}{n!}$, the third sum contains $\binom{n}{3}$ identical terms $\frac{(n-3)!}{n!}$, and so on. Therefore

$$P_n = P\left(\bigcup_{i=1}^{n} A_i\right) = \binom{n}{1}\frac{(n-1)!}{n!} - \binom{n}{2}\frac{(n-2)!}{n!} + \binom{n}{3}\frac{(n-3)!}{n!} + \ldots + (-1)^{n+1}\frac{1}{n!}$$
$$= 1 - \frac{1}{2!} + \frac{1}{3!} - \frac{1}{4!} + \ldots + (-1)^{n+1}\frac{1}{n!} = \sum_{i=1}^{n} (-1)^{i+1}\frac{1}{i!}.$$

A table of the values P_n is as follows:

The number of letters n	Probability P_n that at least one boy gets the correct letter
2	.5
3	.666667
4	.625000
5	.633333
6	.631944
7	.632143
8	.632118

It can be shown that P_n is approaching $1 - \frac{1}{e} = .632121$ (where $e = 2.718282$ is the base of the natural logarithms).

EXERCISES (for Chapters VI and VII)

(Answers in Appendix to exercises marked)*

1. A coin is flipped, then 1 card is selected from an ordinary deck of 52 cards, then an ordinary die is rolled. In how many distinct ways can these three procedures be accomplished?

*2. (a) How many license plates can be made if each plate contains three different letters and three different digits? (Assume that 26 letters are available.)
   (b) Solve the problem if three letters are followed by three digits in such a way that the first letter may not be X, the first digit may not be zero, but there is no condition that all letters and all digits have to be distinct.

3. A digit combination lock contains four disks on a common axis. Each disk is divided into ten sectors with the digits 0 through 9 on the sectors. The lock can open only if each of the disks occupies a certain position with respect to the body of the lock. Find the probability that the lock will open for an arbitrary combination of the digits.

*4. A "crown and anchor" wheel used in gambling can stop in 18 equally likely positions numbered 1 to 18. A customer placed bets on all the positions divisible by 3. What is the probability that he will win?

5. A cube whose faces are colored is split into 1000 small cubes of equal size. The cubes thus obtained are mixed thoroughly. Find the probability that a cube drawn at random will have exactly two colored faces.

6. A pair of fair dice is thrown. Find the probability that the maximum of two numbers is
   (a) greater than 4.
   (b) equal to 4.

7. An empty train coach has nine separated compartments. Five passengers enter the train at night, and each of them selects a compartment at random. What is the probability that at least two of them will find themselves in the same compartment in the morning?

*8. Six equally beautiful girls decided to attend a five o'clock tea dance and sat together around the same table. For the first dance, four young gentlemen starting running toward their table, each of them obviously having selected one of the girls well ahead of time. Assuming that the men did not consult each other and that they will stick to their decisions, what is the probability that no one of them will have to return sadly frustrated without a partner for the dance?

9. How many different necklaces can be made from six beads of different colors, always using all of them?

10. The 26 letters of the alphabet are written in a row by a monkey. What is the probability that x and y will be adjacent?

11. For an experiment in extrasensory perception, four cards from an ordinary deck are placed face down in a row before the subject. What is the probability that he names each card correctly if he guesses at random?

*12. Four boys and four girls attend a party. They sit along a straight table. What is the probability that
   (a) four girls sit together?
   (b) the boys and girls alternate?

13. Four boys and four girls sit around a circular table. What is the probability that
   (a) the four boys sit together?
   (b) the boys and girls alternate?

14. Four men and their wives attend a party. If all of them are seated at random around a circular table, find the probability that the appropriate couples sit together.

*15. A party of 12 men, of whom A and B are 2, stand in a line. Find the probability that
   (a) A and B are next to one another.
   (b) there are exactly four men between A and B.
   (c) there are no more than four men between A and B.

16. The Security Council of the United Nations numbers 15 members, with Great Britain, China, France, Russia and the U.S. as permanent members. If at a meeting the members sit down at random, find the probability that the British and French delegates are next to each other but that the Russian and American delegates are not, if
   (a) the delegates sit in a row.
   (b) the delegates sit around a circular table.

17. There are 10 people at a meeting.
   (a) In how many ways can a committee of three be selected?
   (b) In how many ways can a committee consisting of a chairman, a secretary, and a treasurer be selected? (It is known that the selection will be in the order indicated above.)

*18. A student studies 12 problems, from which the professor will choose (randomly) 6 for an exam. If the student can solve 8 out of the 12 problems, what is the probability that (s)he can solve at least 5 problems during the exam?

19. Five women out of eleven at a party are single. If a young man will introduce himself at random to six of them, what is the probability that
(a) exactly three will be single?
(b) at least two will be single?

20. A lot contains 10 items, of which 4 are defective. If a random sample containing five items is taken, find the probability that the sample will contain
(a) exactly three defective items.
(b) at least three nondefective items.

*21. A student is to answer 10 out of 13 questions on an exam.
(a) How many choices has he?
(b) How many if he must answer the first two?
(c) How many if he must answer exactly three of the first five questions?

22. You need five eggs to make omelets for breakfast. You find a dozen eggs in the refrigerator but do not realize that three of these eggs are rotten. What is the probability that of the five eggs you choose
(a) none is rotten?
(b) exactly one is rotten?
(c) exactly two are rotten?

*23. A domino piece selected at random is not a double. Find the probability that a second piece, also selected at random, will match the first one. (See Example 6 of this chapter.)

24. Two of ten tickets are prizewinners. Find the probability that among five tickets taken at random
(a) one is a prizewinner.
(b) two are prizewinners.
(c) at least one is a prizewinner.

*25. In a lottery there are 90 numbers, of which 5 win. By agreement, one can bet any sum on any one of the 90 numbers, on any set of 2, 3, 4 or 5 numbers. What is the probability of winning in each of the indicated cases? (In order to win, all the numbers selected for a bet must be winning.)

*26. The manager of a music store ordered 100 cheap transistor radios. He does not trust the manufacturer, and therefore he decided to test the quality: he will accept the whole lot only if out of five randomly selected radios no more than one will be defective. What is the probability that he will accept the order if there are actually six defective radios in the whole lot?

27. n persons are seated in an auditorium that can accommodate $n + k$ people. Find the probability that $m \leq n$ given seats are occupied. (Assume that people select their seats at random.)

*28. (a) One straight line is determined by two points in a plane. Three lines are determined by three noncolinear points. Six lines are determined by four points, no three of which are colinear. How many lines are determined by n points, no three of which are colinear?
(b) A triangle has no diagonals; a quadrilateral has two diagonals; a pentagon has five diagonals. How many diagonals does a polygon of n sides have?

29. Three tickets are selected at random from among five tickets worth one dollar each, three tickets worth three dollars each and two tickets worth five dollars each. Find the probability that
(a) at least two of them have the same price.
(b) all three of them cost seven dollars.

*30. A box contains four red, four white, and five green balls. Three balls are drawn from the box together. Find the chance that they may be
(a) all of different colors.
(b) all of the same color.

31. A supermarket shelf has 3 brands of canned tuna on it: 10 cans of the first kind, 5 of the second kind and 3 of the third. Suppose that six cans are chosen randomly. What is the probability that two cans of each brand are picked?

32. A box contains 20 jars of peach preserves, 15 of strawberry and 18 of raspberry. If four jars are selected from the box at random, find the probability that
(a) all four are peach preserves.
(b) exactly three are peach preserves.
(c) at least three are peach preserves.

*33. An urn contains 20 balls, of which 5 are red, 5 are white, 5 are blue and 5 are green. Suppose a sample of five balls is chosen (without replacement).
(a) What is the probability that the sample contains three balls of one color and two of another?
(b) What is the probability that the sample contains at least one ball of each color?

34. Four cards are drawn out of a well shuffled deck of fifty-two cards. What is the probability of getting exactly two spades and exactly two aces among them?

35. Thirteen toys are to be distributed among three children so that the oldest one gets five toys and the other two get four toys each. In how many ways can you do it?

36. Find the probability that when a bridge hand is dealt, each player has exactly one ace.

*37. Peter and Sam decide to flip a coin k times, Peter betting a dollar on heads and Sam a dollar on tails. What is the probability that they will break even? (Discuss with respect to k.)

38. The numbers 2, 4, 6, 7, 8, 10, 11, 12, 13 and 17 are written, respectively, on 10 indistinguishable cards. Two cards are selected at random from the ten. Find the probability that the fraction formed with them is reducible.

*39. Determine the probability that a randomly selected integer N with at most six digits gives a number ending with a 1 as a result of
(a) squaring.
(b) raising to the fourth power.

*40. You ask your friend to write down nine as a sum of four nonnegative integers. He just writes the required numbers at random. What is the probability that they will contain a five? (Consider partitions in different order as different.)

41. You tell your friend that all the numbers in problem 40 must be positive. What is the probability now?

*42. In a town of n inhabitants, a person tells a rumor to a second person, who in turns repeats it to a third person, and so on. At each step the recipient of the rumor is chosen at random from the $(n - 1)$ people available. Find the probability that the rumor will be told r times without
(a) returning to the originator.
(b) being repeated to any person.

43. Eight couples attend a dancing party. The highlight of the party is a midnight dance for which every woman gets a partner selected at random. Find numerically the probability that no original couple will dance together.

44. Ten men checked their hats in the cloakroom. The attendant hopelessly confused the hats, and decided, therefore, to hand them out at random. What is the probability that no man gets his own hat?

45. Using the binomial theorem, calculate:
(a) $(2x + 1)^4$.
(b) $\left(3x^2 - \frac{y}{2}\right)^5$.
(c) $(t - 2u)^8$.
(d) $(a^2 + b^4)^7$.
(e) $(3 - 6t)^6$.
(f) $(2 + 4w)^3$.

46. Use the binomial formula for approximation of the following:
(a) $(.993)^7$. (b) $(1.004)^{15}$. (c) $(.995)^4$.
(d) $(.998)^{20}$. (e) $(1.001)^{30}$. (f) $(.989)^2$.

47. Prove that $\sum_{k=0}^{n} \binom{n}{k}(-1)^k = 0$. (See Example 17 of Chapter VI.)

48. Prove that $\sum_{i=0}^{n} \binom{2n}{2i} = 2^{2n-1}$.

49. Prove that $\sum_{i=0}^{n} \binom{n}{i}^2 = \binom{2n}{n}$.

50. Prove that $\sum_{i=0}^{n} \binom{2n}{2i} = \sum_{i=0}^{n-1} \binom{2n}{2i + 1}$.

51. Using the multinomial theorem, find
(a) $(a + 2 - c)^3$.
(b) $(x - y + 1)^4$.
(c) $(2t + u - 1)^3$.

52. Letters of the word LATENT are written on six cards. The cards are shuffled, and then drawn one at a time at random. What is the probability that the order of the letters will create the word TALENT?

CONDITIONAL PROBABILITY AND INDEPENDENCE

Many probabilities concurring preuaile much.

Thomas Granger, 1620

In the previous chapters, we investigated probabilities of events related to an experiment under the assumption that we have no prior information about the outcome of an actual trial. However, very often we have some incomplete and distorted information which still leaves some uncertainty about the outcome, but which eliminates some possibilities and influences the likelihood of others. In such situations we speak about *conditional probabilities.* (As a matter of fact, we always have some information about a trial before its realization, even though perhaps very incomplete. Therefore some authors consider all probabilities to be conditional. We will not follow this point of view in our treatment.)

For example, imagine the following experiment: A card is drawn at random from a well shuffled deck of bridge cards. What is the probability that it is a red face card? (A red card means hearts or diamonds, a face card means a jack, queen or king.) If we denote this event by A, then obviously

$$P(A) = \frac{6}{52} = \frac{3}{26}.$$

Imagine now that you are told that the drawn card is a heart (denote this event by B). Thus there are now only 13 possible outcomes instead of the original 52, and of these 3 are favorable. Therefore, if we denote the conditional probability that A occurs given that B has occurred $P(A/B)$, then obviously

$$P(A/B) = \frac{3}{13} = \frac{6}{26}.$$

We can see that the conditional probability of A given B considerably increased, compared with $P(A)$. Can we conclude that there is any relation between ordinary and conditional probabilities? We can easily see that

$$P(B) = \frac{13}{52}, \quad P(A \cap B) = \frac{3}{52}$$

and it looks like that in this case:

$$P(A/B) = \frac{3}{13} = \frac{\frac{3}{52}}{\frac{13}{52}} = \frac{P(A \cap B)}{P(B)}.$$

Or consider the following example. If a symmetrical die is rolled, what is the probability that the outcome will be less than four (event A), given the fact that it is an even number (event B)? Obviously,

$$P(A) = \frac{3}{6}, \quad P(B) = \frac{3}{6}, \quad P(A \cap B) = \frac{1}{6} \quad \text{and} \quad P(A/B) = \frac{1}{3} = \frac{P(A \cap B)}{P(B)}$$

In this case also, we can establish the above mentioned relation. As a matter of fact, we can easily see that this relation is very natural and realistic. Perhaps we should remember for a while the classical definition of probability that we abandoned in Chapter IV. If n is the total number of equally likely outcomes, and m_A is the total number of outcomes favorable to A, then we defined

$$P(A) = \frac{m_A}{n}.$$

Now, what is the probability of A if we know that B has happened? In this case there are only m_B (assume $m_B > 0$) possible equally likely outcomes, out of which $m_{A \cap B}$ are favorable to A. Hence

$$P(A/B) = \frac{m_{A \cap B}}{m_B} = \frac{\frac{m_{A \cap B}}{n}}{\frac{m_B}{n}} = \frac{P(A \cap B)}{P(B)}.$$

If we look closer at this string of equalities, we notice that the first one indicates that the conditional probability of A given B is in fact ordinary probability of A, if we restrict our sample space to all the outcomes that form the set B. If we know that B has happened, then nothing else can happen and we can take B as a "new" sample space. The conditional probability is in this sense a relative probability with respect to the event which has happened. We have used the classical definition of probability only to motivate the general definition of conditional probabilities that comes next, and also to give the reader some intuitive feeling. We can now discard again the classical definition of probability, but we should keep in mind the above derived conclusion about the probability of A given that B has happened.

These examples lead us to define also, in the general case, the probability of A under the condition that B has happened as the fraction of $P(A \cap B)$ and $P(B)$. It is obvious that we have to restrict ourselves to the cases when $P(B) > 0$. *Throughout this chapter we always assume that the probabilities of the conditions are strictly positive.*

DEFINITION 1: If A,B are two events and $P(B) > 0$, then the conditional probability that A occurs given that B has occurred, written $P(A/B)$ (read, the probability of A given B), is defined as

$$P(A/B) = \frac{P(A \cap B)}{P(B)}.$$

COMMENT. We can see that Definition 1 implies

$$P(A \cap B) = P(B)P(A/B) = P(A)P(B/A).$$

This fact is very important and will be often used and generalized for the case of an intersection of finitely many sets.

EXAMPLE 1

It is known that 40% of the families in a certain suburban community have a mortgage on their house, about 60% of families have a mortgage on the car and 20% of families have both mortgages. In order to conduct a poll, some families are chosen at random for an interview. If the selected family has a house mortgage (event B), what is the probability that it has also a car mortgage (event A)?

SOLUTION. From the given data, we get

$$P(A) = .6, \quad P(B) = .4, \quad P(A \cap B) = .2.$$

The required probability is actually

$$P(A/B) = \frac{P(A \cap B)}{P(B)} = \frac{.2}{.4} = \frac{1}{2}.$$

We can also find the probability that a family has a house mortgage, given the fact that it has a car mortgage:

$$P(B/A) = \frac{P(A \cap B)}{P(A)} = \frac{.2}{.6} = \frac{1}{3}.$$

EXAMPLE 2

A die is loaded so that the probability of any number of points is directly proportional to this number. The die is rolled and it is known that the outcome is an even number (event B). What is the probability that it is less than four (event A)? Find also $P(B/A)$.

SOLUTION. If p_i is the probability that we get i points, then

$$p_i = ip_1.$$

Further, $\sum_{i=1}^{6} p_i = p_1 + 2p_1 + 3p_1 + 4p_1 + 5p_1 + 6p_1 = 21/p_1.$

Therefore $$p_1 = \frac{1}{21}.$$

$$P(B) = P(\{2,4,6\}) = p_2 + p_4 + p_6 = \frac{1}{21}(2 + 4 + 6) = \frac{12}{21}.$$

$$P(A \cap B) = P(\{2\}) = p_2 = \frac{2}{21}.$$

The probability that we get less than four points, given the fact that the outcome is an even number, is therefore

$$P(A/B) = \frac{P(A \cap B)}{P(B)} = \frac{\frac{2}{21}}{\frac{12}{21}} = \frac{2}{12} = \frac{1}{6}.$$

Further, $$P(A) = P(\{1,2,3\}) = \frac{1}{21}(1 + 2 + 3) = \frac{6}{21}.$$

The probability that we get an even number of points, given the fact that the outcome is less than four, is therefore

$$P(B/A) = \frac{P(A \cap B)}{P(A)} = \frac{\frac{2}{21}}{\frac{6}{21}} = \frac{1}{3}.$$

→ **THEOREM 1:** Let (Ω, P) be a probability measure space. For any fixed $B \subset \Omega$, the function $P(A/B)$ defined for all $A \subset \Omega$ is a probability measure.

Proof: We have to show that the axioms A1–A3 are satisfied.
A1: We know that $P(A \cap B) \geq 0$, $P(B) > 0$, hence

$$P(A/B) = \frac{P(A \cap B)}{P(B)} \geq 0.$$

A2: $P(\Omega/A) = \dfrac{P(A \cap \Omega)}{P(A)} = \dfrac{P(A)}{P(A)} = 1.$

A3: Let A and C be disjoint sets, then

$$P(A \cup C/B) = \frac{P((A \cup C) \cap B)}{P(B)} = \frac{P((A \cap B) \cup (C \cap B))}{P(B)} = \frac{P(A \cap B)}{P(B)} + \frac{P(C \cap B)}{P(B)} = P(A/B) + P(C/B).$$

q.e.d.

EXAMPLE 3

An ordinary deck of 52 cards has been dealt to 4 players, each getting 13 cards. Our man, even before he touched his cards, noticed that the player on his right had the ace of hearts. The first question he asks is: "What is the probability that the man on my right has all four aces in his hand?"

SOLUTION. Let the events A and B be as follows:

$A = \{\text{the observed hand contains the ace of hearts}\}$

$B = \{\text{the observed hand contains all four aces}\}$

Then $P(B/A) = \dfrac{\binom{48}{9}}{\binom{51}{12}} = \dfrac{12.11.10}{51.50.49} = .010564.$

If the man does not have the chance to observe his neighbor's hand, then

$$P(B) = \frac{\binom{48}{9}}{\binom{52}{13}} = \frac{13.12.11.10}{52.51.50.49} = .002641.$$

Comparing the numerical values, we easily see that in this case $P(B/A) > P(B)$.

EXAMPLE 4

Two symmetrical dice are rolled, and the maximum on the dice is six (event B). What is the probability that the sum is at least eight (event A)?

COMMENT. The maximum of two numbers, denoted $\max(x,y)$ is the bigger of the two numbers. If $x = y$, then $\max(x,y) = x = y$.

SOLUTION. We have to find $P(A/B)$. It is quite easy to find straightforwardly, but using the consequences of Theorem 1 is even simpler. Because $P(A/B)$ is a probability measure, we can write

$$P(A/B) = 1 - P(A'/B).$$

Now $B = \{(1,6), (2,6), (3,6), (4,6), (5,6), (6,6), (6,5), (6,4), (6,3), (6,2), (6,1)\}$

$A' \cap B = \{(1,6), (6,1)\}$

$$P(B) = \frac{11}{36}, P(A' \cap B) = \frac{2}{36}$$

$$P(A/B) = 1 - P(A'/B) = 1 - \frac{P(A' \cap B)}{P(B)} = 1 - \frac{\frac{2}{36}}{\frac{11}{36}} = \frac{9}{11}.$$

However, if we wish, we can evaluate $P(A/B)$ directly.

$$A \cap B = \{(2,6), (3,6), (4,6), (5,6), (6,6), (6,5), (6,4), (6,3), (6,2)\}$$

$$P(A/B) = \frac{P(A \cap B)}{P(B)} = \frac{\frac{9}{36}}{\frac{11}{36}} = \frac{9}{11}.$$

We can also calculate the conditional probability that the maximum is six, given the fact that the sum is at least eight. This is equal to $P(B/A)$.

Now $$P(A) = \frac{15}{36} \quad \text{(express } A \text{ as a set)}$$

and therefore $$P(B/A) = \frac{P(A \cap B)}{P(A)} = \frac{\frac{9}{36}}{\frac{15}{36}} = \frac{9}{15}.$$

We are now going to derive several important theorems about conditional probabilities. These theorems are very useful in practical applications, but their theoretical formulations look perhaps somewhat complicated, abstract and puzzling. For this reason, we begin with an analysis of a situation that might happen in anybody's life. Only after we have acquired some feeling for this type of problem shall we give a general theoretical formulation of the corresponding theorems. Unfortunately, an exact formulation of these theorems looks rather dry and forbidding. Let us start, therefore, with something tantalizing, in the hope that it will keep the reader moving. More formal examples follow after the proof of Theorem 3.

EXAMPLE 5

Ernie is enjoying the Halloween dancing party organized by the Student's Union very much. Perhaps he is so happy because the drinks are very good tonight, but more likely it is Carol and Alice who are the source of his exuberant mood. He knows that sooner or later he will ask one of the girls to spend a quiet evening just for two with him, but he cannot decide whom he should ask. Carol is very attractive and beautiful; however, he is aware that she is a little bit uptight and inaccessible. If he asks her, he can imagine that they will have a tremendous time together, but there is some chance, say three in five, that everything will go wrong. He is more sure about Alice. She is not as exciting for him, but she is really a very nice and warm person, and as far as he is concerned, 90 per cent of the time they have real fun together. To make a decision between the girls is virtually beyond Ernie's mental powers, and he decides to roll a die. If a one or a two appears he will ask Alice, but otherwise he will ask Carol. He is quite confident that he will not be immediately rebuffed by either girl, but he also knows that if the evening deteriorates later on he will have lost his opportunity with the other girl. He feels that his illustrious moral code of the cavalier would not allow it. Given this conflicting situation, what is the chance that Ernie will enjoy a really exciting evening?

SOLUTION. We have to be a little formal to solve this informal problem. According to the outcome on the die, Ernie can spend his evening either with Alice (event A) or with Carol (event C). The probabilities of choosing the girls are as follows:

$$P(A) = \frac{1}{3}, \qquad P(C) = \frac{2}{3}.$$

In either case, he can have either an exciting evening and fun (event E) or not. The probability of everything going wrong with Carol is "three in five," and therefore the conditional probability of an exciting evening with Carol is

$$P(E/C) = \frac{2}{5} = \frac{4}{10}.$$

The conditional probability of having an exciting evening given that he has chosen Alice is

$$P(E/A) = .9 = \frac{9}{10}.$$

He can have an exciting evening either with Carol or with Alice, but not with both of them at the same time—at least as far as we know. Therefore the events C and A are mutually exclusive, and because he will certainly ask one of the girls, they are also exhaustive. Therefore

$$E = (E \cap A) \cup (E \cap C)$$

and

$$P(E) = P(E \cap A) + P(E \cap C).$$

(Can you interpret the last line in words?)

According to the comment after Definition 1, we can write

$$P(E \cap A) = P(A) \cdot P(E/A) = \frac{1}{3} \cdot \frac{9}{10},$$

$$P(E \cap C) = P(C) \cdot P(E/C) = \frac{2}{3} \cdot \frac{4}{10}.$$

And finally we get

$$P(E) = \frac{1}{3} \cdot \frac{9}{10} + \frac{2}{3} \cdot \frac{4}{10} = \frac{9 + 8}{30} = \frac{17}{30}.$$

Therefore Ernie has $\frac{17}{30}$ chance of spending an exciting evening with a girl. The idea which we used to solve this problem is sometimes referred to as the *formula of total probability,* and we shall derive it in a general form in Theorem 2.

EXAMPLE 6

(See Example 5.) Imagine now that not only has Ernie had a tremendously exciting evening with a girl yesterday, but all he remembers about it this morning is that he perhaps behaved a little too benevolently and excessively towards a nearby bottle of 132 proof Demerara rum. He really does not remember with which of the two girls he spent the last evening, and his exuberant cavalier mood is irreversibly gone when he discovers that his car keys have disappeared. Now he has to phone Carol or Alice. Which one should he call first; which one is more likely to have spent last evening with him?

SOLUTION. We have to find the conditional probability that Ernie was with Carol given the fact that he has had an exciting evening; that is, we need $P(C/E)$.

Now
$$P(C/E) = \frac{P(C \cap E)}{P(E)} = \frac{P(C)P(E/C)}{P(E)}$$

as follows from the comment just after Definition 1. From the previous example, we know that $P(E) = \frac{17}{30}$, $P(C) = \frac{2}{3}$ and $P(E/C) = \frac{4}{10}$, therefore

$$P(C/E) = \frac{P(C)P(E/C)}{P(E)} = \frac{\frac{2}{3} \cdot \frac{4}{10}}{\frac{17}{30}} = \frac{8}{17}.$$

Similarly, we can calculate the conditional probability that he was with Alice given the fact that he had an exciting evening.

$$P(A/E) = \frac{P(A \cap E)}{P(E)} = \frac{P(A)P(E/A)}{P(E)} = \frac{\frac{1}{3} \cdot \frac{9}{10}}{\frac{17}{30}} = \frac{9}{17}.$$

We can see that Alice is more likely to have spent the evening with Ernie, even though originally he gave much more preference to Carol. We can observe the following changes in the probabilities before and after the event.

$$P(C) = \frac{2}{3} = \frac{34}{51}. \qquad P(C/E) = \frac{24}{51}.$$

$$P(A) = \frac{1}{3} = \frac{17}{51}. \qquad P(A/E) = \frac{27}{51}.$$

We can now schematically express our calculations in the form of a *tree.* There are two possibilities in the beginning: the selection of either Carol (C) or Alice (A), with the corresponding probabilities $\frac{2}{3}$ and $\frac{1}{3}$. After that, there are again two possibilities: either an exciting evening (E) or a failure (F), and again we can find the corresponding probabilities.

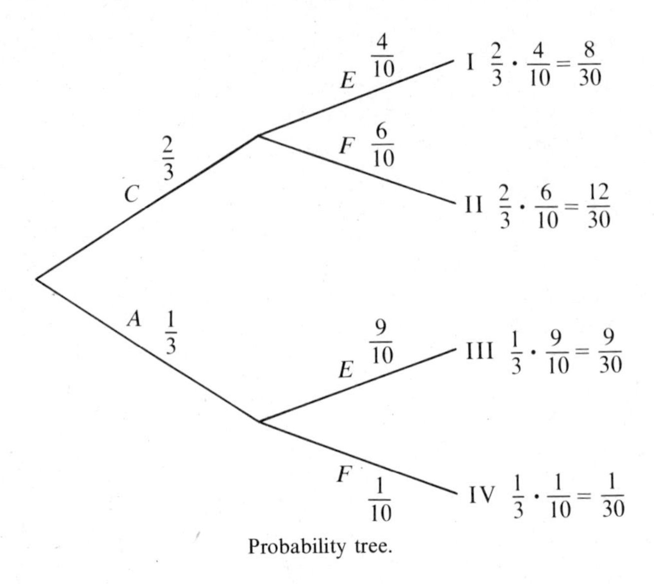

Probability tree.

There are four branches I, II, III and IV, and the probability of each one is the product of the probabilities along the branch. The event E occurs in case I or III, therefore

$$P(E) = \frac{8}{30} + \frac{9}{30} = \frac{17}{30}.$$

Further

$$P(C/E) = \frac{P(\text{branch I})}{P(E)} = \frac{8}{17},$$

$$P(A/E) = \frac{P(\text{branch III})}{P(E)} = \frac{9}{17}.$$

This schematic approach does not replace the previous calculations; it makes them better organized, systematic and clear. We shall have many opportunities to use this method of solving examples involving conditional probabilities. Now we shall give the previously mentioned general theorems, the proofs of which are based on the same ideas as the solutions of Examples 5 and 6.

→ **THEOREM 2:** *Formula of Total Probability:* Let (Ω, P) be a probability measure space and $B \subset \Omega$. Further, let $A_1, A_2, \ldots, A_n$ be such that $A_j \subset \Omega$, $A_i \cap A_j = \phi$ for $i \neq j$ and $\bigcup_{j=1}^{n} A_j = \Omega$; i.e., the sets A_j, $j = 1, 2, \ldots, n$ form a partition of Ω. Then

$$P(B) = \sum_{j=1}^{n} P(A_j)P(B/A_j).$$

Proof: Using the distributive law, and the fact that the sets A_j form a partition of Ω, we get

$$B = B \cap \Omega = B \cap \bigcup_{j=1}^{n} A_j = \bigcup_{j=1}^{n} (B \cap A_j).$$

The sets $(B \cap A_j)$ for $j = 1,2, \ldots, n$ are pairwise disjoint. We can see from the following diagram that B is expressed as a union of pairwise disjoint sets $(B \cap A_j)$.

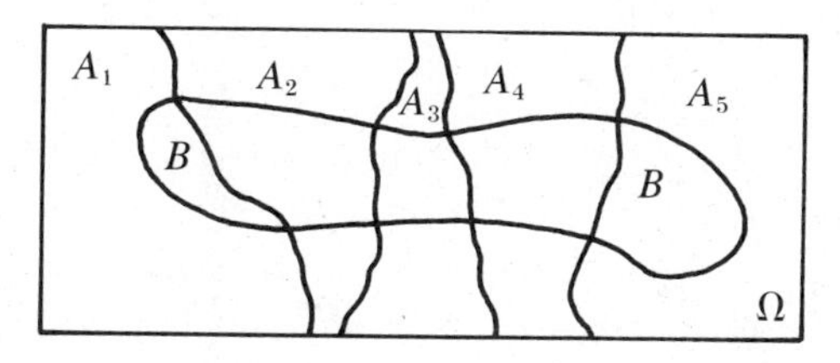

Therefore, by Theorem 6 in Chapter V, we get

$$P(B) = P\left(\bigcup_{j=1}^{n} (B \cap A_j)\right) = \sum_{j=1}^{n} P(B \cap A_j).$$

However, according to the comment after Definition 1, page 78,

$$P(B \cap A_j) = P(A_j)P(B/A_j).$$

And therefore,

$$P(B) = \sum_{j=1}^{n} P(B \cap A_j) = \sum_{j=1}^{n} P(A_j)P(B/A_j).$$

q.e.d.

COMMENT 1. This theorem gives us a method for calculating $P(B)$ by splitting B into several parts $(B \cap A_j)$, evaluating each of the probabilities $P(B \cap A_j)$ and then adding them together in order to get $P(B)$.

COMMENT 2. In Example 5, we assigned the events as follows:

$$A_1 = A, \qquad A_2 = C, \qquad B = E.$$

Obviously, A_1 and A_2 (Alice or Carol) form a partition of Ω. (Ernie must ask one of the girls but he cannot ask both of them.) We calculated $P(E)$ according to Theorem 2.

THEOREM 3: *Bayes' Theorem:* Let (Ω,P) be a probability measure space and $B \subset \Omega$. Further, let $A_j \subset \Omega$, $j = 1,2, \ldots, n$ be a partition of Ω. Then for every $i = 1,2, \ldots, n$

$$P(A_i/B) = \frac{P(A_i)P(B/A_i)}{\sum_{j=1}^{n} P(A_j)P(B/A_j)}.$$

Proof: From Definition 1, we get

$$P(A_i/B) = \frac{P(B \cap A_i)}{P(B)}$$

and from the comment after Definition 1, we get

$$P(A_i \cap B) = P(A_i)P(B/A_i).$$

If we use the formula of total probability (Theorem 2), we get

$$P(A_i/B) = \frac{P(A_i)P(B/A_i)}{\sum_{j=1}^{n} P(A_j)P(B/A_j)}.$$

q.e.d.

COMMENT 1. The probabilities $P(C/E)$ and $P(A/E)$ in Example 6 were calculated using Bayes' theorem.

COMMENT 2. The probabilities $P(A_i)$ $i = 1,2, \ldots, n$ are the probabilities of A_i's *before* B happened, and therefore we call them *à priori* probabilities. However $P(A_i/B)$ $i = 1,2, \ldots, n$ are the probabilities of A_i's *after* it had become known that B happened and therefore they are called *à posteriori* probabilities. In Examples 6 and 7, $P(A)$ and $P(C)$ are à priori probabilities of Alice and Carol being selected before anything happened while $P(A/E)$ and $P(C/E)$ are à posteriori probabilities that Alice or Carol had been selected given what happened the previous evening. Examples 5 and 6 might appear rather shallow and unimportant. Let us consider a serious problem related to industrial quality control.

EXAMPLE 7

Three machines A_1, A_2, A_3 produce 50%, 30% and 20% of the total number of items in a factory. The percentage of defective output of these machines is 3%, 4% and 5%, respectively.
(a) If an item is selected at random, find the probability that it is defective.
(b) Suppose that the item selected at random proved to be defective. What is the probability that it was produced by machine A_1?

SOLUTION. (a) Let us denote events as follows:

B—the selected item is defective

A_1—the selected item has been produced by machine A_1

A_2—the selected item has been produced by machine A_2

A_3—the selected item has been produced by machine A_3

Obviously, A_1, A_2, A_3 is a partition (an item must be produced by machine A_1 or A_2 or A_3 but cannot be produced by two machines at once). We can use the formula of total probability:

$$P(A_1) = .5, \quad P(A_2) = .3, \quad P(A_3) = .2, \quad P(B/A_1) = .03, \quad P(B/A_2) = .04,$$
$$P(B/A_3) = .05.$$

$$P(B) = \sum_{j=1}^{3} P(A_j)P(B/A_j) = (.5)(.03) + (.3)(.04) + (.2)(.05) = .037.$$

(b) In this case we can use Bayes' theorem. We know that $P(A_1) = .5$ and $P(B/A_1) = .03$, hence

$$P(A_1/B) = \frac{P(A_1)P(B/A_1)}{\sum_{j=1}^{3} P(A_j)P(B/A_j)} = \frac{(.5)(.03)}{.037} = \frac{15}{37}.$$

Also in this case we can use a probability tree for the solution.

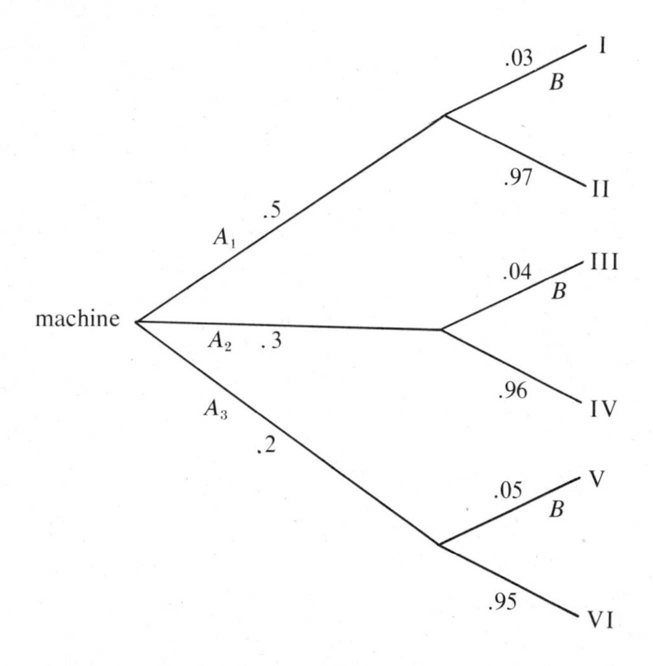

$$P(\text{the selected item is defective}) = P(\text{I}) + P(\text{III}) + P(\text{V}) = (.5)(.03) + (.3)(.04) + (.2)(.05) = .037.$$

Similarly $P(\text{item produced by } A_1/\text{it is defective}) = \dfrac{P(\text{I})}{.037} = \dfrac{(.5)(.03)}{.037} = \dfrac{15}{37}$

EXAMPLE 8

Telecommunications in the presence of a noise. A telegraphic communication system transmits the signals dot or dash. However, there is a random noise present in the transmission channel that causes some distortion of the messages. Assume that an average of two fifths of the dots and one third of the dashes is changed. Suppose that the ratio between the transmitted dots and dashes is 5:3. What is the probability that a received signal will be the same as the transmitted signal if
(a) the received signal is a dot?
(b) the received signal is a dash?

SOLUTION. Let us denote the events as follows:

A—a dot is received

B—a dash is received

H_1—the transmitted signal was a dot

H_2—the transmitted signal was a dash

By assumption, $P(H_1) : P(H_2) = 5{:}3$, hence $P(H_1) = \frac{5}{8}$, $P(H_2) = \frac{3}{8}$.

Further, $P(A|H_1) = \frac{3}{5}$, $P(A|H_2) = \frac{1}{3}$, $P(B|H_1) = \frac{2}{5}$, $P(B|H_2) = \frac{2}{3}$.

From the total probability formula,

$$P(A) = \frac{5}{8}\cdot\frac{3}{5}+\frac{3}{8}\cdot\frac{1}{3}=\frac{1}{2}. \qquad P(B) = \frac{5}{8}\cdot\frac{2}{5}+\frac{3}{8}\cdot\frac{2}{3}=\frac{1}{2}.$$

The required probabilities are

$$P(H_1|A) = \frac{P(H_1)P(A|H_1)}{P(A)} = \frac{\frac{5}{8}\cdot\frac{3}{5}}{\frac{1}{2}} = \frac{3}{4},$$

$$P(H_2|B) = \frac{P(H_2)P(B|H_2)}{P(B)} = \frac{\frac{3}{8}\cdot\frac{2}{3}}{\frac{1}{2}} = \frac{1}{2}.$$

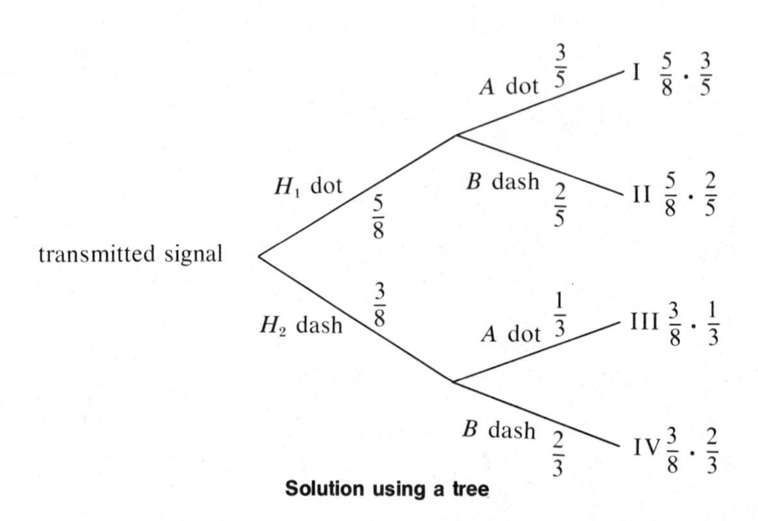

Solution using a tree

$$P(A) = P(\mathrm{I}) + P(\mathrm{III}) = \frac{5}{8}\cdot\frac{3}{5}+\frac{3}{8}\cdot\frac{1}{3}=\frac{1}{2}. \qquad P(B) = P(\mathrm{II}) + P(\mathrm{IV}) = \frac{5}{8}\cdot\frac{2}{5}+\frac{3}{8}\cdot\frac{2}{3}=\frac{1}{2}.$$

$$P(H_1|A) = \frac{P(\mathrm{I})}{P(A)} = \frac{\frac{5}{8}\cdot\frac{3}{5}}{\frac{1}{2}} = \frac{3}{4}. \qquad P(H_2|B) = \frac{P(\mathrm{IV})}{P(B)} = \frac{\frac{3}{8}\cdot\frac{2}{3}}{\frac{1}{2}} = \frac{1}{2}.$$

There are many more solved examples of this type in the next chapter.

Some people have a rather critical view of Bayes' theorem and its practical applications. Consider again Example 6. It is quite reasonable to speak about the à priori probabilities of each girl being selected *before anything happens.* However *after* a selection has been made, we are sure about the selection (at least in principle if we do not know the actual outcome), all the uncertainty, chance or randomness is gone and it makes no sense to speak about the probability that a certain girl was selected. Bayes' theorem gives in such a case only some sorts of pseudo-probabilities.

Sometimes à priori probabilities are not even known and are rather arbitrarily and subjectively guessed. The reasons used for such guesses are usually based on vague principles of "incomplete information or indifference" or on anticipation of the "most adverse situation." These reasons are frequently quite nebulous and irrational. Therefore, such a practice is also considered very objectionable because à priori probabilities have a strong influence on the final solution (see the discussion at the end of Example 13b, Chapter IX).

However, many probabilists and statisticians take a completely opposite point of view regarding Bayes' theorem. They recognize the practical fact that even though a selection has been made, there still can be some uncertainty in the individual's mind related, perhaps, to incomplete or distorted information about the outcome of an experiment. In such a case, Bayes' theorem gives the best estimation of the probabilities of the events which have or have not happened in the past. There is an apparent schism among probabilists and statisticians as far as practical applications of Bayes' theorem are concerned. However, the theorem itself is being utilized in an increasing number of areas, and we shall devote proper attention to it. (See the section on decision under uncertainty, page 289.)

THEOREM 4: *General Rule of Multiplication:* Let $E_1, E_2, \ldots, E_n$ be any sets, $E_i \subset \Omega$, $i = 1,2, \ldots, n$ such that $P(E_1 \cap E_2 \cap \ldots \cap E_{n-1}) > 0$. Then

$$P(E_1 \cap E_2 \cap \ldots \cap E_n) = P(E_1)P(E_2/E_1) \cdot P(E_3/E_1 \cap E_2) \ldots P(E_n/E_1 \cap E_2 \cap \ldots \cap E_{n-1}).$$

Proof: $P(E_1) \cdot P(E_2/E_1)P(E_3/E_1 \cap E_2) \ldots P(E_n/E_1 \cap E_2 \cap \ldots \cap E_{n-1})$

$$= P(E_1) \cdot \frac{P(E_1 \cap E_2)}{P(E_1)} \cdot \frac{P(E_1 \cap E_2 \cap E_3)}{P(E_1 \cap E_2)} \cdots \frac{P(E_1 \cap E_2 \cap \ldots \cap E_n)}{P(E_1 \cap E_2 \cap \ldots \cap E_{n-1})} = P(E_1 \cap E_2 \cap \ldots \cap E_n)$$

q.e.d.

EXAMPLE 9

There are six semiconductors in a box; four of them are in working condition and two are defective. Three semiconductors are selected at random, one after another. Find the probability that the first and third will be nondefective, and the second will be defective.

SOLUTION. Let us denote

E_1—the first item is nondefective

D_2—the second item is defective

E_3—the third item is nondefective

Then the asked probability $P(E_1 \cap D_2 \cap E_3)$ can be found as

$$P(E_1 \cap D_2 \cap E_3) = P(E_1)P(D_2/E_1)P(E_3/E_1 \cap D_2).$$

Now

$$P(E_1) = \frac{4}{6} = \frac{2}{3},$$

$$P(D_2/E_1) = \frac{2}{5},$$

$$P(E_3/E_1 \cap D_2) = \frac{3}{4}.$$

Hence $$P(E_1 \cap D_2 \cap E_3) = \frac{2}{3} \cdot \frac{2}{5} \cdot \frac{3}{4} = \frac{1}{5}.$$

It is quite expedient to use a probability tree if we want to capture systematically all the possibilities offered in this example. Denote by G the selection of a nondefective (good) item and by D the selection of a defective item.

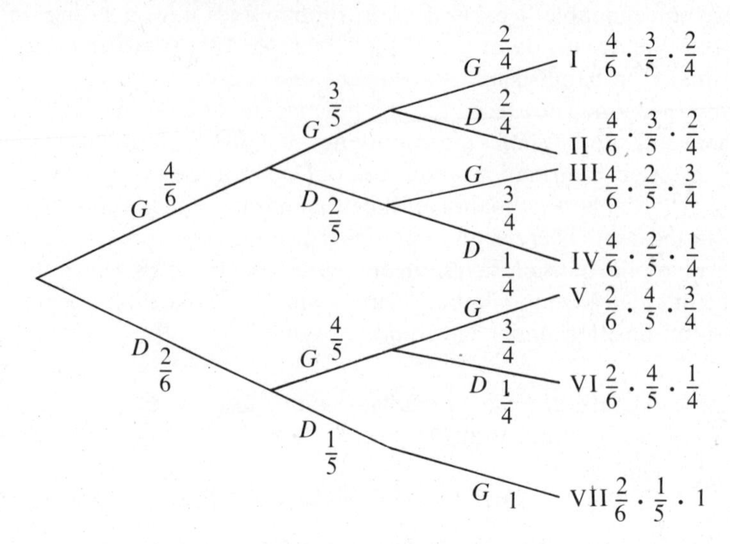

The required probability is the probability of branch III:

$$P = \frac{4}{6} \cdot \frac{2}{5} \cdot \frac{3}{4} = \frac{1}{5}.$$

Let us calculate some other probabilities.

$$\begin{aligned} P(\text{second item defective}) &= P(\text{III}) + P(\text{IV}) + P(\text{VII}) \\ &= \frac{4}{6} \cdot \frac{2}{5} \cdot \frac{3}{4} + \frac{4}{6} \cdot \frac{2}{5} \cdot \frac{1}{4} + \frac{2}{6} \cdot \frac{1}{5} \cdot 1 = \frac{3+1+1}{15} = \frac{4}{15}. \end{aligned}$$

$$\begin{aligned} P(\text{last two items good/more good than defective items obtained}) &= \frac{P(\text{I}) + P(\text{V})}{\frac{4}{5}} \\ &= \frac{\frac{4}{6} \cdot \frac{3}{5} \cdot \frac{2}{4} + \frac{2}{6} \cdot \frac{4}{5} \cdot \frac{3}{4}}{\frac{4}{5}} \\ &= \frac{1}{2} \end{aligned}$$

because

$$\begin{aligned} P(\text{more good than defective items obtained}) &= P(\text{I}) + P(\text{II}) + P(\text{III}) + P(\text{V}) \\ &= \frac{4}{6} \cdot \frac{3}{5} \cdot \frac{2}{4} + \frac{4}{6} \cdot \frac{3}{5} \cdot \frac{2}{4} + \frac{4}{6} \cdot \frac{2}{5} \cdot \frac{3}{4} + \frac{2}{6} \cdot \frac{4}{5} \cdot \frac{3}{4} \\ &= \frac{4}{5}. \end{aligned}$$

COMMENT. This probability can be calculated differently:

$$P(\text{more good than defective items}) = P(3\text{ good}) + P(2\text{ good}) = \frac{\binom{4}{3}\binom{2}{0} + \binom{4}{2}\binom{2}{1}}{\binom{6}{3}}$$

$$= \frac{4 + 12}{20} = \frac{4}{5}.$$

Thus we can see that the same probability can be calculated using different methods. To choose the best and most elegant method for a given problem is a component of the art of solving problems.

The next notion we are going to investigate is that of the *independence* of two events. Imagine the following experiment. Your friend selects one card at random from a well shuffled deck of bridge cards. Denote the event that the selected card will be a face card (jack, queen or king) by A. There are 12 face cards in the deck, therefore

$$P(A) = \frac{12}{52} = \frac{3}{13}.$$

Now imagine that your friend looks at the card, and tells you that it is a heart. Denote this fact by the letter B. There are 13 heart cards, and 3 of them are face cards. Therefore

$$P(A/B) = \frac{3}{13}.$$

We immediately notice that the probability of A does not change whether we do or do not know if the event B has happened; that is,

$$P(A) = P(A/B).$$

The occurrence of the event B does not change the probability of occurrence of the event A. In this case, we are inclined to say that A is independent of B. However, this definition looks somewhat nonsymmetrical. Could we say that if A is independent of B, that is, if

$$P(A) = P(A/B) = \frac{P(A \cap B)}{P(B)}$$

then necessarily B is independent of A? If A is independent of B, then from the above relation, we get

$$P(A \cap B) = P(A)P(B)$$

and therefore

$$P(B/A) = \frac{P(A \cap B)}{P(A)} = \frac{P(A)P(B)}{P(A)} = P(B).$$

Hence B is necessarily independent of A. Independence of the events A and B is therefore symmetrical, even though the condition given above might not look so. However, there is another problem with the above condition for independence: it is applicable only when $P(A)$

> 0 and $P(B) > 0$. We can see that in this case the above condition is equivalent to the following condition:

$$P(A \cap B) = P(A)P(B).$$

We can immediately notice three advantages of the second condition:
(a) If $P(A)P(B) \neq 0$, then both conditions are equivalent.
(b) The latter condition is applicable, even if $P(A)P(B) = 0$.
(c) The latter condition is, at first glance, symmetrical.
For these reasons, we shall accept the latter condition as a definition of independence—however, we should not forget its original intuitive meaning.

DEFINITION 2: Let A and B be two sets of a probability measure space (Ω, P). We say that A and B are independent if

$$P(A \cap B) = P(A)P(B)$$

Otherwise A and B are called dependent.

EXAMPLE 10

Let a fair coin be tossed thrice, and let us denote the events A, B, C as follows:

A—the first toss is a head

B—the second toss is a head

C—exactly 2 heads are tossed in a row

Then

$$A = \{HHH, HHT, HTH, HTT\} \text{ and } P(A) = \frac{4}{8} = \frac{1}{2}.$$

$$B = \{HHH, HHT, THH, THT\} \text{ and } P(B) = \frac{4}{8} = \frac{1}{2}.$$

$$C = \{HHT, THH\} \text{ and } P(C) = \frac{2}{8} = \frac{1}{4}.$$

Further,

$$A \cap B = \{HHH, HHT\} \text{ and } P(A \cap B) = \frac{2}{8} = \frac{1}{4}.$$

$$A \cap C = \{HHT\} \text{ and } P(A \cap C) = \frac{1}{8}.$$

$$B \cap C = \{HHT, THH\} \text{ and } P(B \cap C) = \frac{2}{8} = \frac{1}{4}.$$

Comparing the probabilities:

$$P(A)P(B) = \frac{1}{2} \cdot \frac{1}{2} = \frac{1}{4} = P(A \cap B) \text{ and hence } A \text{ and } B \text{ are independent,}$$

$$P(A)P(C) = \frac{1}{2} \cdot \frac{1}{4} = \frac{1}{8} = P(A \cap C) \text{ and hence } A \text{ and } C \text{ are independent,}$$

while $P(B)P(C) = \frac{1}{2} \cdot \frac{1}{4} \neq \frac{1}{4} = P(B \cap C)$ and hence B and C are dependent.

EXAMPLE 11

Two men shoot at a target. Mr. A hits the target with probability $\frac{1}{4}$, Mr. B hits the target with probability $\frac{2}{5}$. Each shoots once at the target. What is the probability that the target will be hit?

SOLUTION. Let us denote the events as follows:

A—Mr. A hits the target

B—Mr. B hits the target

Then we are seeking the probability $P(A \cup B)$,

$$P(A \cup B) = P(A) + P(B) - P(A \cap B).$$

Now
$$P(A) = \frac{1}{4}, \qquad P(B) = \frac{2}{5}.$$

Further, a success or a failure of Mr. A does not influence the skill of Mr. B and vice versa. Hence we may reasonably assume that the events A and B are independent (see the discussion on page 97).

Therefore
$$P(A \cap B) = P(A) \cdot P(B) = \frac{1}{4} \cdot \frac{2}{5}$$

Finally
$$P(A \cup B) = \frac{1}{4} + \frac{2}{5} - \frac{1}{4} \cdot \frac{2}{5} = \frac{11}{20}$$

THEOREM 5: If A and B are independent sets, then so are the pairs of sets A, B', A', B and A', B'. ←

Proof: A and B are independent, hence

$$P(A \cap B) = P(A)P(B), \quad A' \cap B = B - A = B - A \cap B \quad \text{and} \quad A \cap B \subset B.$$

Using Theorem 3 from Chapter V, we get

$$P(A' \cap B) = P(B) - P(A \cap B) = P(B) - P(A)P(B) = P(B)[1 - P(A)] = P(A')P(B)$$

hence A' and B are independent.

Similarly,
$$P(A \cap B') = P(A - A \cap B) = P(A) - P(A \cap B) = P(A) - P(A)P(B)$$
$$= P(A)[1 - P(B)] = P(A)P(B')$$

hence A and B' are independent.

Now A and B are independent, hence so are A' and B, as follows from the first step; but then so are A' and B', as follows from the second step. q.e.d.

Imagine that we are given three events A, B and C which are *pairwise independent,* i.e.,

$$P(A \cap B) = P(A)P(B), \quad P(A \cap C) = P(A)P(C) \quad \text{and} \quad P(B \cap C) = P(B)P(C).$$

Can we conclude that all three events are independent, taken all three together? No, it is not always true, as shown by the following adaptation of a very famous example.

EXAMPLE 12

(According to Bernstein.) Consider a four faced symmetrical tetrahedron colored red on one face, green on another, blue on a third and combined red, green and blue on the fourth. Suppose the tetrahedron is to be thrown once; let A be the event that the face on which it lands contains red, B that it contains green and C that it contains blue. Consider the independence of A, B and C.

SOLUTION. If we abbreviate red by r, blue by b and green by g, then a sample space is

$$\Omega = \{r,b,g,rbg\}$$

where rbg stands for the multicolored face. The events A, B and C are as follows:

$$A = \{r,rbg\} \qquad P(A) = \frac{1}{2}$$

$$B = \{g,rbg\} \qquad P(B) = \frac{1}{2}$$

$$C = \{b,rbg\} \qquad P(C) = \frac{1}{2}$$

Further, we have $$A \cap B = A \cap C = B \cap C = \{rbg\},$$

hence $$P(A \cap B) = P(A \cap C) = P(B \cap C) = \frac{1}{4}\cdot$$

Hence
$$P(A \cap B) = P(A)P(B) \quad \text{and } A,B \text{ are independent,}$$
$$P(A \cap C) = P(A)P(C) \quad \text{and } A,C \text{ are independent,}$$
$$P(B \cap C) = P(B)P(C) \quad \text{and } B,C \text{ are independent.}$$

We can see that A, B and C are pairwise independent. However,

$$A \cap B \cap C = \{rbg\}, \text{ hence}$$

$$P(A/B \cap C) = \frac{P(A \cap B \cap C)}{P(B \cap C)} = \frac{\frac{1}{4}}{\frac{1}{4}} = 1 \neq \frac{1}{2} = P(A).$$

Therefore A is not independent of B and C taken together. This is intuitively clear. If we know that both B and C have occurred, then the face on which the tetrahedron landed must contain green and blue, hence it must be the combined face rbg also containing red. The simultaneous occurrence of B and C implies the occurrence of A. Therefore the events A, B, C are pairwise independent, but they are not independent taken all three together.

This example suggests the following definition.

DEFINITION 3: Let A, B and C be sets of the probability measure space (Ω, P). We say that A, B and C are

(a) pairwise independent if
$P(A \cap B) = P(A)P(B)$, $P(A \cap C) = P(A)P(C)$, $P(B \cap C) = P(B)P(C)$,

(b) mutually independent if they are pairwise independent, and furthermore
$P(A \cap B \cap C) = P(A)P(B)P(C)$.

COMMENT 1. Mutual independence always implies pairwise independence, but not vice versa, as follows from Example 12.

COMMENT 2. Some authors use the expression "complete independence" or "stochastic independence" instead of mutual independence. Some use independence without any modifier.

WARNING: There are two important properties of events in probability theory. Events A and B may be *disjoint* or *exclusive* in the sense that

$$A \cap B = \phi.$$

However, events A and B may be *independent* in the sense that

$$P(A \cap B) = P(A)P(B).$$

These two properties are not related at all and should not be confused. Unfortunately, they are a source of frequent mistakes, mix-ups and chaos, expecially because of the expressions "mutually exclusive" and "mutually independent."

EXAMPLE 13

Toss a fair coin three times so that all $2^3 = 8$ outcomes are equally likely. Let the events A, B, C and D be as follows:

A means that the first toss results in a head,

B means that the second toss results in a head,

C means that the third toss results in a head,

D means that the first toss results in a tail.

Investigate the independence of A, B and C. Are A and D independent or mutually exclusive?

SOLUTION. Obviously $P(A) = P(B) = P(C) = P(D) = \frac{1}{2}$,

$$P(A \cap B) = P(A \cap C) = P(B \cap C) = \frac{1}{4},$$

and finally $$P(A \cap B \cap C) = \frac{1}{8}.$$

Therefore the events A, B and C are both pairwise and mutually independent.

However A, B and C are not mutually exclusive. (Why?) Furthermore, $A \cap D = \phi$, hence A and D are mutually exclusive. However,

$$P(A \cap D) = P(\phi) = 0 \neq \frac{1}{2} \cdot \frac{1}{2} = P(A)P(D),$$

hence A and D are not independent.

→ **THEOREM 6.** Let A, B and C be mutually independent events. Then the events
(a) A and $(B \cap C)$ are independent,
(b) A and $(B \cup C)$ are independent,
(c) A', B and C are mutually independent,
(d) A', B' and C' are mutually independent.

COMMENT. Theorem 6 contains several examples of independent events which can be expressed in terms of A, B and C. The reader will easily construct many other examples.

Proof:

(a) $$P(A \cap (B \cap C)) = P(A \cap B \cap C) = P(A)P(B)P(C) = P(A)[P(B)P(C)] = P(A)P(B \cap C).$$

Thus the events A and $(B \cap C)$ are independent.

(b) $$\begin{aligned} P(A \cap (B \cup C)) &= P((A \cap B) \cup (A \cap C)) = P(A \cap B) + P(A \cap C) - P(A \cap B \cap C) \\ &= P(A)P(B) + P(A)P(C) - P(A)P(B)P(C) \\ &= P(A)[P(B) + P(C) - P(B \cap C)] = P(A)P(B \cup C). \end{aligned}$$

Thus A and $(B \cup C)$ are independent.

(c) A', B and C are pairwise independent as follows from Theorem 5. For the mutual independence,

$$\begin{aligned} P(A' \cap B \cap C) &= P(B \cap C - A \cap B \cap C) = P(B \cap C) - P(A \cap B \cap C) = P(B)P(C) - P(A)P(B)P(C) \\ &= P(B)P(C)[1 - P(A)] = P(A')P(B)P(C). \end{aligned}$$

Thus A', B and C are mutually independent.

(d) A, B, C are mutually independent (m.i.) and by (c) also A', B, C are m.i.; however B, A', C are also m.i. and by (c) also B', A', C are m.i.; however C, B', A' are also m.i. and by (c) also C', B', A' are m.i.; and therefore A', B', C' are m.i.

q.e.d.

DEFINITION 4: The sets $A_1, A_2, \ldots, A_n$ of a probability measure space (Ω, P) are called mutually independent if for all combinations $1 \leq i < j < k < \ldots \leq n$, the following conditions are satisfied:

$$P(A_i \cap A_j) = P(A_i)P(A_j)$$

$$P(A_i \cap A_j \cap A_k) = P(A_i)P(A_j)P(A_k)$$

$$\vdots$$

$$P(A_1 \cap A_2 \cap \ldots \cap A_n) = P(A_1)P(A_2) \ldots P(A_n).$$

EXAMPLE 14

Mr. A and Mr. B shoot at a target. The probability that Mr. A hits the target is $\frac{1}{4}$ while the probability that Mr. B hits the target is $\frac{1}{3}$. If Mr. A can fire only twice, how many times must Mr. B fire so that there is at least a 90% probability that the target will be hit?

SOLUTION. Let us denote the events as follows:

D—no hit by Mr. A in his two shots

D_n—no hit by Mr. B in his first n shots

E—at least one hit in the whole action

A—hit by Mr. A in one shot

B—hit by Mr. B in one shot

Then

$$P(A) = \frac{1}{4}, \qquad P(B) = \frac{1}{3}.$$

If we assume that successive shots of Mr. A and Mr. B are independent, then

$$P(D) = [P(A')]^2 = \left(\frac{3}{4}\right)^2,$$

$$P(D_n) = [P(B')]^n = [1 - P(B)]^n = \left(\frac{2}{3}\right)^n.$$

The complement of E means "no hit in the whole action" and is equivalent to

$$(D \cap D_n)$$

that is,

$$E' = D \cap D_n.$$

Assuming independence of D and D_n, we get

$$P(E') = P(D) \cdot P(D_n) = \left(\frac{3}{4}\right)^2\left(\frac{2}{3}\right)^n.$$

Our condition requires $P(E) \geq 0.9$, which is equivalent to $P(E') \leq 0.1$. Hence the condition for the number of shots of Mr. B is as follows:

$$\left(\frac{3}{4}\right)^2\left(\frac{2}{3}\right)^n \leq .1.$$

We can solve this inequality by logarithms or by substituting 1,2,3 etc. for n. In either case, we can find that the minimal number of shots for Mr. B is $n = 5$ in order that there is at least a 90% chance for a hit.

Here we have to make a very important comment of a philosophical nature. In the previous example, we assumed that the events A: "Mr. A hits the target" and B: "Mr. B hits the target" are independent. The same assumption was made about the events "Mr. A hits the target on the first shot," "Mr. A hits the target on the second shot," and so on. From these assumptions, we concluded that $P(A \cap B) = P(A)P(B)$. It might seem that at this point we are a bit inconsistent by *assuming* independence instead of *verifying* the requirements of the definition. We should realize, however, that we are actually using the word "independent" in two different meanings. We are studying a situation in the real physical world, and we try to describe it by a mathematical probabilistic model. Events of the real world are then interpreted as subsets—events of the sample space. The fact of *actual independence* of real physical events (which may be nebulous, obscured or even questionable and doubtful) is *interpreted* in the probabilistic model by the *formal independence* corresponding to the multiplication of probabilities: $P(A \cap B) = P(A)P(B)$. How far this interpretation is reasonable and correct is beyond the reach of this text. Nevertheless, we shall make the assumption that all events related to shooting, successive repeated trials, and so forth, are independent unless it is explicitly stated otherwise or unless the physical situation clearly requires a model with dependent events.

DEFINITION 5: Partitions $\mathsf{A}_1, \mathsf{A}_2, \ldots, \mathsf{A}_k$ of the sample space Ω are called mutually independent if for every selection of a set $A_1 \in \mathsf{A}_1$, a set $A_2 \in \mathsf{A}_2, \ldots,$ and a set $A_k \in \mathsf{A}_k$, the sets $A_1, A_2, \ldots, A_k$ are mutually independent.

EXERCISES

Exercises for this chapter follow after the next chapter.

SELECTED EXAMPLES OF CONDITIONAL PROBABILITIES AND INDEPENDENCE

This chapter contains selected examples of conditional probabilities, the formula of total probability and Bayes' theorem, the multiplication rule, pairwise and mutual independence, as introduced in the preceding chapter. Many more examples of this type, but of a considerably more complicated structure, can be found in Chapter XV, which deals with practical applications: the safety of blood transfusions, population genetics and cases of disputed paternity, decision making under uncertainty, and so on.

EXAMPLE 1

Let A and B be mutually exclusive and independent sets. Then

$$\min(P(A),P(B)) = 0.$$

COMMENT. For any two real numbers x and y, $min(x,y)$ is the smaller of the numbers x and y.

SOLUTION. Since A and B are mutually exclusive, we have $A \cap B = \phi$. Hence

$$P(A \cap B) = P(\phi).$$

On the other hand, A and B are independent, hence

$$P(A)P(B) = P(A \cap B) = 0.$$

Therefore, either $\quad P(A) = 0 \quad \text{or} \quad P(B) = 0$

and thus $\quad \min(P(A),P(B)) = 0.$

EXAMPLE 2

If A, B, C are exhaustive and mutually independent sets with $P(A) < 1$ and $P(B) < 1$, then $P(C) = 1$.

SOLUTION. Using the properties of A, B, C, we have

$$\begin{aligned} 1 &= P(A \cup B \cup C) \\ &= P(A) + P(B) + P(C) - P(A)P(B) - P(A)P(C) - P(B)P(C) + P(A)P(B)P(C). \end{aligned}$$

Hence $\quad P(C)[1 - P(A) - P(B) + P(A)P(B)] = 1 - P(A) - P(B) + P(A)P(B).$

We know that $1 - P(A) - P(B) + P(A)P(B) = (1 - P(A))(1 - P(B)) > 0$

hence
$$P(C) = \frac{1 - P(A) - P(B) + P(A)P(B)}{1 - P(A) - P(B) + P(A)P(B)} = 1.$$

This example can be generalized to four or more events.

EXAMPLE 3

The problem of Chevalier de Méré. "What is more likely: throwing at least one ace with four dice or at least one double ace in 24 throws of a pair of dice?" ("Ace" means the number 1.) For the history of this example and its place in the development of probability theory, see Chapter I, page 1.

SOLUTION. Let us find the probabilities of complements; let q_1 be the probability of not getting a 6 in four tosses of a die and q_2 the probability of not getting a pair of 6's in 24 tosses of a pair of dice. The probability of not getting a 6 in one toss is $\frac{5}{6}$ and because the successive tosses are independent, we have

$$q_1 = \left(\frac{5}{6}\right)^4.$$

Similarly, the probability of not getting two 6's in one toss of a pair of dice is $\frac{35}{36}$ and because the successive tosses are independent,

$$q_2 = \left(\frac{35}{36}\right)^{24}.$$

If p_1 and p_2 are the desired probabilities, then

$$p_1 = 1 - q_1 = 1 - \left(\frac{5}{6}\right)^4 = .518,$$

$$p_2 = 1 - q_2 = 1 - \left(\frac{35}{36}\right)^{24} = .491.$$

Hence
$$p_1 > .5 > p_2.$$

This also explains the gambling adventures of Chevalier de Méré. Betting at equal odds on getting at least one ace with four dice must bring a definite profit in a long run. However, betting at equal odds on getting at least one double ace in 24 throws of a pair of dice must bring a loss.

EXAMPLE 4

Let A be the event that a family has children of both sexes, and let B be the event that the family has at most one boy. Investigate separately the independence of A and B for families with two and three children (assume that the sexes of children are equally likely).

SOLUTION 1. In the case of two children,

$$\Omega = \{bb, bg, gb, gg\}$$

is a sample space with an equally likely probability measure.

$$A = \{bg, gb\} \qquad P(A) = \frac{1}{2}.$$

$$B = \{bg, gb, gg\} \qquad P(B) = \frac{3}{4}.$$

$$A \cap B = \{bg, gb\} \qquad P(A \cap B) = \frac{1}{2}.$$

Since $P(A \cap B) \neq P(A)P(B)$, A and B are dependent.

SOLUTION 2. In the case of three children,

$$\Omega = \{bbb, bbg, bgb, gbb, ggb, gbg, bgg, ggg\}$$

is a sample space with an equally likely probability measure.

$$A = \{bbg, bgb, gbb, ggb, gbg, bgg\} \qquad P(A) = \frac{6}{8} = \frac{3}{4}.$$

$$B = \{bgg, gbg, ggb, ggg\} \qquad P(B) = \frac{4}{8} = \frac{1}{2}.$$

$$A \cap B = \{bgg, gbg, ggb\} \qquad P(A \cap B) = \frac{3}{8}.$$

Since $P(A \cap B) = P(A)P(B)$, the events A and B are independent.

EXAMPLE 5

A certain type of missile hits its target with probability .3. How many missiles should be fired so that there is at least an 80% probability of hitting the target? (Assume the independence of successive shots.)

SOLUTION. The probability of a missile missing its target is .7. Hence the probability that n (independent) missiles miss the target is $(.7)^n$. We seek the smallest n such that

$$1 - (.7)^n \geq .8$$

Solving this inequality, we find that the least number of missiles is $n = 5$.

EXAMPLE 6

Five masked robbers attempted to hold up a bank during a rush hour when ninety-five men were waiting in the bank's main hall. However, an attentive clerk managed to push the alarm button, and the police arrived immediately. The policemen blocked all the entrances to the hall, and the robbers, realizing that they were trapped, decided on a quick action: they threw away their masks, and disappeared amidst the chaotic crowd. Thus the police ended up with 100 men in the hall, all of them claiming that they were innocent and indicating that someone else was a robber. The investigators decided to use a lie detector. However, it is well known that such a machine is not absolutely reliable. If a guilty man is monitored by a lie detector, he will be discovered with the probability 90%, while an innocent man will be accused (unjustly) of the crime with the probability 5%. The first man is selected at random from the crowd and questioned while monitored by the device.

(a) What is the probability that he will be accused?
(b) If he is accused of the crime, what is the probability that he is innocent?

SOLUTION. (a) Denote by G the event that the man is guilty and by I that he is innocent. Then G and I form a partition of the sample space and further,

$$P(G) = \frac{5}{100} = .05 \qquad P(I) = \frac{95}{100} = .95.$$

If we denote by A the fact that he is accused after the questioning, and by N that he is not accused, then

$$P(A/G) = .9 \qquad P(N/G) = .1$$
$$P(A/I) = .05 \qquad P(N/I) = .95.$$

Using the formula of total probability, we get

$$P(A) = P(G)P(A/G) + P(I)P(A/I) = (.05) \times (.9) + (.95) \times (.05)$$
$$= \frac{5}{100} \cdot \frac{90}{100} + \frac{95}{100} \cdot \frac{5}{100} = \frac{90 + 95}{2000} = \frac{185}{2000} = .0925.$$

So the man will be accused with the probability .0925.

(b) We are seeking $P(I/A)$, and by Bayes' theorem:

$$P(I/A) = \frac{P(I)P(A/I)}{P(A)} = \frac{\frac{95}{100} \cdot \frac{5}{100}}{\frac{185}{2000}} = \frac{95}{185} = .5135.$$

Even if the man is accused after the monitoring by the lie detector, it is still more likely that he is innocent than guilty, and the lie detector test in this case does not seem to be a very reliable method of questioning.

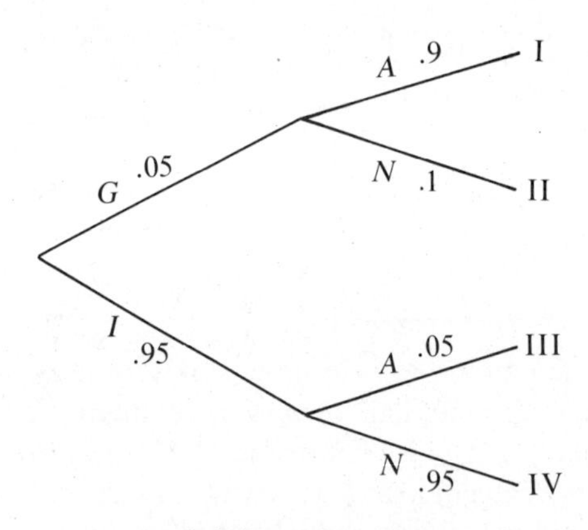

Solution using a tree.

(a) $$P(A) = P(\mathrm{I}) + P(\mathrm{III}) = (.05) \times (.9) + (.95) \times (.05) = .0925.$$

(b) $$P(I/A) = \frac{P(\mathrm{III})}{P(\mathrm{I}) + P(\mathrm{III})} = \frac{(.95) \times (.05)}{.0925} = .5135.$$

EXAMPLE 7

The hypothetical and renowned match, Army versus Navy, is approaching again, and Jim argues with his friends about the chances of both teams. After some heated discussions, they agree that if the game should be held at West Point, then Army could win with probability .6. However, if the game should be held at Annapolis then Navy will beat Army with probability .8. Unfortunately, nobody knows yet where the game will be held except that there will be a draw between the cities giving both teams equal chances for their favorite locations. (a) What is the probability that Navy will beat Army? (Disregard the possibility of a tie.)

SOLUTION. Denote the event of the game being held at West Point by W and at Annapolis by A. Further, let N mean that Navy defeats Army. Then we get the following probability tree.

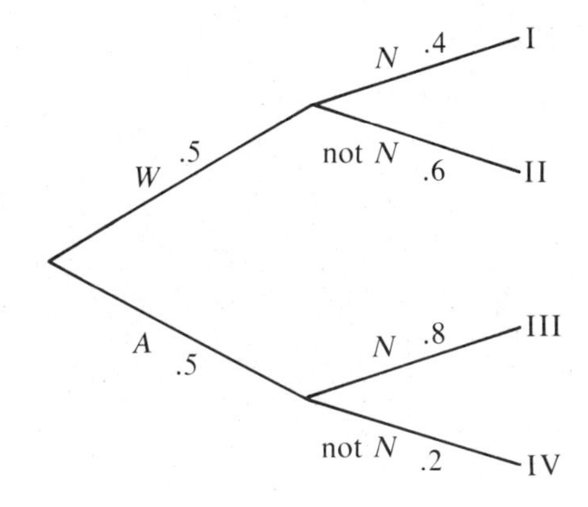

$$P(N) = P(\text{I}) + P(\text{III}) = (.5) \times (.4) + (.5) \times (.8) = .6.$$

Thus Navy will defeat Army with the probability .6.

(b) Imagine now that Jim's interest in the game was considerably cooled by his examinations. When he recuperates from exam shock and returns to his senses, he is told by a friend that Navy really did defeat Army. Unfortunately, Jim cannot find out where the game was held. What is the probability that it was at Annapolis?

SOLUTION. Using the previous notation and the probability tree, we are seeking

$$P(A/N) = \frac{P(\text{III})}{P(\text{I}) + P(\text{III})} = \frac{(.5) \times (.8)}{.6} = \frac{2}{3}.$$

EXAMPLE 8

During your visit to Las Vegas, you are told that one of the three slot machines in a parlor pays off with probability $\frac{1}{2}$, while each of the other two pays off with probability $\frac{1}{3}$. Unfortunately you do not know which is which. You play twice a randomly selected slot machine, losing the first time and winning the second time. What is the probability that you have chosen the favorable machine? Is this evidence supporting a guess that you are playing the favorable machine?

SOLUTION. Let us denote the events A, B_1, B_2 as follows:

A—you lose the first time, win the second time

B_1—you are playing the favorable machine

B_2—you are playing an unfavorable machine

Then the desired probability is $P(B_1/A)$. We know that

$$P(B_1) = \frac{1}{3}, \qquad P(B_2) = \frac{2}{3},$$

$$P(A/B_1) = \frac{1}{2} \cdot \frac{1}{2} = \frac{1}{4}, \qquad P(A/B_2) = \frac{2}{3} \cdot \frac{1}{3} = \frac{2}{9}.$$

Using Bayes' theorem,

$$P(B_1/A) = \frac{\frac{1}{3} \cdot \frac{1}{4}}{\frac{1}{3} \cdot \frac{1}{4} + \frac{2}{3} \cdot \frac{2}{9}} = \frac{9}{25}$$

This probability is greater than $\frac{1}{3} = P(B_1)$; thus it is now more likely that you have selected the favorable machine than it was before you started playing.

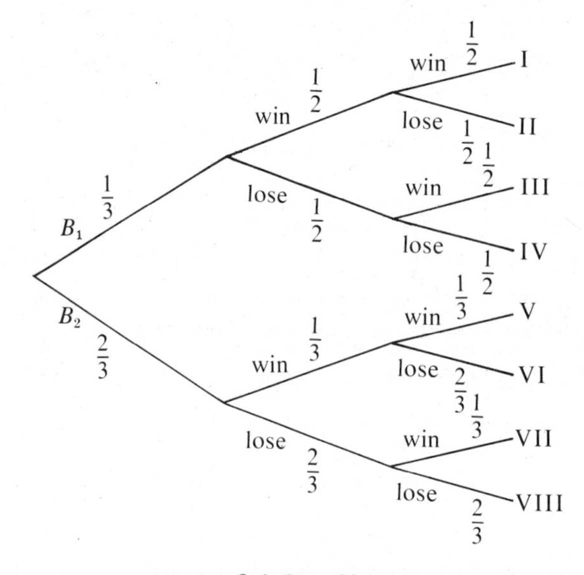

Solution with a tree

$$P(\text{lose first time, win second time}) = P(\text{III}) + P(\text{VII}) = \frac{1}{3} \cdot \frac{1}{2} \cdot \frac{1}{2} + \frac{2}{3} \cdot \frac{2}{3} \cdot \frac{1}{3} = \frac{25}{108}.$$

$$P(\text{playing the favorable machine/lose first, win second time}) = \frac{P(\text{III})}{P(\text{III}) + P(\text{VII})}$$

$$= \frac{\frac{1}{3} \cdot \frac{1}{2} \cdot \frac{1}{2}}{\frac{25}{108}} = \frac{\frac{9}{108}}{\frac{25}{108}} = \frac{9}{25}.$$

EXAMPLE 9

Find $P(A)$, $P(B)$ and $P(C)$ if you know that the following four conditions must be satisfied:

(a) A,B,C are mutually independent,

(b) $P(A \cap B) = \frac{1}{6}$,

(c) $P(B \cap C') = \frac{1}{2} P(A)$,

(d) $P(B) = \frac{1}{3P(C)}$.

SOLUTION. From (a) and (d) we get

$$P(B \cap C) = P(B)P(C) = \frac{1}{3}.$$

From (c), $P(B) = P(B \cap C') + P(B \cap C) = \frac{1}{2}P(A) + \frac{1}{3}$.

From (b), $P(A \cap B) = P(A)P(B) = \frac{1}{6}$.

Hence
$$P(A)\left[\frac{1}{2}P(A) + \frac{1}{3}\right] = \frac{1}{6}.$$

$$[P(A)]^2 + \frac{2}{3}P(A) - \frac{1}{3} = 0.$$

Solving the quadratic equation, we get

$$x_{1,2} = \frac{-\frac{2}{3} \pm \sqrt{\frac{4}{9} + \frac{4}{3}}}{2} = \frac{-\frac{2}{3} \pm \frac{4}{3}}{2} = \begin{cases} \frac{1}{3} \\ -1 \end{cases}$$

However, the solution $x_2 = -1$ is not acceptable for $P(A)$; hence we get

$$P(A) = \frac{1}{3},$$

$$P(B) = \frac{1}{6}\Big/\frac{1}{3} = \frac{1}{2},$$

$$P(C) = \frac{1}{3}\Big/\frac{1}{2} = \frac{2}{3}.$$

EXAMPLE 10

An antiaircraft gun can fire three shots at an enemy plane. The probability of a hit with the first shot is .4, with the second it is .5 and with the third it is .7. If the airplane is hit just once, it can still fly away with probability .6; if it is hit twice it can be salvaged with probability .2;

but if it is hit thrice, then it is inevitably downed. What is the probability that the plane will be destroyed? (Consider the successive shots as independent.)

SOLUTION. Let us denote the events as follows:

A_i – the plane is hit exactly i times, $i = 0,1,2,3$

H_i – the plane is hit by the i-th shot, $i = 1,2,3$

M_i – the plane is missed by the i-th shot, $i = 1,2,3$

D – the plane is destroyed

We can use the formula of total probability to calculate the required probability.

$$P(D) = \sum_{i=1}^{3} P(A_i)P(D/A_i).$$

$$P(A_0) = P(M_1 \cap M_2 \cap M_3) = P(M_1)P(M_2)P(M_3) = (.6) \times (.5) \times (.3) = .09.$$

$$P(A_1) = P(H_1 \cap M_2 \cap M_3) + P(M_1 \cap H_2 \cap M_3) + P(M_1 \cap M_2 \cap H_3)$$
$$= (.4) \times (.5) \times (.3) + (.6) \times (.5) \times (.3) + (.6) \times (.5) \times (.7) = .36.$$

$$P(A_2) = P(H_1 \cap H_2 \cap M_3) + P(H_1 \cap M_2 \cap H_3) + P(M_1 \cap H_2 \cap H_3)$$
$$= (.4) \times (.5) \times (.3) + (.4) \times (.5) \times (.7) + (.6) \times (.5) \times (.7) = .41.$$

$$P(A_3) = P(H_1 \cap H_2 \cap H_3) = (.4) \times (.5) \times (.7) = .14.$$

$$P(D/A_1) = .4, \qquad P(D/A_2) = .8, \qquad P(D/A_3) = 1.$$

Using the formula of total probability:

$$P(D) = \sum_{i=1}^{3} P(A_i)P(D/A_i) = (.36) \times (.4) + (.41) \times (.8) + (.14) = .612.$$

Set up the probability tree and use it to obtain a solution.

EXAMPLE 11

Two shipments of cars, shipment A and shipment B, were sent by the Marshal Flivvers Company to two dealers in downtown San Francisco at the same time. Shipment A consisted of 10 small cars and 5 expensive cars while shipment B consisted of 5 small cars and 5 expensive cars. During their transit there was some confusion, and two cars, one from each shipment, got exchanged with each other. On arrival of the two shipments, one car was selected at random from shipment B for tests. Find the probability that

(a) The selected car was expensive.
(b) The car transferred from A to B was small given that the selected car was expensive.
(c) The car transferred from B to A was expensive given that the selected car was expensive.
(d) The car transferred from A to B was small given that the car transferred from B to A was small.
(e) The selected car was expensive given that the car transferred from A to B was expensive.
(f) The car transferred from A to B was small given that the car transferred from B to A was expensive and the selected car was expensive.

SOLUTION. Let us denote events as follows:

C—the car transferred from A to B is small

C'—the car transferred from A to B is expensive

D—the car transferred from B to A is small

D'—the car transferred from B to A is expensive

H—the selected car is expensive

Then obviously,

$$P(C) = \frac{10}{15} = \frac{2}{3}, \qquad P(C') = \frac{1}{3}.$$

$$P(D) = \frac{1}{2}, \qquad P(D') = \frac{1}{2}.$$

Thus we have 4 cases of exchange for which the probabilities are

$$P(C \cap D) = P(C)P(D) = \frac{2}{3} \cdot \frac{1}{2} = \frac{1}{3}, \qquad P(C' \cap D) = \frac{1}{3} \cdot \frac{1}{2} = \frac{1}{6},$$

$$P(C \cap D') = \frac{2}{3} \cdot \frac{1}{2} = \frac{1}{3}, \qquad P(C' \cap D') = \frac{1}{3} \cdot \frac{1}{2} = \frac{1}{6}.$$

(a) $P(H) = P(H \cap C \cap D) + P(H \cap C \cap D') + P(H \cap C' \cap D) + P(H \cap C' \cap D')$

$$= P(C \cap D)P(H|C \cap D) + P(C \cap D')P(H|C \cap D') + P(C' \cap D)P(H|C' \cap D) + P(C' \cap D')P(H|C' \cap D')$$

$$= \frac{5}{10} \cdot \frac{1}{3} + \frac{4}{10} \cdot \frac{1}{3} + \frac{6}{10} \cdot \frac{1}{6} + \frac{5}{10} \cdot \frac{1}{6} = \frac{29}{60} = .483.$$

(b) $$P(C|H) = \frac{P(C \cap H \cap D) + P(C \cap H \cap D')}{P(H)} = \frac{\frac{5}{10} \cdot \frac{1}{3} + \frac{4}{10} \cdot \frac{1}{3}}{\frac{29}{60}} = \frac{18}{29} = .621.$$

(c) $$P(D'|H) = \frac{P(H \cap D' \cap C) + P(H \cap D' \cap C')}{P(H)} = \frac{\frac{4}{10} \cdot \frac{1}{3} + \frac{5}{10} \cdot \frac{1}{6}}{\frac{29}{60}} = \frac{18}{29} = .448.$$

(d) $$P(C|D) = \frac{P(C \cap D)}{P(D)} = \frac{\frac{1}{3}}{\frac{1}{2}} = \frac{2}{3} = .667.$$

(e) $$P(H|C') = \frac{P(H \cap C')}{P(C')} = \frac{P(H \cap C' \cap D) + P(H \cap C' \cap D')}{P(C')}$$

$$= \frac{\frac{6}{10} \cdot \frac{1}{6} + \frac{5}{10} \cdot \frac{1}{6}}{\frac{1}{3}} = \frac{11}{20} = .55.$$

(f) $$P(C|D' \cap H) = \frac{P(C \cap D' \cap H)}{P(D' \cap H)} = \frac{\frac{4}{10}\cdot\frac{1}{3}}{\frac{29}{60}\cdot\frac{13}{29}} = \frac{8}{13} = .615.$$

Set up a probability tree to obtain the solution; it is really very expedient.

EXAMPLE 12

Maximization of Odds. The Sultan said to Ali Baba, "Here are two urns, a white balls and b black balls. Divide all the balls between the urns in any way you wish. Then I will make the urns indistinguishable. You will choose an urn and draw a ball from it. If it is a white ball, I will spare your life." How can Ali Baba maximize his chance to survive?

SOLUTION. Consider the urn in which Ali Baba puts fewer, say c, white balls. Then $c \leq a - c$, and assume that he puts d black balls into this urn. The other urn contains $(a - c)$ white and $(b - d)$ black balls. Let $p(c,d)$ denote the probability that Ali Baba will draw a white ball from a randomly chosen urn. By the formula of total probability,

$$p(c,d) = \frac{1}{2}\left(\frac{c}{c+d} + \frac{a-c}{a+b-c-d}\right).$$

Let us investigate the following difference:

$$p(c,0) - p(c,d) = \frac{1}{2}\left(1 + \frac{a-c}{a+b-c}\right) - \frac{1}{2}\left(\frac{c}{c+d} + \frac{a-c}{a+b-c-d}\right).$$

Taking the first and the third term together and the second and the fourth together, we get

$$p(c,0) - p(c,d) = \frac{1}{2}\left(\frac{(a+b-c-d)[(c+d)-c]}{(a+b-c-d)(c+d)} + \frac{(a-c)[(a+b-c-d)-(a+b-c)]}{(a+b-c-d)(a+b-c)}\right)$$

$$= \frac{1}{2}\frac{d}{a+b-c-d}\left(\frac{a+b}{c+d} - 1 - \frac{a-c}{a+b-c}\right).$$

The bracket expression will reach its minimum if $d = b$. Therefore

$$p(c,0) - p(c,d) \geq \frac{1}{2}\frac{d}{a+b-c-d}\left(\frac{a+b}{c+b} - 1 - \frac{a-c}{a+b-c}\right)$$

$$= \frac{1}{2}\frac{d}{a+b-c-d}\frac{(a+b-c)[(a+b)-(c+b)] - (a-c)(c+b)}{(a+b-c)(c+b)}$$

$$= \frac{1}{2}\frac{d}{a+b-c-d}\frac{(a-c)[(a+b-c)-(c+b)]}{(a+b-c)(c+b)}$$

$$= \frac{1}{2}\frac{d}{(a+b-c-d)}\frac{(a-c)(a-2c)}{(a+b-c)(c+b)} \geq 0$$

because every bracket is nonnegative according to our assumption. Hence

$$p(c,0) \geq p(c,d),$$

and, therefore, the urn containing fewer white balls should not contain any black balls. Further, for any natural number $c \geq 1$,

$$\frac{a-1}{a+b-1} \geq \frac{a-c}{a+b-c},$$

hence

$$\frac{1}{2}\left(1+\frac{a-1}{a+b-1}\right) \geq \frac{1}{2}\left(1+\frac{a-c}{a+b-c}\right)$$

and this means that $p(1,0) \geq p(c,0)$.

Therefore, Ali Baba should put one white and no black balls into one urn and the remaining balls into the other urn in order to maximize his chances for survival. This answer is intuitively plausible. Imagine that the first urn really contains a single white ball and the rest of the balls are in the second urn. If Ali Baba wants to change this distribution, he has to put a ball from the second to the first urn (the given distribution being at least as good as all the balls in one urn). If he transfers a white ball, his chances for survival from the first urn do not change, but they decrease for the second urn. If he transfers a black ball, his chances from the first urn decrease dramatically (from 100% to 50%), while they increase less for the second urn.

EXAMPLE 13a

Decision Making under Uncertainty. Two young brothers have identical piggy banks for their savings. The older boy has two quarters and three dimes in his bank while the younger one has two quarters and one dime. One day the banks got hopelessly mixed, and the boys could not tell them apart except by examining their contents. However, they were afraid to ask their father for the keys, therefore they started shaking one of the banks in order to get one or two coins out and determine whose it was. They can perform the following experiments in order to guess:

(a) Get one coin out and see if it is a quarter or a dime.
(b) Same as (a), but replace the coin, then get one more coin.
(c) Get two coins without replacing the first one.

Using the à posteriori probabilities, help the boys decide in each possible case whose bank is more likely to be in their hands.

SOLUTION. Denote the events as follows:

O—the bank of the older brother is being investigated

Y—the bank of the younger brother is being investigated

Q—a quarter drops out of the bank

D—a dime drops out of the bank.

The brothers are completely ignorant of whose bank is being investigated. It seems plausible, therefore, to assume that $P(O) = (Y) = \frac{1}{2}$.

(a) If a quarter drops out, then the probability that it is the younger brother's bank is

$$P(Y/Q) = \frac{P(Y)P(Q/Y)}{P(Y)P(Q/Y) + P(O)P(Q/O)} = \frac{\frac{1}{2}\cdot\frac{2}{3}}{\frac{1}{2}\cdot\frac{2}{3} + \frac{1}{2}\cdot\frac{2}{5}} = \frac{5}{8}\cdot$$

The probability that it is the older brother's bank is

$$P(O/Q) = 1 - P(Y/Q) = 1 - \frac{5}{8} = \frac{3}{8}\cdot$$

In this case, it is more likely that they have the younger brother's bank. Similarly, we shall calculate the remaining probabilities:

$$P(Y/D) = \frac{\frac{1}{2}\cdot\frac{1}{3}}{\frac{1}{2}\cdot\frac{1}{3} + \frac{1}{2}\cdot\frac{3}{5}} = \frac{5}{14}, \qquad P(O/D) = 1 - \frac{5}{14} = \frac{9}{14}\cdot$$

The following is the decision table for this experiment

Coin	The more likely bank
Q	Y
D	O

(b) There are four possible outcomes if two coins (with replacement) are taken out, namely QD, DQ, QQ, DD (in the order indicated).

$$P(Y/QD) = \frac{P(Y)P(Q/Y)P(D/Y \cap Q)}{P(Y)P(Q/Y)P(D/Y \cap Q) + P(O)P(Q/O)P(D/O \cap Q)}$$

$$= \frac{\frac{1}{2}\cdot\frac{2}{3}\cdot\frac{1}{3}}{\frac{1}{2}\cdot\frac{2}{3}\cdot\frac{1}{3} + \frac{1}{2}\cdot\frac{2}{5}\cdot\frac{3}{5}} = \frac{25}{52}\cdot$$

$$P(O/QD) = 1 - \frac{25}{52} = \frac{27}{52}\cdot$$

$$P(Y/DQ) = \frac{\frac{1}{2}\cdot\frac{1}{3}\cdot\frac{2}{3}}{\frac{1}{2}\cdot\frac{1}{3}\cdot\frac{2}{3} + \frac{1}{2}\cdot\frac{2}{5}\cdot\frac{3}{5}} = \frac{25}{52} \qquad P(O/DQ) = \frac{27}{52}\cdot$$

$$P(Y/QQ) = \frac{\frac{1}{2}\cdot\frac{2}{3}\cdot\frac{2}{3}}{\frac{1}{2}\cdot\frac{2}{3}\cdot\frac{2}{3} + \frac{1}{2}\cdot\frac{2}{5}\cdot\frac{2}{5}} = \frac{25}{34} \qquad P(O/QQ) = \frac{9}{34}\cdot$$

$$P(Y/DD) = \frac{\frac{1}{2}\cdot\frac{1}{3}\cdot\frac{1}{3}}{\frac{1}{2}\cdot\frac{1}{3}\cdot\frac{1}{3}+\frac{1}{2}\cdot\frac{3}{5}\cdot\frac{3}{5}} = \frac{25}{106} \quad P(O/DD) = \frac{81}{106}.$$

The decision table for this experiment is as follows:

Coins	The more likely bank
QD	O
DQ	O
QQ	Y
DD	O

(c) There are four possible outcomes as in the previous example, but the probabilities are different because the first coin is not replaced.

$$P(Y/QD) = \frac{\frac{1}{2}\cdot\frac{2}{3}\cdot\frac{1}{2}}{\frac{1}{2}\cdot\frac{2}{3}\cdot\frac{1}{2}+\frac{1}{2}\cdot\frac{2}{5}\cdot\frac{3}{4}} = \frac{10}{19} \quad P(O/QD) = \frac{9}{19}.$$

$$P(Y/DQ) = \frac{\frac{1}{2}\cdot\frac{1}{3}\cdot 1}{\frac{1}{2}\cdot\frac{1}{3}+\frac{1}{2}\cdot\frac{3}{5}\cdot\frac{2}{4}} = \frac{10}{19} \quad P(O/DQ) = \frac{9}{19}.$$

$$P(Y/QQ) = \frac{\frac{1}{2}\cdot\frac{2}{3}\cdot\frac{1}{2}}{\frac{1}{2}\cdot\frac{2}{3}\cdot\frac{1}{2}+\frac{1}{2}\cdot\frac{2}{5}\cdot\frac{1}{4}} = \frac{10}{13} \quad P(O/QQ) = \frac{3}{13}.$$

$$P(Y/DD) = 0. \qquad P(O/DD) = 1.$$

The decision table for this experiment is as follows:

Coins	The more likely bank
QD	Y
DQ	Y
QQ	Y
DD	O

EXAMPLE 13b

In the previous example we derived a decision rule for each experiment, telling us which bank should be considered more likely given an outcome of an experiment. Let us calculate the probability of the correct decision for each experiment.

SOLUTION.

(a) $$P(\text{correct decision}) = P(Y)P(Q/Y) + P(O)P(D/O) = \frac{1}{2}\cdot\frac{2}{3} + \frac{1}{2}\cdot\frac{3}{5} = .63\overline{3}.$$

(See the decision table.)

(b) $$P(\text{correct decision}) = P(Y)P(QQ/Y) + P(O)[P(QD/O) + P(DQ/O) + P(DD/O)]$$
$$= \frac{1}{2}\left(\frac{2}{3}\right)^2 + \frac{1}{2}\left[\frac{2}{5}\cdot\frac{3}{5} + \frac{3}{5}\cdot\frac{2}{5} + \frac{3}{5}\cdot\frac{3}{5}\right] = .64\overline{2}.$$

(c) $$P(\text{correct decision}) = P(Y)[P(QD/Y) + P(DQ/Y) + P(QQ/Y)] + P(O)P(DD/O)$$
$$= \frac{1}{2}\left[\frac{2}{3}\cdot\frac{1}{2} + \frac{1}{3}\cdot\frac{2}{2} + \frac{2}{3}\cdot\frac{1}{2}\right] + \frac{1}{2}\cdot\frac{3}{5}\cdot\frac{2}{4} = .650.$$

Therefore, the third experiment has the highest probability of a correct decision that is apparently obtained by the greatest amount of information about the piggy bank in the boys' hands.

Some people might object to this extensive use of à posteriori probabilities on the grounds mentioned in the previous chapter: First of all, it is rather arbitrary to say that the probability of either bank being investigated is $\frac{1}{2}$ just because we are completely ignorant of which one it is. Somebody else could be convinced that the older brother's bank is more likely to be investigated, and assign the à priori probabilities as $P(O) = \frac{2}{3}$, $P(Y) = \frac{1}{3}$ or perhaps $P(O) = \frac{3}{4}$, $P(Y) = \frac{1}{4}$. Furthermore, it makes little sense to speak about the probability that the bank in the hand could be the older brother's bank. Either it is or it is not, and we could easily find out by getting the key to open it. However, regardless of these objections, the above mentioned ideas are widely used in decision making under uncertainty and we shall have an opportunity to investigate them in detail in the practical applications in Chapter XV.

EXAMPLE 14

Polya's Scheme of Contagious Diseases. The following scheme was proposed by G. Polya as a model describing the phenomenon of spreading a contagious disease. An urn contains initially m red and n green balls. Balls are drawn randomly, and replaced after each drawing. At the same time, a number k of balls of the same color as the ball drawn are added to the urn. The occurrence of a red ball in a certain draw increases the probability of getting a red ball on the next draw. This is in a way similar to the spreading of a contagious disease: each occurrence of such a disease increases the probability of its further occurrences. Polya's urn scheme can be considered as a model of this spreading, though not a perfect one. The number k can be negative, and in this case k balls are removed from the urn. We would like to find the probability $P(i)$ that, in N drawings, exactly i red balls are obtained. Obviously, there must be $j = N - i$ green balls drawn.

SOLUTION. First of all, we can see that the actual contents of the urn at any stage depend only on the numbers of red and green balls previously drawn, but not on the order in which they were drawn. Let us calculate the probability P that among N balls first i draws will result in red balls and remaining $j = N - i$ draws will yield green balls. Using the formula for multiplication of conditional probabilities, we get

$$P = \left(\frac{m}{m+n}\right)\left(\frac{m+k}{m+n+k}\right) \cdots \left(\frac{m+(i-1)k}{m+n+(i-1)k}\right)\left(\frac{n}{m+n+ik}\right)\left(\frac{n+k}{m+n+(i+1)k}\right) \cdots$$
$$\left(\frac{n+(j-1)k}{m+n+(N-1)k}\right).$$

Furthermore, it is intuitively clear that the probability of drawing i red and j green balls *in any order* is the same and is therefore equal to P. However, there are $\binom{N}{i}$ ways to get i red balls among N balls. Therefore

$$P(i) = \binom{N}{i} P.$$

It is quite expedient to use general binomial coefficients in order to express $P(i)$. We can define for any real number x and for any natural number l

$$\binom{x}{l} = \frac{x(x-1)\ldots(x-l+1)}{l!}.$$

Then obviously,

$$P(i) = \frac{N!\left(\frac{m}{k}\right)\left(\frac{m}{k}+1\right)\cdots\left(\frac{m}{k}+(i-1)\right)\left(\frac{n}{k}\right)\left(\frac{n}{k}+1\right)\cdots\left(\frac{n}{k}+(j-1)\right)}{i!(N-i)!\left(\frac{m+n}{k}\right)\left(\frac{m+n}{k}+1\right)\cdots\left(\frac{m+n}{k}+(N-1)\right)}$$
$$= \frac{\binom{\frac{m}{k}+i-1}{i}\binom{\frac{n}{k}+j-1}{j}}{\binom{\frac{m+n}{k}+N-1}{N}}.$$

Now we can write $(N-i)$ instead of j, $j = N - i$ and therefore

$$P(i) = \frac{\binom{\frac{m}{k}+i-1}{i}\binom{\frac{n}{k}+N-i-1}{N-i}}{\binom{\frac{m+n}{k}+N-1}{N}}.$$

EXAMPLE 15

Behavior Conditioning. Effects of behavior conditioning in the process of learning are often studied through experiments on laboratory animals. A rat is released at the entrance to a Y maze.

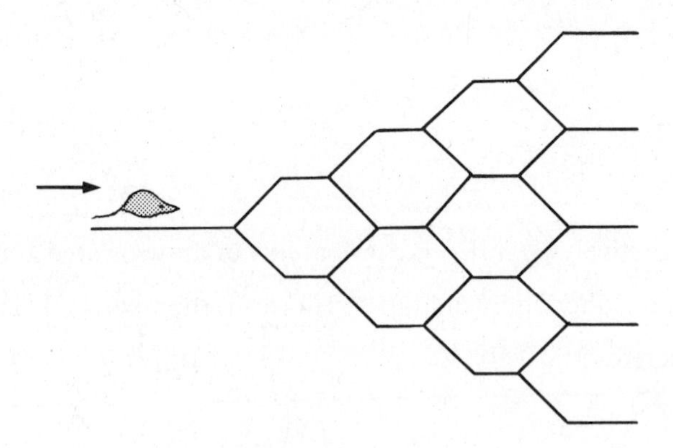

After each right turn, the animal gets a reward (food) and after each left turn the animal is punished (falls into water, gets an electric shock, or something similar). In the beginning, the animal will go to the left as equally likely as to the right. However, after a reward for going to the right, the animal will turn right on the next intersection with the probability .6. After a punishment for going to the left, the animal will turn left only with the probability .3. Investigate the probability $P(n)$ of the animal going right on the n-th turn.

SOLUTION. We describe the rat's behavior using a probability tree:

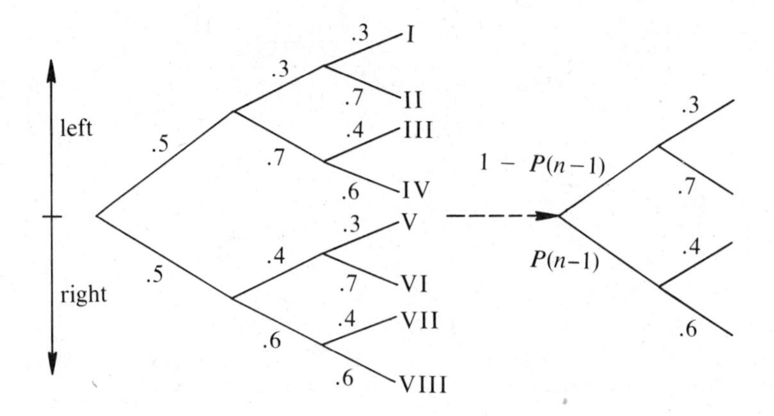

Then $P(3) = .5 \times .3 \times .7 + .5 \times .7 \times .6 + .5 \times .4 \times .7 + .5 \times .6 \times .6 = .635$ is the probability of turning right on the third turn. In general, if the rat turned right on the $(n - 1)$ turn (and this happens with the probability $P(n - 1)$), then it will turn right on the n-th turn with the probability .6. However, if the animal turned left on the $(n - 1)$ turn (and this happens with the probability $1 - P(n - 1)$), then it will turn right on the n-th turn with the probability .7. Therefore

$$P(n) = P(n - 1) \times .6 + (1 - P(n - 1)) \times .7$$

or

$$P(n) = .7 - (.1)P(n - 1).$$

Using this formula recursively, we get

$$\begin{aligned} P(n) &= .7 - (.1)P(n - 1) = .7 - (.1)[.7 - (.1)P(n - 2)] \\ &= .7(1 - (.1) + (.1)^2 - (.1)^3 \ . \ . \ . \ (-.1)^{n-2}) + (-.1)^{n-1}P(1). \end{aligned}$$

Using the formula
$$1 + q + q^2 + \ldots + q^m = \frac{1 - q^{m+1}}{1 - q}$$

for the summation of a geometric series, we get

$$P(n) = .7\frac{1 - (-.1)^{n-1}}{1.1} + (-.1)^{n-1}P(1).$$

Hence
$$P(n) = .\overline{63}(1 - (-.1)^{n-1}) + .5(-.1)^{n-1}.$$

This formula gives the probability that the animal will go right on the n-th turn. With increasing n, this probability approaches $.\overline{63}$.

EXAMPLE 16

Let A and B be independent events. Let the simultaneous realizations of A and B imply C and the simultaneous realizations of A' and B' imply C'. Show that

$$P(A) \cdot P(C) \leq P(A \cap C).$$

SOLUTION. The Venn diagram of A, B and C looks somewhat like the following sketch:

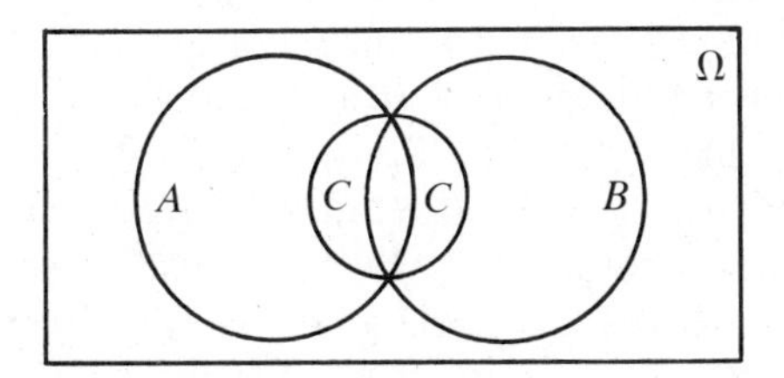

We can express the conditions concerning the implications as follows:

$$C \supset A \cap B. \qquad C' \supset A' \cap B'.$$

Therefore

$$\begin{aligned} P(A \cap C) &= P(A \cap B \cap C) + P(A \cap B' \cap C) = P(A \cap B) + P(A \cap B' \cap C) \\ &= P(A)P(B) + P(A \cap B' \cap C) \geq P(A)P(B) + P(A)P(A \cap B' \cap C) \\ &= P(A)[P(B) + P(A \cap B' \cap C)]. \end{aligned}$$

$$C' \supset A' \cap B' \text{ implies } C \subset A \cup B,$$

therefore if we intersect both sides by $B' \cap C$, we get

$$C \cap B' \subset A \cap B' \cap C.$$

This is equivalent to the fact that

$$C - B = C \cap B' = A \cap B' \cap C.$$

We can write $P(B) + P(A \cap B' \cap C) = P(B) + P(C - B) = P(B \cup C)$.

Therefore $P(A \cap C) \geq P(A)[P(B) + P(A \cap B' \cap C)] = P(A)P(B \cup C) \geq P(A)P(C)$.

EXAMPLE 17

Crazy Game of Sucker Bets. Using a fair coin, two players play the following game: When a coin is flipped three times, there are eight equally likely possibilities: *HHH, HHT, HTH, THH, HTT, THT, TTH, TTT*. One player selects one of these patterns and the other selects a different one. Now the coin is flipped repeatedly till one of the two chosen triplets appears as a run. The person having chosen this pattern wins the game. For example, if the first player selects *HTH* and the other selects *HHT*, then the series of tosses *TTHHHT* is won by the second player. Analyze the game.

SOLUTION. At first glance, it appears that both players have equal chances to win the game because all the patterns are equally likely to appear. A majority of people would be willing to play the game with equal stakes. However, the real situation is more complicated.

Let us consider, first, that the players select patterns of two coins, that is *HH, HT, TH, TT*. If the players select *HH* and *HT*, they have an equal chance $\frac{1}{2}$ to win, because if an *H* appears on a certain toss, the next toss will equally likely be *H* or *T*. There are four distinct patterns and therefore $\binom{4}{2} = 6$ ways that a selection can be made. Let us consider other possibilities. For the same reason as above, the selections *TT* and *TH* give both players equal odds to win. *H* and *T* have the same chance to appear on any toss, therefore, if one pattern can be obtained from another by replacement of *H* by *T* and vice versa, then both of them must have equal chances to win like *HH* and *TT* or *HT* and *TH*. However, the pattern *TH* is much more likely to win than *HH*. Any appearance of *HH* is necessarily preceded by *TH*, except if *HH* appears immediately in the beginning. That is, any game that does not start with *HH* will be necessarily won by the player who has *TH*. There are three patterns for the beginning giving a victory of *TH*, and one when *TH* loses. Therefore, the chances of *TH* winning over *HH* are $\frac{3}{4}$. The same is true for the patterns *HT* and *TT*. The following is the matrix of the above derived conditional probabilities that the player *B* wins given the selections of player *A* and *B*.

B \ A	HH	HT	TH	TT
HH	–	$\frac{1}{2}$	$\frac{1}{4}$	$\frac{1}{2}$
HT	$\frac{1}{2}$	–	$\frac{1}{2}$	$\frac{3}{4}$
TH	$\frac{3}{4}$	$\frac{1}{2}$	–	$\frac{1}{2}$
TT	$\frac{1}{2}$	$\frac{1}{4}$	$\frac{1}{2}$	–

Thus if *A* selects his pattern first, then *B* can always find a pattern for himself that gives him a chance of $\frac{1}{2}$ or better to win.

The situation is much more interesting when we investigate triplets. For the reasons given above (symmetry of H and T), the following pairs of triplets have an equal chance to win: HHH and TTT; HHT and TTH; HTH and THT. Other probabilities are harder to calculate. For example, THH will always precede HHH unless the latter appears in the beginning of the game. There are seven openings in favor of THH and only one in favor of HHH, therefore THH beats HHH with the probability $\frac{7}{8}$. We shall not calculate the remaining probabilities, but examine the accompanying table of conditional probabilities that B wins given the selections of both players A and B.

B \ A	HHH	HHT	HTH	HTT	THH	THT	TTH	TTT
HHH		$\frac{1}{2}$	$\frac{2}{5}$	$\frac{2}{5}$	$\frac{1}{8}$	$\frac{5}{12}$	$\frac{3}{10}$	$\frac{1}{2}$
HHT	$\frac{1}{2}$		$\frac{2}{3}$	$\frac{2}{3}$	$\frac{1}{4}$	$\frac{5}{8}$	$\frac{1}{2}$	$\frac{7}{10}$
HTH	$\frac{3}{5}$	$\frac{1}{3}$		$\frac{1}{2}$	$\frac{1}{2}$	$\frac{1}{2}$	$\frac{3}{8}$	$\frac{7}{12}$
HTT	$\frac{3}{5}$	$\frac{1}{3}$	$\frac{1}{2}$		$\frac{1}{2}$	$\frac{1}{2}$	$\frac{3}{4}$	$\frac{7}{8}$
THH	$\frac{7}{8}$	$\frac{3}{4}$	$\frac{1}{2}$	$\frac{1}{2}$		$\frac{1}{2}$	$\frac{1}{3}$	$\frac{3}{5}$
THT	$\frac{7}{12}$	$\frac{3}{8}$	$\frac{1}{2}$	$\frac{1}{2}$	$\frac{1}{2}$		$\frac{1}{3}$	$\frac{3}{5}$
TTH	$\frac{7}{10}$	$\frac{1}{2}$	$\frac{5}{8}$	$\frac{1}{4}$	$\frac{2}{3}$	$\frac{2}{3}$		$\frac{1}{2}$
TTT	$\frac{1}{2}$	$\frac{3}{10}$	$\frac{5}{12}$	$\frac{1}{8}$	$\frac{2}{5}$	$\frac{2}{5}$	$\frac{1}{2}$	

From this table we can discover a surprising fact. Given any selection of the player A, the player B can always find a pattern which will assure him a win with a probability of at least $\frac{2}{3}$ but sometimes as high as $\frac{7}{8}$. Therefore even though at first glance the game appears to be fair, it actually gives a much better chance to win to the player who makes his selections later. For any selection of the first player, there always exists a selection for the second player that is "better" in the sense that it gives a better chance to win. It is a rather striking observation, when we realize that both players are choosing from the same set of patterns.

EXAMPLE 18

Coding Paradox. An alphabet consists of four letters, A, B, C and D. For the sake of security, these letters are coded into strings of 0's and 1's before any message is transmitted. Following is the table of probabilities of occurrences of the four letters in messages, and also the corresponding codes.

Letter	Probability	Code
A	$\frac{1}{2}$	0
B	$\frac{1}{4}$	10
C	$\frac{1}{8}$	110
D	$\frac{1}{8}$	111

If a coded symbol is intercepted by an enemy, what is the probability that it is a 0?

SOLUTION 1. The intercepted symbols may originate from coding of A, B, C or D. The following are conditional probabilities of getting a 0 if a symbol is picked up at random from the code of the given letter.

$$P(0/A) = 1, \quad P(0/B) = \frac{1}{2}, \quad P(0/C) = \frac{1}{3}, \quad P(0/D) = 0.$$

Therefore the probability of getting 0 can be calculated using the formula of total probability as follows:

$$P(0) = P(A)P(0/A) + P(B)P(0/B) + P(C)P(0/C) + P(D)P(0/D)$$

$$= \frac{1}{2} \cdot 1 + \frac{1}{4} \cdot \frac{1}{2} + \frac{1}{8} \cdot \frac{1}{3} + \frac{1}{8} \cdot 0 = \frac{2}{3} \cdot$$

SOLUTION 2. Imagine a long message of, say, 800 letters. Then there will be about 400 A's giving 400 0's, 200 B's giving 200 0's and 200 1's, 100 C's giving 200 1's and 100 0's and 100 D's giving 300 1's. We can see that we may expect about 700 0's and 700 1's and therefore

$$P(0) = \frac{1}{2} \cdot$$

DISCUSSION. With two different solutions of the same problem, we can ask: Which one is correct and which is false? It looks as though Solution 1 should be correct because it was obtained using an "approved" theoretical formula. However, the situation is more complicated. Both solutions are correct and false. The first solution answers correctly the question: If the enemy intercepts a whole coded letter but due to distortion is able to recognize only one of its symbols, what is the probability that it will be 0? The second solution is the correct answer for the question: If the enemy attempts to intercept the whole sequence of symbols representing a coded message but due to distortions is able to recognize only one symbol, what is the probability that it will be 0? In this sense, both solutions are correct but they solve different practical problems (and it seems that the second one is more plausible). At the same time, both solutions are incorrect because they try to solve a question which is inaccurately and nebulously formulated in the first place. It is not clear at all what the underlying physical experiment is, what a corresponding sample space should be and how we can express investigated events. This is a frequent situation in applications of probability and our example serves as a warning against abuses of probability. When we investigate a practical problem in the *real world,* we can use different *probabilistic models* for its description. Some models might be more accurate and better express the properties of the investigated phenomenon, some models might be poorer. However, only actual performance of the experiment can *verify* which model is more suitable.

EXERCISES

(Answers in Appendix to exercises marked)*

*1. Four machines A, B, C, D produce 40%, 30%, 20% and 10% of the total production of a factory. Percentages of defective items of these machines are 3%, 4%, 5% and 7%, respectively. If an item selected at random is defective, find the probability that it was produced by machine B.

2. A college entrance board views students as capable or incapable. The examination given to prospective students is passed by a capable student with probability .9 and with probability .4 by an incapable student. If the probability is .55 that a student taking the exam is capable, what is the probability that a student who has passed is capable?

*3. A certain man is a frequent visitor at the race track. When he uses his own "scientific method," he wins 10% of the time. When he follows the suggestions of a tout, he wins 40% of the time, and when he follows his wife's suggestions, he wins 80% of the time. Like many husbands, he uses his own method 60% of the time, he listens to the tout 30% of the time and he listens to his wife 10% of the time. Today he won. What method do you think he used, i.e., what method is most probable?

4. Four persons, called North, South, East and West, are each dealt 13 cards from an ordinary deck of 52 cards.
   (a) If South has exactly one ace, what is the probability that his partner has the other three aces? (Partners are $N - S$ or $E - W$.)
   (b) If North and South together have 10 hearts, what is the probability that either East or West has the other 3 hearts?

5. In a certain college, 30% of the boys and 20% of the girls are studying statistics. The girls constitute 60% of the student body. If a student is selected at random and is studying statistics, determine the probability that the student is a girl.

*6. We are given two urns with these contents:

   Urn A contains four red marbles, two white marbles and seven blue marbles.

   Urn B contains two red marbles and six white marbles.

   A fair die is tossed; if 2 or 5 appears, a marble is chosen from B, otherwise a marble is chosen from A. Find the probability that (a) a red marble is chosen, (b) a white marble is chosen, (c) a blue marble is chosen.

7. Refer to the preceding problem. (a) If a red marble is chosen, what is the probability that it came from urn A? (b) If a white marble is chosen, what is the probability that a 5 appeared on the die?

8. There are five bolts in a container. Three are defective. The bolts are tested one after the other until the three defective ones are discovered. What is the probability that the process stopped on the
   (a) third test?
   (b) fourth test?

9. Refer to the preceding problem. If a process stopped on the fourth test, what is the probability that the first bolt is nondefective?

*10. Suppose that the reliability of a chest x-ray test for the detection of tuberculosis is specified as follows: of people with tuberculosis, 90% of the x-ray examinations detect the disease, but 10% go undetected. Of people free of tuberculosis, 99% of the x-rays are judged as free of the disease, but 1% are diagnosed as showing tuberculosis. From a large population of which only 0.1% have tuberculosis, one person is selected at random, given a chest x-ray and the radiologist reports the presence of tuberculosis. What is the probability that the person has tuberculosis?

11. A fair coin is flipped. If a head appears, th symmetrical tetrahedron die (with 1,2. points on its faces) is rolled; if a tail then an ordinary cubic die is roll the probability that the number the die will be more than two

*

12. Three friends, Paul, Bob and Bill, arrive at an airport at the same time as three other people. All six came to board a plane which has only three vacant seats. They agree to decide who makes the trip by drawing straws.
    (a) What is the probability that Paul and Bob will go but not Bill?
    (b) What is the probability that all three friends made the trip?
    (c) What is the probability that at least one of them will make the trip?

13. The SCUBAPRO Company makes sharkskin wet suits in three different sizes: small, medium and large. Each size comprises the following percentage of the total production: S–40%, M–40%, L–20%. The nylon linings of these suits are red or blue, with the percentages of red for different sizes being: S–30%, M–25%, and L–20%. If you pick up a suit at random,
    (a) what is the probability that it has a blue lining?
    (b) if it has a blue lining, what is the probability that it is a large size suit?

*14. Scuba divers spear-fishing in Puget Sound are usually after ling cod, rock cod, or sea bass. Assume that at certain times of the year no other species of fish are encountered by the divers, and that these three kinds of fish represent 20%, 30% and 50% of the fish population. A novice diver will hit the elusive ling cod with his spear-gun with the probability .2. He will score on the spooky rock cod with the probability .3 and he will connect on a slow sea bass with the probability .4.
    (a) What is the probability that he will get a fish on his first shot?
    (b) If he gets a fish on the first shot, what is the probability that it will be a delicious rock cod?
    (c) If he gets a fish on the first shot, what kind is it most likely to be?
    (d) Imagine that he has a chance to shoot at three fish. What is the probability that he will get a sea bass and a rock cod on his three shots?

15. The probability that a doctor can diagnose a certain rare disease correctly is .3. If the disease is diagnosed correctly, the probability that it will be cured is .8. If it is not diagnosed correctly, the patient still has a probability $\frac{1}{3}$ of being cured in the course of nature. Of four patients who consult the doctor, what is the probability that at least half of them will be cured? (Assume that patients can be cured independently of each other, and also that they consult the doctor without knowing that they suffer from the disease.)

16. There are four red and five blue marbles in a box. We select three marbles from the box one after another, without replacement. Calculate the probability that the third marble will be red given that the second marble was blue (having forgotten the color of the first marble).

17. Two numbers X and Y are selected as follows: First, X is selected at random from the set $\{1,2,3,4\}$. Then Y is selected at random from the set $\{1, \ldots, X\}$ where X is the previously selected number.
    (a) Find the probability that $Y = 1$.
    (b) If $Y = 1$, find the probability that X was 2.

18. Imagine a box of three drawers: a top, a middle and a bottom drawer. In each drawer there are two boxes, each containing one coin. In the top drawer both coins are of gold, in the middle drawer both are of silver and in the bottom there is one gold and one silver coin. A drawer is selected at random and presented to you. You do not know which it is but upon opening one of its boxes, you find a silver coin. What is the probability that the other coin will be gold?

*19. According to the latest reports, some of the UFO's are piloted by Ookpiks. Reliable observers confirmed beyond all doubt the unlimited mental capabilities and some of the interests of these interterrestrial intelligences: about 60% of Ookpiks were sighted while enjoying themselves cutting the peculiar figures of a jitterbug dance. Unverified rumors claimed that only 20% of Ookpiks neither wear little pink hats nor dance jitterbug. However, it was more than exceedingly documented in the opening address of the last Intragalaxial Congress on UFO's that 70% of Ookpiks are either passionate jitterbug dancers or wear little pink hats, but not both. Recently we received a confidential communication that our Everready Units succeeded in catching one of the Ookpiks at random, and it was discovered that every morning he put on a little pink hat. According to usually well informed circles, some of our leading scientists are working presently on the conditional probability that this Ookpik enjoys dancing jitterbug. Can you help them?

20. A warehouse contains 300 cases of high quality thread, 500 cases of medium quality thread and 200 cases of low quality thread. A flood removes all identification labels. The owners decide to test the thread to determine the quality. In a given test, the high, medium and low quality threads break 10%, 20% and 30% of the time, respectively. A case is selected, two spools are selected from it at random, and both are tested. If neither breaks, what is the probability that the case contains thread of high quality?

*21. There are two lots of items; it is known that 90% of the items of one lot satisfy technical standards and 20% of the items of the other lot are defective. Suppose that an item from a lot selected at random turns out to be good. Find the probability that a second item of the same lot will be defective, if the first item is returned to the lot after it has been checked.

22. A box contains three coins, two of them fair and one two-headed. A coin is selected at random and tossed. If heads appears, the coin is tossed again; if tails appears, then another coin is selected from the two remaining coins and tossed.
   (a) Find the probability that heads appears twice.
   (b) If the same coin is tossed twice, find the probability that it is the two-headed coin.
   (c) Find the probability that tails appears twice.

23. Three dice are rolled. Find the probability of obtaining more than 14 dots.

*24. Three dice are tossed and the sum thus obtained is fourteen. What is the probability that two dice showed 6 each?

25. Three dice are rolled. Given that the total is more than 14 what is the probability that the second die shows less than 6 dots?

26. This is the problem of four liars. One person (a) out of four receives information in the form of a "yes" or "no" and he transmits it to the second person (b). The second person transmits it to the third (c), the third to the fourth (d) and the fourth communicates the received information to the sender. Given the fact that only one person in three tells the truth, find the probability that the first liar told the truth if the fourth told the truth.

27. Three hunters shoot simultaneously at a wild boar, which is killed by one bullet. Find the probability that the boar is killed by the first, second or third hunter, respectively, if the probabilities of their hitting the boar are 0.2, 0.4 and 0.6, respectively.

*28. A territorial defense unit has deployed three earth-to-earth missiles of type A, and a very high number of missiles of type B. The probability of destruction of a target by a type A missile is .4, while the probability is only .3 for type B missiles. If a strategically important object is to be destroyed, first all missiles A are fired and then missiles B. How many missiles of type B should be used so that there is at least a 95% chance that the object will be destroyed?

29. Find the probability that a person rolling a fair die will get his second 6 on the fourth roll.

30. You keep rolling a fair die. How many times should you roll it so that the probability of getting at least one 6 is greater than .9?

31. Suppose that airplane engines operate independently and fail in flight with the probability of $\frac{1}{5}$. Assuming that a plane makes a successful flight if at least half of its engines run (regardless of their location), determine whether a four-engine plane or a two-engine plane has the higher probability of a successful flight.

32. Two students independently attempt to solve a certain problem. The probability that A solves it is $\frac{1}{10}$ and the probability that B solves it is $\frac{1}{3}$. What is the probability that the problem will be solved?

33. A certain professor of statistics does not trust the belt on his pants: he knows that it may snap open during a lecture with the probability .01, and he can lose his cover in front of the class. He trusts his suspenders even less because he knows that they may fail with the probability .02. However, he believes that the belt and the suspenders are independent means of support of his dignity and he always wears both of them. What is the probability that both of them will let go simultaneously, thus inadvertently causing an exposure in front of the class? What is the probability that at least one of them will fail?

34. A team wins with probability .5, loses with probability .3 and ties with probability .2. The team plays twice and the outcomes of these games are independent. Find the probability that the team wins at least once.

35. A movie producer is filming a difficult sequence with a stuntman. It is known that 40% of the time something fails and the shot will be useless, even though this may be discovered only later in the studio. How many times will he have to shoot this sequence if he wants to be at least 99% sure that at least one shot is good?

*36. The probability that A hits a target is $\frac{1}{3}$, that B hits it is $\frac{1}{2}$ and that C hits it is $\frac{2}{3}$. Each of them shoots once at the target. What is the probability that there will be exactly one hit? If there is one hit only, what is the probability that it is B's success?

37. Using an ordinary deck of 52 cards, we shuffle the hearts, diamonds, spades and clubs separately, and from each suit we deal 2 cards. What is the probability that among the 8 cards dealt there are exactly 2 kings?

38. Over a long period of time, three fourths of all students passed a course on finite probability. If three students Agamemnon, Brunhilde and

Clytemnestra take the examination, find the probability that
(a) A passes, B passes, C fails.
(b) exactly two students pass.
(c) at least two students pass.

39. A survey of 200 students of the University of Kodiak Island is made to determine whether they prefer beer, wine, hard liquor or do not drink at all. The students are also classified according to sex. The results are summarized as follows:

	Beer	Wine	Hard Liquor	Teetotaler
Female	33	12	12	18
Male	51	20	53	1

(a) Given that a randomly selected student is a female, what is the probability that she is a teetotaler?
(b) Given that a randomly selected student is a teetotaler, what is the probability that the student is a female?
(c) For which of the four drinking categories, if any, is membership in that category independent of the student's sex?

40. A coin is biased so that the probability of getting heads is .506. If the coin is tossed four times, find the probability of getting at least three heads.

*41. In a three game hockey play-off between Toronto and Montreal, Toronto is figured to have a 58% chance of winning the first game. Winning a game boosts the morale of either team so that its chances of winning its next game are improved 5%. What is the probability that Toronto wins the series? (Disregard ties.)

42. Let A be the event that a family has children of both sexes, and let B be the event that the family has at most one boy. Investigate the independence of A and B if the family has four children. (See Example 4, page 100.)

43. To control the quality of a manufacturing process, each item produced passes through three inspections. Of four items, A, B, C and D, it is known that A passed only inspection 1, B passed only inspection 2, C passed only inspection 3 and D passed all three inspections. One of the four items is selected at random. Let:
$E_1 = \{\text{item passed inspection 1}\}$
$E_2 = \{\text{item passed inspection 2}\}$
$E_3 = \{\text{item passed inspection 3}\}$
Investigate the pairwise independence of events (E_1,E_2), (E_1,E_3), (E_2,E_3) and then investigate the mutual independence of all three events (E_1,E_2,E_3).

44. Toss two fair coins and let the events A, B and C be as follows:
A means that the first toss is a head,
B means that the second toss is a head,
C means that there is a match (either both heads or both tails).
Investigate the independence of A, B and C.

45. An urn contains ten green and six red balls. A ball is drawn at random and its color is recorded. Then the ball is replaced and five more balls of the same color are added. This procedure is repeated three times. What is the probability of drawing exactly two red balls?

*46. There are 15 tennis balls in a box, of which 9 are new. For the first game, three balls are selected at random and, after play, they are returned to the box. For the second game, three balls are also selected at random. Find the probability that all the balls taken for the second game will be new.

47. The manager of an upland bird hunting preserve discovered at the spring count that only seven pheasants and five partridges survived harsh winter conditions. Therefore, he set the following rules for the beginning of the new season: There will be only one hunter allowed each day on the preserve and the daily bag limit will be one bird. Moreover the hunter, if successful, will have to release on the same day three live birds of the same kind. Joe Doe succeeded in booking the first three days for himself and he got his bird each day. Find the probability that he brought home two pheasants and one partridge (assume that the birds on the preserve are taken at random).

*48. The owner of a corner grocery store decided to stock only two brands of Scotch: Johnnie Walker and Black & White. At the same time, he felt that the number of bottles displayed of each kind should correspond to the actual demand of his customers. Therefore, he put six bottles of JW and eight bottles of B & W on the self-service shelf, and decided that for each bottle bought by a customer he would immediately put back two bottles of the same brand. However, after a while he noticed that customers just grabbed at random one of the displayed bottles. Find the probability that among ten bottles of Scotch sold on the first day, there were six bottles of JW and four bottles of B & W. (Assume that each customer bought only one bottle.)

49. Show that if A and B_1 are independent, A and B_2 are independent and $B_1 \cap B_2$ has probability 0, then A and $B_2 \cup B_2$ are also independent.

50. Let A and B be events with $P(A) = \frac{1}{4}$, $P(A \cup B) = \frac{1}{3}$ and $P(B) = p$.

(a) Find p if A and B are mutually exclusive.
(b) Find p if A and B are independent.
(c) Find p if A is a subset of B.

51. Can any event be independent of itself?

52. Of the three events E, F and G, we know that E and F are independent and $G \subseteq E$. Does it follow that G and F are independent? Defend your answer.

53. Let A and B be exhaustive and independent events such that $P(A) < 1$. Prove that $P(B) = 1$.

54. Let A,B,C and D be exhaustive and mutually independent events such that $P(A) < 1$, $P(B) < 1$, $P(C) < 1$. Prove that $P(D) = 1$.

55. Let $A \subset \Omega$, $P(A) = 1$. Prove that A is independent of any other event $B \subset \Omega$.

56. Let $P(A/B) = P(A/B')$. Show that A and B are independent.

57. Prove that $P(A/B) \geq 1 - \dfrac{P(A')}{P(B)}$.

58. Let $A \subset B \subset \Omega$, $P(A) > 0$. Find $P(B/A)$.

*59. BINGO. A BINGO card consists of five columns, each of them headed by a letter of the word BINGO. The first column contains five randomly chosen numbers from among the numbers 1 to 15, the second column contains five randomly chosen numbers from among 16 to 30, the third (exceptional) column contains only four numbers from among 31 to 45 and the last two columns each contain five numbers from 46 to 60 and 61 to 75, respectively.

B	I	N	G	O
14	23	42	55	65
8	19	31	56	63
9	18	F	51	72
12	17	44	54	70
5	20	36	49	62

A typical BINGO card

A BINGO blower contains 75 numbered balls, the first fifteen of them (1 to 15) with the letter B, the second fifteen of them (16 to 30) with the letter I, the third fifteen of them (31 to 45) with the letter N, and so on. The first five balls drawn from the blower (at random) constitute the word BINGO. What is the probability that you have all the drawn numbers on your card?

RANDOM VARIABLES

In a state that is not stable, But chaungynge ay and variable.

The Romaunt of the Rose, 1400 (Anon.)

In previous chapters, we investigated sample spaces and probability measures. In the process of evaluation of the probability of an event, we were often interested not in the sample points as such but in some of their common properties: e.g., when tossing a coin five times, we were interested in the number of heads; when choosing four eggs for an omelet from a dozen containing two rotten ones, we were interested in the number of rotten eggs among four chosen; similarly, we were interested in the number of aces in a hand for bridge, in the total number of points on three dice, in the number of black balls drawn from an urn containing balls of three different colors, and so on.

When an experiment is performed, we are not always interested in the outcome for its own sake but very often we are concerned about some *numerical value* associated with it, with some *evaluation* of the outcome. In tossing a coin five times, the outcome might be *HTTHH* but we could be interested just in the fact that we got three heads, regardless of the order in which they appeared, regardless of the location of tails among them, regardless of whether or not all the heads came one after another and regardless of all the other properties of the actual outcome. When one is told that he can keep for himself all the coins displaying heads, then that's all that counts. The actual outcome of an experiment is sometimes very complicated but we are often interested only in one particular property of the outcome, which can be expressed simply by a *number,* and we disregard all its other features.

Imagine that in a game of blackjack you have been dealt the 2, 3, 4, 5 and 6 of hearts. It does not matter that you have all cards of the same suit—it might be a good sign for bridge but it does not interest you now, you do not really care how many cards you have in your hand, you do not even notice that you have a straight flush for poker. The only thing that matters is that you have a total of 20, which is so close to 21 that you will probably beat the dealer and win some money. Right now only the total number is important; everything else is irrelevant. The actual outcome consisting of five cards, the 2, 3, 4, 5, and 6 of hearts, is labeled with the number 20; it is enumerated, evaluated, *the number 20 is uniquely assigned to the outcome* as the only relevant characteristic for a game of blackjack. And we may say that this is a typical approach of the human intellect in its observation of reality. The observed phenomenon is often so complex, involved, perplexed and complicated that the human mind tries to grasp just one of its attributes which is important for its purposes at the moment, and which usually can be expressed by a number. All the other features and characteristics are disregarded even though they may appear relevant in other circumstances.

Let us consider several examples from everyday life. A tree in the forestry reserve is selected, cut down and its diameter at the base measured in inches for an estimation of the yield of lumber—all its other properties, such as the color of the bark, the number of branches, the shape of the eggs of the woodpecker nesting in the tree, etc., are unimportant to the lumber company. A student on campus is selected at random and his height is measured for purposes of an anthropometric research project, all other complex features of his person being disregarded. The very same student may be selected by chance in the next survey conducted by the Student Union to investigate the academic standing of its members; this time the relevant characteristic might be his grade point average and everything else, including his height, might be disregarded. Let us mention briefly a few more situations: the number of phone calls going through a switchboard; the cargo weight on a flight from Paris to Rome; the number of loves in one's life during the last five years; the length of a queue at a box office; the lifetime of a lightbulb, and so on. In all the cases there is an experiment performed (selection of a student; observation of a switchboard; selection of a tree), and a real number is assigned to the outcome (weight, number of calls, diameter, length of a queue, lifetime).

This situation is very common in everyday life; however, we should start our analysis with an elementary problem. Let us study a simple case of tossing coins. The experiment consists of tossing three fair coins, and the number assigned to an outcome is the number of heads which appear. An appropriate sample space Ω and the number $X(\omega)$ assigned to an outcome ω might be as follows:

Outcomes in the sample space Ω	Assigned number $X(\omega)$
$\omega_1 = TTT$	0
$\omega_2 = TTH$	1
$\omega_3 = THT$	1
$\omega_4 = HTT$	1
$\omega_5 = HHT$	2
$\omega_6 = HTH$	2
$\omega_7 = THH$	2
$\omega_8 = HHH$	3

For every point ω_i in the sample space $\omega_i \in \Omega$, we have defined a real number $X(\omega_i)$. For example, $X(HHT) = 2$, $X(HHH) = 3$, etc. Hence $X(\omega)$ is a function on Ω and we shall call it a *random variable.*

DEFINITION 1: Any real valued function X defined on a sample space Ω is called a random variable.

COMMENT 1. The expression "random variable" will often be abbreviated as r.v.

COMMENT 2. Sometimes we shall say "a r.v. defined on a probability measure space (Ω,P)" but we shall still mean the same thing, that is, a r.v. defined on Ω.

COMMENT 3. Our simple definition of a random variable is meaningful only in the case of finite (and later on countable) sample spaces. Its extension to other more complicated spaces requires an incomparably higher level of mathematical sophistication, and is very frequently abused.

COMMENT 4. At this moment, we are facing an important question. Why do we select a sample space Ω in such a way that a r.v. is a *function* on Ω; why do we not simply consider the actual *values* of the random variable as outcomes of the experiment? In terms of the previous example, why do we select $\Omega_1 = \{HHH, HHT, HTH, THH, TTH, THT, HTT, TTT\}$? We are interested in the number of heads only, so why do we not simply select $\Omega_2 = \{0,1,2,3\}$? It is clear that Ω_2 is perfectly acceptable but it is suitable for very limited purposes only. We certainly can investigate questions related to the number of heads or the number of tails using Ω_2. But we cannot investigate runs, for example. (A run is a sequence of the same symbols limited on both ends by different symbols or by a blank.) Let Y, for example, mean that number of runs. Then we cannot find the probability that $Y = 1$ using Ω_2, because the event $Y = 1$ cannot be expressed as a subset of Ω_2. But using Ω_1, we have $Y = 1$ for $\{HHT, THH, TTH, HTT\}$. It follows that we have to select a sample space that is sufficiently rich and suitable for investigation of a sufficiently large and interesting class of random variables.

EXAMPLE 1

Draw a hand of 13 cards out of an ordinary deck of 52 cards. Then a sample space has $\binom{52}{13}$ possible outcomes. For each hand ω, let $X(\omega)$ be the number of aces in the hand ω. Then the random variable $X(\omega)$ can attain values 0, 1, 2, 3 or 4.

EXAMPLE 2

Toss two dice and let Ω be the set of all pairs $\omega = (x,y)$, where x is the score on the first die and y is the score on the second die. For any outcome ω, let $X(\omega)$ be the total. Then Ω has 36 points and $X(\omega)$ can attain values 2,3, . . . ,12. For example $X(3,5) = 8$, $X(2,4) = 6$, etc.

EXAMPLE 3

You make an even money bet at roulette, putting five dollars on even numbers. (You win \$5 if an even number comes up; you lose \$5 if an odd number or 0 or 00 appears.) The sample space is $\Omega = \{00,0,1,2,3\ .\ .\ .\ ,34,35,36\}$. For each $\omega \in \Omega$, let $X(\omega)$ be your payoff. Then $X(\omega) = +5$ if ω is even, $X(\omega) = -5$ if ω is odd and $X(0) = -5$, $X(00) = -5$.

EXAMPLE 4

A lot contains 100 items, of which 7 are defective. A quality control inspector selects 5 items at random for testing. For any possible sample ω, let the r.v. $X(\omega)$ be the number of defective items in the sample. The r.v. X can attain 0, 1, 2, 3, 4 and 5.

EXAMPLE 5

A champion shoots five rounds of ammunition in the rapid fire discipline at a regulation target. Each time, either he hits (H) or misses (M) the bull's eye. A sample space consists of all quintuples of letters H and M. Let the r.v. X be the number of hits; then, for example,

$$X(HMMHM) = 2, \qquad X(MHHHH) = 4.$$

The r.v. can attain values 0, 1, 2, 3, 4 and 5.

If a r.v. X is given, then we can perform some numerical calculations with the values of X, or we can even take a composite function, for example:

$$Y(\omega) = X(\omega) + 1.$$

$$Z(\omega) = \frac{1}{2}X^2(\omega).$$

$$S(\omega) = \sin X(\omega) + 12, \text{ etc.}$$

Obviously, the composite functions $Z(\omega)$, $Y(\omega)$, $S(\omega)$, etc., are also functions on Ω and hence random variables. Investigate once more the example of tossing three coins where $X(\omega)$ means the number of heads.

Ω	$X(\omega_i)$	$Y(\omega_i) = X(\omega_i) + 1$	$Z(\omega_i) = \frac{1}{2}X^2(\omega_i)$
$\omega_1 = TTT$	0	1	0

Ω	$X(\omega_i)$	$Y(\omega_i) = X(\omega_i) + 1$	$Z(\omega_i) = \frac{1}{2}X^2(\omega_i)$
$\omega_2 = TTH$	1	2	$\frac{1}{2}$
$\omega_3 = THT$	1	2	$\frac{1}{2}$
$\omega_4 = HTT$	1	2	$\frac{1}{2}$
$\omega_5 = HHT$	2	3	2
$\omega_6 = HTH$	2	3	2
$\omega_7 = THH$	2	3	2
$\omega_8 = HHH$	3	4	$4\frac{1}{2}$

Thus we have a method for creating an unlimited number of random variables. Let $X(\omega)$ be a r.v. on Ω, that is, a function mapping Ω into the set of real numbers. Further, let F be a function mapping the set of all real numbers into itself. Then the composite function $F(X(\omega))$ is again a function mapping Ω into the set of real numbers and so, therefore, a random variable. We can take any function F and we get a r.v., for example:

$$Y(\omega) = \log(X(\omega) + 1)^{12} - \tan\frac{X(\omega)}{2}.$$

$$Z(\omega) = 6X^3(\omega) - 4X^2(\omega) + 17X(\omega) - 1.$$

The simplest random variable is the so-called *indicator function.* Let $A \subset \Omega$ be a subset of Ω and let us define:

$$I_A(\omega) = 1 \quad \text{if} \quad \omega \in A.$$

$$I_A(\omega) = 0 \quad \text{if} \quad \omega \notin A.$$

Then $I_A(\omega)$ is a random variable called an indicator function. For example, if Ω is the set of all outcomes of a toss of the die, $\Omega = \{1,2,3,4,5,6\}$, and $A \subset \Omega$ is the set of even outcomes, $A = \{2,4,6\}$, then

$$I_A(1) = 0. \qquad I_A(4) = 1.$$

$$I_A(2) = 1. \qquad I_A(5) = 0.$$

$$I_A(3) = 0. \qquad I_A(6) = 1.$$

Thus $I_A(\omega)$ indicates whether the outcome ω is odd (by 0) or even (by 1).

PROBABILITY DISTRIBUTION OF A RANDOM VARIABLE

Let $X(\omega)$ be a random variable on a sample space $\Omega = \{\omega_1,\omega_2,\omega_3, \ldots ,\omega_n\}$ of a probability measure space (Ω,P). Not all the values $X(\omega_1)$, $X(\omega_2)$, . . . , $X(\omega_n)$ are necessarily different. Let $\{x_1,x_2, \ldots ,x_k\}$ be the set of all possible distinct values that $X(\omega)$ can attain (obviously $k \leq n$). Define:

$$A_i = \{\omega: \omega \in \Omega, X(\omega) = x_i\} \qquad i = 1,2, \ldots ,k.$$

Obviously, the class $\{A_1,A_2, \ldots ,A_k\}$ is a partition of Ω. (Remember that a partition of a sample space Ω is any class of sets $\{A_1,A_2, \ldots ,A_k\}$ that are pairwise disjoint, i.e., $A_i \cap A_j = \Phi$ for $i \neq j$ and exhaustive, i.e., $\bigcup_{i=1}^{k} A_i = \Omega$).

DEFINITION 2: Let X be a random variable on a sample space Ω and let $\{x_1,x_2, \ldots ,x_k\}$ be the set of all possible distinct values that X can attain. Then the class of sets $\{A_1,A_2, \ldots ,A_k\}$ where $A_i = \{\omega: \omega \in \Omega, X(\omega) = x_i\}$ is called the partition induced by X.

It is very important to know the probability with which the r.v. X, defined on a probability measure space (Ω,P), attains its particular values. Let us denote by $f(x_i)$ the probability that X attains the value x_i. Once we have the partition of Ω induced by X, it is very easy to find these probabilities.

EXAMPLE 6

Consider tossing a coin three times, and let $X(\omega)$ be the number of heads. Then $X(\omega)$ can attain values $x_1 = 3$, $x_2 = 2$, $x_3 = 1$, $x_4 = 0$. The corresponding partition is as follows:

$$\Omega = \{HHH, HHT, HTH, THH, TTH, THT, HTT, TTT\}.$$

$$A_1 = \{\omega: X(\omega) = x_1 = 3\} = \{HHH\}.$$

$$A_2 = \{\omega: X(\omega) = x_2 = 2\} = \{HHT, HTH, THH\}.$$

$$A_3 = \{\omega: X(\omega) = x_3 = 1\} = \{HTT, THT, TTH\}.$$

$$A_4 = \{\omega: X(\omega) = x_4 = 0\} = \{TTT\}.$$

Obviously, $A_i \cap A_j = \Phi$ for $i \neq j$ and $\bigcup_{i=1}^{4} A_i = \Omega$ so $\{A_1,A_2,A_3,A_4\}$ is a partition of Ω.

Let us investigate now the r.v. Y, which means the number of runs. Consider a sequence of two symbols, say a and b,

aaabbabaabbb

Any uninterrupted string of symbols of the same kind, limited on both ends by symbols of a different kind or by a blank space, is called a *run.* Thus, in the given sequence, we have six runs as follows:

aaa; *bb*; *a*; *b*; *aa*; *bbb*

Given the sample space for tossing a coin three times, the number of runs Y can be 1,2 or 3 and the partition induced by Y is as follows:

$$B_1 = \{\omega: Y(\omega) = 1\} = \{HHH, TTT\}.$$

$$B_2 = \{\omega: Y(\omega) = 2\} = \{HHT, HTT, TTH, THH\}.$$

$$B_3 = \{\omega: Y(\omega) = 3\} = \{HTH, THT\}.$$

For the r.v. X, we get the following set of probabilities:

$$f(x_1) = P\{\omega : X(\omega) = x_1 = 3\} = P(A_1) = \frac{1}{8} = f(3).$$

$$f(x_2) = P\{\omega : X(\omega) = x_2 = 2\} = P(A_2) = \frac{3}{8} = f(2).$$

$$f(x_3) = P\{\omega : X(\omega) = x_3 = 1\} = P(A_3) = \frac{3}{8} = f(1).$$

$$f(x_4) = P\{\omega : X(\omega) = x_4 = 0\} = P(A_4) = \frac{1}{8} = f(0).$$

For the r.v. Y, the probabilities are:

$$g(1) = P\{\omega : Y(\omega) = 1\} = P(B_1) = \frac{2}{8}.$$

$$g(2) = P\{\omega : Y(\omega) = 2\} = P(B_2) = \frac{4}{8}.$$

$$g(3) = P\{\omega : Y(\omega) = 3\} = P(B_3) = \frac{2}{8}.$$

Following are graphs of these probabilities.

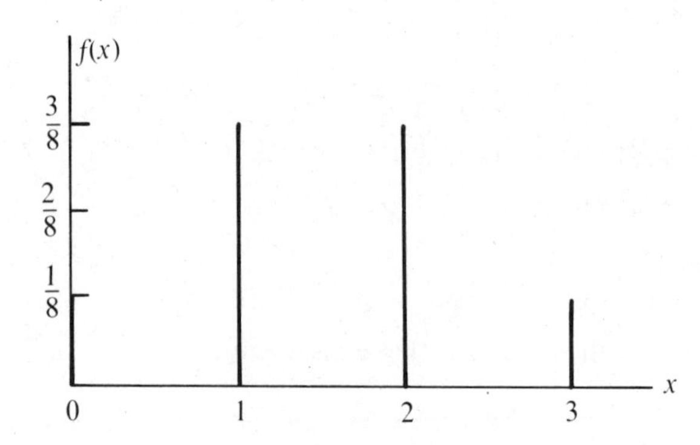

Graph of the probabilities $f(x_i)$

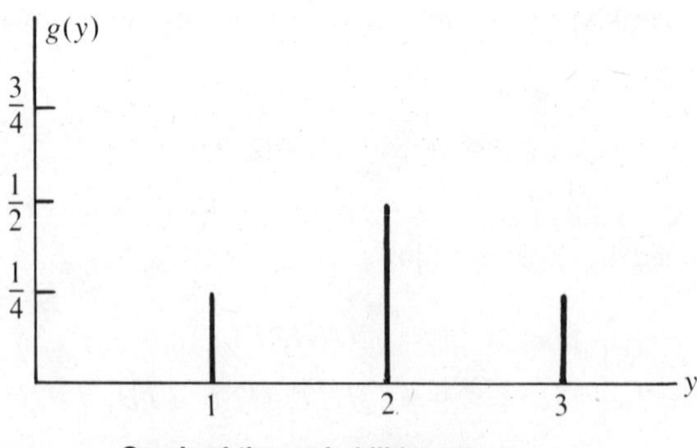

Graph of the probabilities $g(y_j)$

EXAMPLE 7

Roll two symmetrical dice and let the r.v. X be the maximum number on the dice. The sample space Ω has 36 points $\omega = (x,y)$, where $1 \le x,y \le 6$. The r.v. can attain values 1,2,3,4,5 and 6 and the partition induced by X is as follows:

$A_1 = \{\omega\colon X(\omega) = 1\} = \{(1,1)\}.$

$A_2 = \{\omega\colon X(\omega) = 2\} = \{(1,2), (2,1), (2,2)\}.$

$A_3 = \{\omega\colon X(\omega) = 3\} = \{(1,3), (2,3), (3,3), (3,2), (3,1)\}.$

$A_4 = \{\omega\colon X(\omega) = 4\} = \{(1,4), (2,4), (3,4), (4,4), (4,3), (4,2), (4,1)\}.$

$A_5 = \{\omega\colon X(\omega) = 5\} = \{(1,5), (2,5), (3,5), (4,5), (5,5), (5,4), (5,3), (5,2), (5,1)\}.$

$A_6 = \{\omega\colon X(\omega) = 6\} = \{(1,6), (2,6), (3,6), (4,6), (5,6), (6,6), (6,5), (6,4), (6,3), (6,2), (6,1)\}.$

For the maximum on two symmetrical dice, we get the following probabilities:

$$f(1) = P\{\omega : X(\omega) = 1\} = P(A_1) = \frac{1}{36}.$$

$$f(2) = P\{\omega : X(\omega) = 2\} = P(A_2) = \frac{3}{36}.$$

$$f(3) = P\{\omega : X(\omega) = 3\} = P(A_3) = \frac{5}{36}.$$

$$f(4) = P\{\omega : X(\omega) = 4\} = P(A_4) = \frac{7}{36}.$$

$$f(5) = P\{\omega : X(\omega) = 5\} = P(A_5) = \frac{9}{36}.$$

$$f(6) = P\{\omega : X(\omega) = 6\} = P(A_6) = \frac{11}{36}.$$

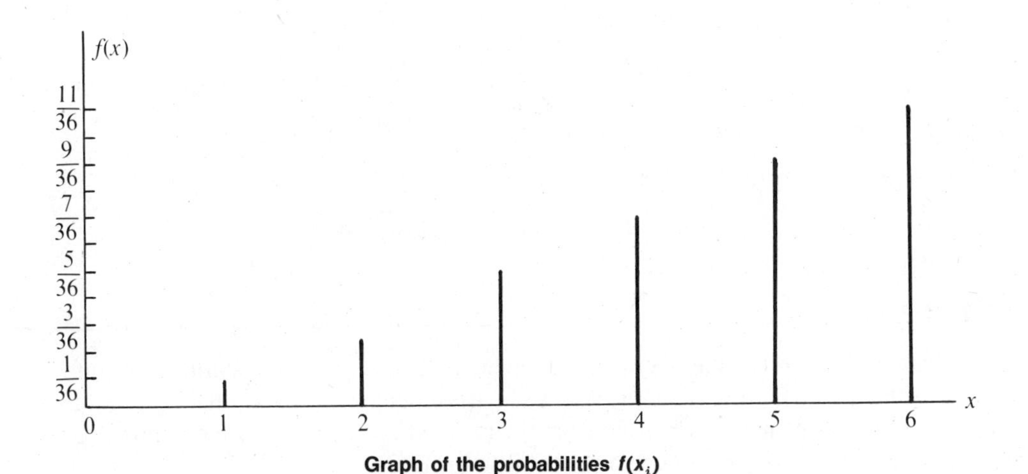

Graph of the probabilities $f(x_i)$

We can imagine $f(x_i)$ as a function defined for all possible distinct values that X can attain. It is quite obvious that this function will be very important for the investigation of random variables.

DEFINITION 3: Let X be a r.v. defined on a probability measure space (Ω,P), and let $\{x_1,x_2, \ldots ,x_k\}$ be the set of all possible distinct values that X can attain. Then the function $f(x_i)$ defined on the set $\{x_1,x_2, \ldots ,x_k\}$ by the relations

$$f(x_i) = P(\{\omega: X(\omega) = x_i\}), \qquad i = 1,2, \ldots ,k$$

is called the probability distribution of X.

COMMENT. Let $\{A_1,A_2, \ldots ,A_k\}$ be the partition induced by X. Then obviously,

$$f(x_i) = P(\{\omega: X(\omega) = x_i\}) = P(A_i).$$

Sometimes we simply write $f(x_i) = P(\{X = x_i\})$.

EXAMPLE 8

A car passes by four traffic lights on its route. Each of them allows it to move ahead or stop with the probability .5. Let the r.v. X denote the number of lights passed by the car before the first stop has occurred. Find the probability distribution of X. Assume that successive lights are independent (even though it is not a realistic assumption for main streets in big cities).

SOLUTION. X can attain the following values:

$$x_1 = 0, \quad x_2 = 1, \quad x_3 = 2, \quad x_4 = 3, \quad x_5 = 4.$$

Let p be the probability of the car being stopped by one set of lights ($p = .5$). The corresponding probabilities $f(x_i) = P\{\omega: X(\omega) = x_i\}$ can be computed as follows:

$$f(x_1) = P\{\omega: X(\omega) = 0\} = p = .5.$$

$$f(x_2) = P\{\omega: X(\omega) = 1\} = (1 - p)p = .25.$$

$$f(x_3) = P\{\omega: X(\omega) = 2\} = (1 - p)^2p = .125.$$

$$f(x_4) = P\{\omega: X(\omega) = 3\} = (1 - p)^3p = .0625.$$

$$f(x_5) = P\{\omega: X(\omega) = 4\} = (1 - p)^4 = .0625.$$

Thus we can get the following table of the probability distribution $f(x_i)$:

x_i	0	1	2	3	4
$f(x_i)$	.5	.25	.125	.0625	.0625

EXAMPLE 9

A man tries to unlock the door of his apartment in the dark. He has four keys, but only one opens the door. He cannot see the keys so he tries them one after another at random. He keeps the keys that do not fit separated from the rest. Let the r.v. X mean the number of trials necessary to open the door. Find the probability distribution of X.

SOLUTION. X attains values 1, 2, 3 or 4. Using a probability tree, we get the following:

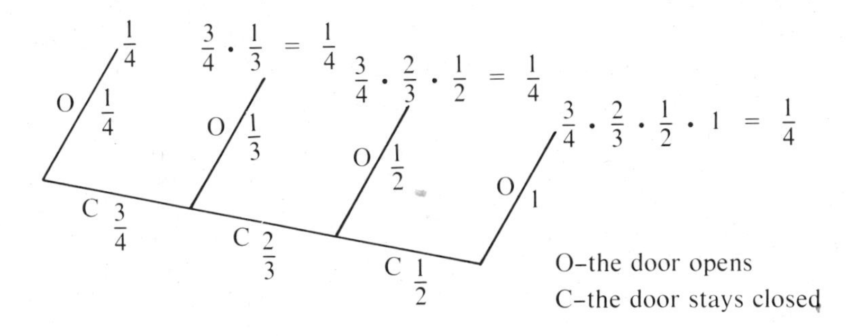

The probability distribution of X:

x_i	1	2	3	4
$f(x_i)$	.25	.25	.25	.25

We can see that the sum of all the probabilities is 1.

EXAMPLE 10

Imagine that the man in the previous example is dead drunk. He tries one key after another, but is unable to keep the keys that do not fit separated from the rest. He has four keys so he gives four tries and decides to sleep in the hall if the door still refuses to open. Let the r.v. Y denote the number of trials. Find the probability distribution of Y.

SOLUTION. The probability that he will open the door in any trial is $p = \frac{1}{4}$, and the probability that he will not is $q = \frac{3}{4}$. Therefore, using a probability tree:

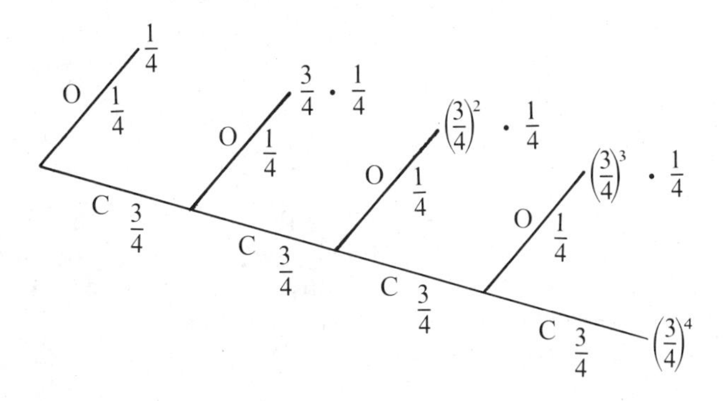

The table of the probability distribution of Y:

y_j	1	2	3	4
$g(y_j)$	.25	.1875	.1406	.4219

It is very likely that the drunkard will sleep in the hall; its probability is $\left(\frac{3}{4}\right)^4 = .3164$.

Notice that in all numerical examples the sum of all probabilities of a distribution is 1. This fact is expressed in the following:

→ **THEOREM 1:** Let $f(x_i)$, $i = 1,2, \ldots, k$ be the probability distribution of a r.v. X. Then

(a) $f(x_i) \geq 0. \qquad i = 1,2, \ldots, k.$

(b) $\sum_{i=1}^{k} f(x_i) = 1.$

Proof: Part (a) is obvious, because $f(x_i) = P(\{\omega: X(\omega) = x_i\}) = P(A_i) \geqq 0$. For part (b), we have

$$\sum_{i=1}^{k} f(x_i) = \sum_{i=1}^{k} P(A_i) = P\left(\bigcup_{i=1}^{k} A_i\right) = P(\Omega) = 1$$

because $\{A_1, A_2, \ldots, A_k\}$ is a partition. q.e.d.

EXAMPLE 11

From a lot of 100 items, of which 10 are defective, a random sample of size 5 is selected for quality control. Let X be the number of defective items in the sample. Find the probability distribution of the r.v. X.

The values of the r.v. X are

$$x_1 = 0, \quad x_2 = 1, \quad x_3 = 2, \quad x_4 = 3, \quad x_5 = 4, \quad x_6 = 5.$$

The probability that the sample contains i defective items is

$$P\{\omega: X(\omega) = i\} = \frac{\binom{10}{i}\binom{90}{5-i}}{\binom{100}{5}}$$

After the numerical calculations, we get

x_i	0	1	2	3	4	5
$f(x_i)$	.5838	.3394	.0702	.0064	.0002	0

EXAMPLE 12

A ski rental shop had five pairs of new and six pairs of used Astro-Norseman men's ski boots No. 10 before the beginning of the season. On the first day of the season, three people rented these boots, and on the second day four pairs of them were rented out (they were all chosen at random). Let the r.v. X denote the number of new pairs of these boots rented on the second day. Find the probability distribution of X. (Consider new boots rented for a day as used.)

SOLUTION. Let the number of new pairs rented on the first day be Y. Then the probability that the r.v. $Y = j$ is equal to

$$P(\{Y = j\}) = \frac{\binom{5}{j}\binom{6}{3-j}}{\binom{11}{3}}, \qquad j = 0,1,2,3.$$

Assume now that j pairs have been rented on the first day, i.e., $Y = j$. The conditional probability that $X = i$ given that $Y = j$ is equal to

$$P(\{X = i/Y = j\}) = \frac{\binom{5-j}{i}\binom{6+j}{4-i}}{\binom{11}{4}}$$

because there are $(5 - j)$ pairs of new boots and $(6 + j)$ pairs of used boots for renting on the second day.

Now, using the formula of total probability, we get

$$f(i) = P(\{X = i\}) = \sum_{j=0}^{3} P(\{Y = j\})P(\{X = i/Y = j\}) = \sum_{j=0}^{3} \frac{\binom{5}{j}\binom{6}{3-j}}{\binom{11}{3}} \frac{\binom{5-j}{i}\binom{6+j}{4-i}}{\binom{11}{4}}.$$

$i = 0,1,2,3,4.$

After some numerical calculations, we get the following probability distribution of the number of pairs of new boots rented out on the second day:

x_i	0	1	2	3	4
$f(x_i)$	.154	.446	.328	.069	.003

In this example, we got the probability distribution of X using the probability distribution of Y. However, in many situations we are interested in the joint behavior of the two r.v.'s. For this purpose, we shall investigate later the joint probability distributions of two or more r.v.'s.

EXAMPLE 13 (VERY DIFFICULT)

Let A and B be any two independent events, and I_A, I_B be their indicator functions (see page 128). Let the r.v. X be defined as $X = I_A + I_B$, and further

$$a = P(\{X = 2\}), \qquad b = P(\{X = 1\}), \qquad c = P(\{X = 0\}).$$

Show that at least one of the numbers a, b, c is at least $\frac{4}{9}$.

SOLUTION. Denote $P(A) = p$, $P(B) = q$, and assume that $1 > p > 0, 1 > q > 0$. Let us investigate the following quadratic polynomial $P(x)$:

$$P(x) = [p(x - 1) + 1][q(x - 1) + 1] = pqx^2 + [(1 - p)q + (1 - q)p]x + (1 - p)(1 - q).$$

From the independence of A and B, we can see that

$$a = pq \neq 0, \qquad b = [(1 - p)q + (1 - q)p] \neq 0, \qquad c = (1 - p)(1 - q) \neq 0,$$

hence

$$P(x) = ax^2 + bx + c.$$

However, the substitution $x = \left(1 - \frac{1}{p}\right)$ or $x = \left(1 - \frac{1}{q}\right)$ yields $P(x) = 0$. The polynomial has real roots and hence its discriminant D must be nonnegative:

$$D = b^2 - 4ac \geq 0.$$

We have the following inequalities and equality:

$$a > 0, \quad b > 0, \quad c > 0, \quad a + b + c = 1, \quad b^2 \geq 4ac.$$

Let us suppose that $a < \frac{4}{9}$, $b < \frac{4}{9}$. Then the last inequality yields $ac \leq \frac{b^2}{4} < \frac{4}{81}$, and the equality yields $a + c = 1 - b > 1 - \frac{4}{9} = \frac{5}{9}$. Hence $c > \left(\frac{5}{9} - a\right) > 0$, and after a substitution into the previous inequality, we get

$$a\left(\frac{5}{9} - a\right) < ac < \frac{4}{81}.$$

Equivalently,

$$-81a^2 + 45a < 4$$

or

$$(9a - 1)(9a - 4) > 0.$$

We have assumed $a < \frac{4}{9}$, thus $(9a - 4) < 0$, hence necessarily $(9a - 1) < 0$ or $a < \frac{1}{9}$.

Therefore, finally,

$$c = 1 - a - b > 1 - \frac{4}{9} - \frac{1}{9} = \frac{4}{9}.$$

We have proved the claim if $1 > p = P(A) > 0$ and $1 > q = P(B) > 0$. However if one of these numbers, say p, is 0, then

$$a = P(\{X = 2\}) = pq = 0, \quad b = P(\{X = 1\}) = P(B), \quad c = P(\{X = 0\}) = P(B')$$

and obviously either b or c is at least $\frac{4}{9}$. Similarly, if one of these numbers, say p, is 1, then

$$a = P(B), \quad b = P(B'), \quad c = 0$$

and either a or b is at least $\frac{4}{9}$.

JOINT PROBABILITY DISTRIBUTION OF RANDOM VARIABLES

Let us consider two random variables X and Y on a probability measure space (Ω,P). Let $S = \{x_1,x_2, \ldots ,x_k\}$ be the set of all possible distinct values that X can attain and let $\{A_1,A_2, \ldots ,A_k\}$ be the partition induced by X. Similarly, let $T = \{y_1,y_2, \ldots ,y_l\}$ be the set of all possible distinct values that Y can attain and let $\{B_1,B_2, \ldots ,B_l\}$ be the corresponding partition. The fundamental task of this section is the investigation of the common, joint behavior of both X and Y.

EXAMPLE 14

Let three coins be tossed, and let X mean the number of heads and Y mean the number of runs (see Example 6). The random variables X and Y are as follows:

ω	$X(\omega)$	$Y(\omega)$
HHH	3	1
HHT	2	2
HTH	2	3
THH	2	2
TTH	1	2
THT	1	3
HTT	1	2
TTT	0	1

$$x_1 = 3 \qquad x_2 = 2 \qquad x_3 = 1 \qquad x_4 = 0$$

$$A_1 = \{HHH\}, \quad A_2 = \{HHT,THH,HTH\}, \quad A_3 = \{HTT,THT,TTH\}, \quad A_4 = \{TTT\}.$$

The values of Y and the induced partition are:

$$y_1 = 1 \qquad y_2 = 2 \qquad y_3 = 3$$

$$B_1 = \{HHH,TTT\}, \quad B_2 = \{HHT,THH,TTH,HTT\}, \quad B_3 = \{THT,HTH\}.$$

Let us investigate now the joint behavior of X and Y. For each value of X and each value of Y, let us write the set of all points for which X and Y are equal to the given numbers at the intersection of the corresponding row and column.

$X \backslash Y$	1	2	3	
3	$\{HHH\}$	ϕ	ϕ	$\{HHH\}$
2	ϕ	$\{HHT,THH\}$	$\{HTH\}$	$\{HHT,THH,HTH\}$
1	ϕ	$\{TTH,HTT\}$	$\{THT\}$	$\{TTH,HTT,THT\}$
0	$\{TTT\}$	ϕ	ϕ	$\{TTT\}$
	$\{HHH,TTT\}$	$\{HHT,THH,TTH,HTT\}$	$\{HTH,THT\}$	Ω

This array describes completely the joint behavior of X and Y. If we want to know, for example, when $X(\omega) = 2$ and $Y(\omega) = 2$, then we look at the intersection of the row containing 2 for X, and the column containing 2 for Y. We get the set $\{HHT,THH\}$, and this is exactly the set of all points for which $X(\omega) = 2$ and $Y(\omega) = 2$. On the far right-hand side, we have the partition induced by X; in the bottom row, we have the partition induced by Y. Every entry in the array is the set-intersection of the corresponding far right set (from the partition induced by X) and the corresponding bottom set (from the partition induced by Y).

In general, we are interested in the set of all outcomes for which $X(\omega) = x_i$ and simultaneously $Y(\omega) = y_j$. This set is

$$\{\omega\colon X(\omega) = x_i \quad \text{and} \quad Y(\omega) = y_j\} = A_i \cap B_j. \qquad i = 1,2, \ldots ,k.$$
$$j = 1,2, \ldots ,l.$$

Furthermore, we are interested in the probabilities that $X(\omega) = x_i$ and at the same time $Y(\omega) = y_j$.

DEFINITION 4: Let X and Y be random variables defined on a probability measure space (Ω,P); let the corresponding sets of all possible and distinct values be S and T. The function $h(x_i, y_j)$ defined on the set $S \times T$ by the relation

$$h(x_i, y_j) = P\{\omega: X(\omega) = x_i, \quad Y(\omega) = y_j\} \qquad \begin{array}{l} i = 1,2, \ldots, k \\ j = 1,2, \ldots, l \end{array}$$

is called the joint probability distribution of X and Y.

COMMENT. Obviously, $h(x_i,y_j) = P(A_i \cap B_j)$, $\quad i = 1,2, \ldots, k$; $\quad j = 1,2, \ldots, l$

We can imagine the joint probability distribution in the form of a table:

$X \backslash Y$	y_1	y_2	...	y_j	...	y_l	f
x_1	$h(x_1,y_1)$	$h(x_1,y_2)$	...	$h(x_1,y_j)$	...	$h(x_1,y_l)$	$f(x_1)$
x_2	$h(x_2,y_1)$	$h(x_2,y_2)$	...	$h(x_2,y_j)$	...	$h(x_2,y_l)$	$f(x_2)$
$\vdots$	$\vdots$	$\vdots$		$\vdots$		$\vdots$	$\vdots$
x_i	$h(x_i,y_1)$	$h(x_i,y_2)$	...	$h(x_i,y_j)$	...	$h(x_i,y_l)$	$f(x_i)$
$\vdots$	$\vdots$	$\vdots$		$\vdots$		$\vdots$	$\vdots$
x_k	$h(x_k,y_1)$	$h(x_k,y_2)$	...	$h(x_k,y_j)$	...	$h(x_k,y_l)$	$f(x_k)$
g	$g(y_1)$	$g(y_2)$	...	$g(y_j)$	...	$g(y_l)$	1

Following is the table of the joint probability distribution of X and Y from Example 14.

$X \backslash Y$	1	2	3	
3	$h(3,1) = \frac{1}{8}$	$h(3,2) = 0$	$h(3,3) = 0$	$f(3) = \frac{1}{8}$
2	$h(2,1) = 0$	$h(2,2) = \frac{2}{8}$	$h(2,3) = \frac{1}{8}$	$f(2) = \frac{3}{8}$
1	$h(1,1) = 0$	$h(1,2) = \frac{2}{8}$	$h(1,3) = \frac{1}{8}$	$f(1) = \frac{3}{8}$
0	$h(0,1) = \frac{1}{8}$	$h(0,2) = 0$	$h(0,3) = 0$	$f(0) = \frac{1}{8}$
	$g(1) = \frac{2}{8}$	$g(2) = \frac{4}{8}$	$g(3) = \frac{2}{8}$	

For example, the entry in the second row, third column $h(2,3) = \frac{1}{8}$ means that the probability of the simultaneous occurrence of $X = 2$ and $Y = 3$ is $\frac{1}{8}$. It is interesting to notice that the marginal sums represent the distribution of X and the distribution of Y as derived in Example 6. We can draw a graph of the joint probability distribution as follows:

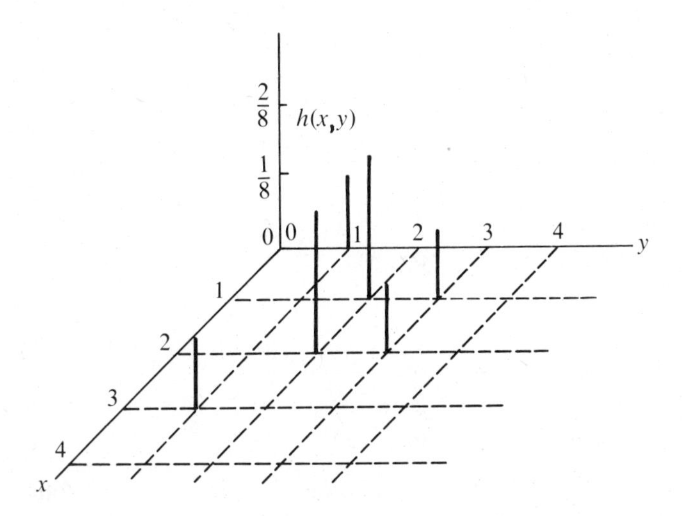

EXAMPLE 15

Let two symmetrical cubic dice be rolled. Let X be the maximum and Y be the total sum thus obtained. Find the joint distribution of X and Y, the distributions of X and Y and compare them.

SOLUTION. The r.v. X can attain values 1,2,3,4,5 and 6. Let f be its probability distribution (see Example 7, including the graph of f). The r.v. Y can attain values $2,3,\ldots,11,12$. Let g be its distribution; let h be the joint distribution.

There are 36 equally likely outcomes (x,y), $1 \leq x,y \leq 6$. We can see that, for example,

$$h(1,2) = P(\{X = 1, \quad Y = 2\}) = P(\{1,1)\}) = \frac{1}{36}\cdot$$

$$h(2,3) = P(\{X = 2, \quad Y = 3\}) = P(\{2,1), (1,2)\}) = \frac{2}{36}\cdot$$

$$h(3,3) = P(\{X = 3, \quad Y = 3\}) = P(\phi) = 0.$$

Similarly, we can find all the remaining probabilities. Following is the table of joint probability distribution of X and Y (all the omitted values are equal to 0).

If we compare the right-hand column with Example 7, we can see that it is the probability distribution of the r.v. X. Notice that each entry in this column is the sum of all the entries in the corresponding row. Similarly, we can see that the bottom row represents the probability distribution of Y, and that each entry is the sum of all the entries in the corresponding column.

$X \backslash Y$	2	3	4	5	6	7	8	9	10	11	12	$f(x)$
1	$\frac{1}{36}$											$\frac{1}{36}$
2		$\frac{2}{36}$	$\frac{1}{36}$									$\frac{3}{36}$
3			$\frac{2}{36}$	$\frac{2}{36}$	$\frac{1}{36}$							$\frac{5}{36}$
4				$\frac{2}{36}$	$\frac{2}{36}$	$\frac{2}{36}$	$\frac{1}{36}$					$\frac{7}{36}$
5					$\frac{2}{36}$	$\frac{2}{36}$	$\frac{2}{36}$	$\frac{2}{36}$	$\frac{1}{36}$			$\frac{9}{36}$
6						$\frac{2}{36}$	$\frac{2}{36}$	$\frac{2}{36}$	$\frac{2}{36}$	$\frac{2}{36}$	$\frac{1}{36}$	$\frac{11}{36}$
$g(y)$	$\frac{1}{36}$	$\frac{2}{36}$	$\frac{3}{36}$	$\frac{4}{36}$	$\frac{5}{36}$	$\frac{6}{36}$	$\frac{5}{36}$	$\frac{4}{36}$	$\frac{3}{36}$	$\frac{2}{36}$	$\frac{1}{36}$	1

The total of all the entries is 1. This observation is expressed in the following theorem:

→ **THEOREM 2:** Let $h(x_i, y_j)$ be the joint probability distribution of the random variables X and Y, $f(x_i)$ the probability distribution of X and $g(y_j)$ the probability distribution of Y, $i = 1,2, \ldots, k$; $j = 1,2, \ldots, l$. Then

(a) $h(x_i, y_j) \geq 0$. $\quad i = 1,2, \ldots, k; \quad j = 1,2, \ldots, l$.

(b) $\sum_{i=1}^{k} h(x_i, y_j) = g(y_j)$. $\quad j = 1,2, \ldots, l$.

(c) $\sum_{j=1}^{l} h(x_i, y_j) = f(x_i)$. $\quad i = 1,2, \ldots, k$.

(d) $\sum_{i=1}^{k} \sum_{j=1}^{l} h(x_i, y_j) = 1$.

COMMENT. The probability distributions $f(x_i)$ and $g(y_j)$ are sometimes called *marginal* probability distributions.

Proof:

(a) $h(x_i, y_j) = P(A_i \cap B_j) \geq 0$.

(b) $\sum_{i=1}^{k} h(x_i, y_j) = \sum_{i=1}^{k} P(A_i \cap B_j) = P\left(\bigcup_{i=1}^{k} (A_i \cap B_j)\right) = P\left(B_j \cap \bigcup_{i=1}^{k} A_i\right) = P(B_j \cap \Omega) = P(B_j) = g(y_j)$.

(c) $\sum_{j=1}^{l} h(x_i, y_j) = \sum_{j=1}^{l} P(A_i \cap B_j) = P\left(\bigcup_{j=1}^{l} (A_i \cap B_j)\right) = P\left(A_i \cap \bigcup_{j=1}^{l} B_j\right) = P(A_i \cap \Omega) = P(A_i) = f(x_i)$.

(d) $\sum_{i=1}^{k} \sum_{j=1}^{l} h(x_i, y_j) = \sum_{i=1}^{k} f(x_i) = 1$. q.e.d.

EXAMPLE 16

A hand of 3 cards is dealt from an ordinary deck of 52 cards. Let the r.v. X be the number of aces and the r.v. Y be the number of spades in the hand. Find the joint probability distributions.

SOLUTION. There are the following numbers of cards in the deck:

Spade Ace	Nonspade Aces	Spade Nonaces	Nonspade Nonaces
1	3	12	36

Following are shown several situations where X and Y attain some particular values:

	SA	NSA	SNA	NSNA
$X = 0, \quad Y = 0$	0	0	0	3
$X = 1, \quad Y = 3$	1	0	2	0
$X = 1, \quad Y = 2$	1	0	1	1
or	0	1	2	0
$X = 2, \quad Y = 1$	1	1	0	1
or	0	2	1	0
and so forth.				

The corresponding probabilities are:

$$h(0,0) = P(\{X = 0, \quad Y = 0\}) = \frac{\binom{36}{3}}{\binom{52}{3}} = .3231.$$

$$h(1,3) = P(\{X = 1, \quad Y = 3\}) = \frac{\binom{1}{1}\binom{12}{2}}{\binom{52}{3}} = .0030.$$

$$h(1,2) = P(\{X = 1, \quad Y = 2\}) = \frac{\binom{1}{1}\binom{12}{1}\binom{36}{1} + \binom{3}{1}\binom{12}{2}}{\binom{52}{3}} = .0285.$$

$$h(2,1) = P(\{X = 2, \quad Y = 1\}) = \frac{\binom{1}{1}\binom{3}{1}\binom{36}{1} + \binom{3}{2}\binom{12}{1}}{\binom{52}{3}} = .0065.$$

Table of the joint probability distribution of X and Y:

$Y \backslash X$	0	1	2	3	$g(y)$
0	.32308	.08552	.00489	.00004	.41353
1	.34208	.08715	.00652	.00014	.43588
2	.10751	.02851	.00163	0	.13765
3	.00995	.00299	0	0	.01294
$f(x)$	.78262	.20416	.01304	.00018	1.0000

EXAMPLE 17

Let X and Y be r.v.'s with the following joint and marginal probability distributions. Find the missing values x,y,z,t,u,v,w,r.

$X \backslash Y$	1	7	8	
2	0	.1	x	y
4	z	t	0	.4
6	u	.3	.1	v
	.1	w	.2	r

SOLUTION. Clearly, $r = 1$.
From the third column, $x + 0 + .1 = .2$, hence $x = .1$.
From the first row, $0 + .1 + x = y$, but $x = .1$, hence $y = .2$.
From the last column, $y + .4 + v = r$ but $r = 1$ and $y = .2$, hence $v = .4$.
From the last row, $.1 + w + .2 = 1$, hence $w = .7$.
From the second column, $.1 + t + .3 = w$ but $w = .7$, hence $t = .3$.
From the second row, $z + t + 0 = .4$ but $t = .3$, hence $z = .1$.
From the first column, $0 + z + u = .1$ but $z = .1$, hence $u = 0$.
Hence the joint and marginal distributions are as follows:

$X \backslash Y$	1	7	8	$f(x)$
2	0	.1	.1	.2
4	.1	.3	0	.4
6	0	.3	.1	.4
$g(y)$	.1	.7	.2	1

INDEPENDENCE OF RANDOM VARIABLES

In Chapter VIII, we investigated the independence of events. There is also a concept of independence for r.v.'s. Let us consider the following case.

EXAMPLE 18

Toss three fair coins so that all eight outcomes are equally likely. Let the r.v. X be 1 or 0, depending on whether the first outcome is H or T, and let the r.v. Y be 1 or 0, depending on whether the second outcome is T or H. The r.v.'s have the following values:

Ω	X	Y
HHH	1	0
HHT	1	0
HTH	1	1
THH	0	0
TTH	0	1
THT	0	0
HTT	1	1
TTT	0	1

Investigate the event that $X = 0$ and $Y = 0$.

By definition, $h(0,0) = P(\{X = 0, Y = 0\}) = P(\{THH, THT\}) = \frac{2}{8} = \frac{1}{4}$.

Further, we see that $f(0) = P(\{X = 0\}) = P(\{THH,THT,TTH,TTT\}) = \frac{1}{2}$.

$$g(0) = P(\{Y = 0\}) = P(\{HHH,HHT,THH,THT\}) = \frac{1}{2}.$$

We can conclude that the events $\{X = 0\}$ and $\{Y = 0\}$ are independent because

$$\{X = 0\} \cap \{Y = 0\} = \{X = 0, \quad Y = 0\}$$

$$P(\{X = 0, \quad Y = 0\} = \frac{1}{4} = \frac{1}{2} \cdot \frac{1}{2} = P(\{X = 0\})P(\{Y = 0\})$$

or in another notation

$$h(0,0) = f(0)g(0).$$

As a matter of fact, we can see that a similar relation remains true for any selection of values of X and Y.

X \ Y	0	1	$f(x)$
0	$\frac{1}{4}$	$\frac{1}{4}$	$\frac{1}{2}$
1	$\frac{1}{4}$	$\frac{1}{4}$	$\frac{1}{2}$
$g(y)$	$\frac{1}{2}$	$\frac{1}{2}$	1

The above joint probability distribution has the remarkable property that each entry is the product of corresponding row and column sums:

$$h(x_i,y_j) = f(x_i)g(y_j).$$

In other words, the two events $\{X = x_i\}$ and $\{Y = y_j\}$ are independent, and it seems convenient to call any two such r.v.'s *independent*. Incidentally, this terminology agrees with our intuition: X is determined only by the outcome of the first toss while Y is determined only by the outcome of the second toss, and it is plausible to say that X and Y are independent.

DEFINITION 5: Let $h(x_i, y_j)$, $f(x_i)$, $g(y_j)$, $i = 1,2, \ldots ,k$; $j = 1,2, \ldots ,l$ be the joint and marginal probability distributions of the r.v.'s X and Y. We say that X and Y are independent if for every $i = 1,2, \ldots ,k$ and every $j = 1,2, \ldots ,l$, we have

$$h(x_i, y_j) = f(x_i)g(y_j).$$

COMMENT 1. The above condition means simply that every entry $h(x_i,y_j)$ of the joint probability distributions is a product of the corresponding row sum $f(x_i)$ and column sum $g(y_j)$.

COMMENT 2. If $\{A_1, A_2, \ldots, A_k\}$ and $\{B_1, B_2, \ldots, B_l\}$ are the partitions induced by X and Y, respectively, then obviously the above condition simply means that for any i and j

$$P(A_i \cap B_j) = P(A_i)P(B_j)$$

or, in other words, any two sets A_i and B_j are independent.

COMMENT 3. If r.v.'s X and Y are not independent, we shall often say that they are *dependent.*

EXAMPLE 19

In the case of tossing three coins, when X means the number of heads and Y the number of runs, the random variables are not independent, because, for example,

$$h(3,1) = \frac{1}{8} \neq \frac{1}{8} \cdot \frac{2}{8} = f(3)g(1). \qquad \text{(See Example 14.)}$$

Similarly, we can see that the r.v.'s in Example 15 and Example 16 are dependent.

EXAMPLE 20

Let the distribution of X and Y be given as follows:

X \ Y	2	3	4	
1	.06	.15	.09	.3
2	.14	.35	.21	.7
	.2	.5	.3	1

Then X and Y are independent, because every entry inside of the array is the product of the corresponding row and column sums.

EXAMPLE 21

Fill in the gaps in the following table, assuming that X and Y are independent random variables.

X \ Y	1	2	3	
4	.06			
5				.3
6			.08	
		.6		

SOLUTION. Let x be the unknown entry $f(4)$, and let y be the unknown entry $g(1)$, thus $f(6) = .7 - x$ and $g(3) = .4 - y$.

X \ Y	1	2	3	$f(x_i)$
4	.06			x
5				.3
6			.08	$(.7 - x)$
$g(y_j)$	y	.6	$(.4 - y)$	

Then we can set up the following equations:

$$xy = .06. \qquad (.7 - x)(.4 - y) = .08.$$

Eliminating x, we get the quadratic equation:

$$70y^2 - 26y + 2.4 = 0.$$

Solving this equation, we get

$$y_1 = .2. \qquad y_2 = .17142857.$$

The corresponding x values are:

$$x_1 = \frac{.06}{.2} = .3. \qquad x_2 = \frac{.06}{.17142857} = .35.$$

Hence the first solution is:

X \ Y	1	2	3	$f(x_i)$
4	.06	.18	.06	.3
5	.06	.18	.06	.3
6	.08	.24	.08	.4
$g(y_j)$	.2	.6	.2	

An approximation of the second solution (after rounding off, which destroys the required property of the matrix) is:

X \ Y	1	2	3	$f(x_i)$
4	.06	.21	.08	.35
5	.05	.18	.07	.3
6	.06	.21	.08	.35
$g(y_j)$	.17	.6	.23	

EXAMPLE 22

Let A and B be two sets $A \subset \Omega$, $B \subset \Omega$, and let the r.v.'s X and Y be their indicator functions, i.e.,

$$X(\omega) = I_A(\omega) \qquad Y(\omega) = I_B(\omega)$$ (See page 128.)

Show that the r.v.'s X and Y are independent if and only if the sets A and B are independent.

SOLUTION. (a) Let X and Y be independent. Obviously

$$\{\omega: X(\omega) = 1 \quad \text{and} \quad Y(\omega) = 1\} = A \cap B.$$

$$\{\omega: X(\omega) = 1\} = A. \qquad \{\omega: Y(\omega) = 1\} = B.$$

Thus
$$P(A \cap B) = P(\{X = 1 \quad \text{and} \quad Y = 1\}) = h(1,1) = f(1)g(1)$$
$$= P(\{X = 1\})P\{Y = 1\}) = P(A)P(B).$$

So A and B are independent.

(b) Let A and B be independent. Then also the pairs of sets A',B; A,B'; A',B' are independent (see Theorem 5, page 93).

$$h(1,1) = P(A \cap B) = P(A)P(B) = f(1)g(1)$$
$$h(1,0) = P(A \cap B') = P(A)P(B') = f(1)g(0)$$
$$h(0,1) = P(A' \cap B) = P(A')P(B) = f(0)g(1)$$
$$h(0,0) = P(A' \cap B') = P(A')P(B') = f(0)g(0)$$

So X and Y are independent.

EXAMPLE 23

Two people are in a shooting competition. The probability that Mr. A hits the target is .6, and the probability for Mr. B is .7. Mr. A shoots twice and Mr. B shoots thrice at the target. Let the r.v.'s X and Y denote the numbers of hits by Mr. A and Mr. B, respectively. Find the joint probability distribution of X and Y. Assume that the performance of Mr. A and Mr. B are independent.

SOLUTION. Let us first find the distribution of X.

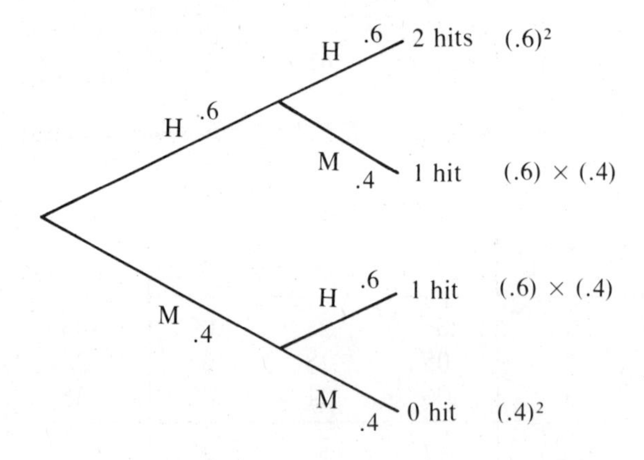

From the above tree we get:

x	0	1	2
$f(x)$	.16	.48	.36

Similarly, we can find the distribution of Y:

y	0	1	2	3
$g(y)$	.027	.189	.441	.343

Assuming the independence of the performances of Mr. A and Mr. B, we can claim that the r.v.'s X and Y are independent. This assumption will give us a fairly adequate mathematical model of the real solution. (See the discussion of independence on page 97.)

Therefore $$h(2,3) = f(2)g(3) = (.36) \times (.343) = .1235.$$
$$h(1,2) = f(1)g(2) = (.48) \times (.441) = .2117.$$

Similarly, we can calculate all probabilities $h(i,j)$:

X \ Y	0	1	2	3	
0	.0043	.0302	.0706	.0549	.16
1	.0130	.0907	.2117	.1646	.48
2	.0097	.0680	.1588	.1235	.36
	.027	.189	.441	.343	

We can generalize the notion of joint distribution to more than two variables.

DEFINITION 6: Let X_i, $i = 1,2, \ldots, n$ be random variables, defined on a probability measure space (Ω,P), and let the corresponding sets of possible distinct values be $S_i = \{x_{i1}, x_{i2}, \ldots, x_{in_i}\}$. Then the function $f_{X_1X_2\ldots X_n}$ defined on the Cartesian product $S_1 \times S_2 \times \ldots \times S_n$ by the relation

$$f_{X_1X_2\ldots X_n}(x_{1i_1}, x_{2i_2}, \ldots, x_{ni_n}) = P\{\omega\colon X_1(\omega) = x_{1i_1}, X_2(\omega) = x_{2i_2}, \ldots, X_n(\omega) = x_{ni_n}\}$$

is called the joint probability distribution of $X_1, X_2, \ldots, X_n$. Random variables $X_1, X_2, \ldots, X_n$ are called mutually independent if

$$f_{X_1X_2\ldots X_n}(x_{1i_1}, x_{2i_2}, \ldots, x_{ni_n}) = f_{X_1}(x_{1i_1})f_{X_2}(x_{2i_2}) \ldots f_{X_n}(x_{ni_n})$$

for any selection of arguments.

EXAMPLE 24

Toss a coin three times and assume that all eight outcomes are equally likely. Let $X_i (i = 1,2,3)$ be equal to 0 if the i-th toss results in H, and be equal to 1 if the i-th toss results in T. Then

$$f_{X_1X_2X_3}(0,0,0) = P(\{HHH\}) = \frac{1}{8} = \frac{1}{2} \cdot \frac{1}{2} \cdot \frac{1}{2} = f_{X_1}(0)f_{X_2}(0)f_{X_3}(0),$$

$$f_{X_1X_2X_3}(1,0,0) = P(\{THH\}) = \frac{1}{8} = \frac{1}{2} \cdot \frac{1}{2} \cdot \frac{1}{2} = f_{X_1}(1) f_{X_2}(0) f_{X_3}(0),$$

$$f_{X_1X_2X_3}(0,1,0) = P(\{HTH\}) = \frac{1}{8} = \frac{1}{2} \cdot \frac{1}{2} \cdot \frac{1}{2} = f_{X_1}(0) f_{X_2}(1) f_{X_3}(0),$$

$$f_{X_1X_2X_3}(0,0,1) = P(\{HHT\}) = \frac{1}{8} = \frac{1}{2} \cdot \frac{1}{2} \cdot \frac{1}{2} = f_{X_1}(0) f_{X_2}(0) f_{X_3}(1),$$

$$f_{X_1X_2X_3}(1,1,0) = P(\{TTH\}) = \frac{1}{8} = \frac{1}{2} \cdot \frac{1}{2} \cdot \frac{1}{2} = f_{X_1}(1) f_{X_2}(1) f_{X_3}(0),$$

$$f_{X_1X_2X_3}(1,0,1) = P(\{THT\}) = \frac{1}{8} = \frac{1}{2} \cdot \frac{1}{2} \cdot \frac{1}{2} = f_{X_1}(1) f_{X_2}(0) f_{X_3}(1),$$

$$f_{X_1X_2X_3}(0,1,1) = P(\{HTT\}) = \frac{1}{8} = \frac{1}{2} \cdot \frac{1}{2} \cdot \frac{1}{2} = f_{X_1}(0) f_{X_2}(1) f_{X_3}(1),$$

$$f_{X_1X_2X_3}(1,1,1) = P(\{TTT\}) = \frac{1}{8} = \frac{1}{2} \cdot \frac{1}{2} \cdot \frac{1}{2} = f_{X_1}(0) f_{X_2}(0) f_{X_3}(0),$$

and therefore X_1, X_2 and X_3 are mutually independent.

EXERCISES

(Answers in Appendix to exercises marked)*

Following are just a few problems. Many more exercises on probability distributions are located at the ends of the next two chapters.

1. Imagine an experiment of tossing a coin four times, and let the sample space Ω be the Cartesian product of the set $\{H,T\}$ with itself four times. Consider the random variables on Ω described by the following: X is the number of heads in ω, Y is the number of tails in ω, Z is the length of the longest run of consecutive tails in ω ($Z = 0$ if no tails are tossed) and W is the number of the toss in which the last tail is tossed ($W = 0$ if no tails are tossed).
   (a) For each $\omega \in \Omega$, give the value of each of the random variables X, Y, Z and W.
   (b) For each $\omega \in \Omega$, give the value of each of the random variables $X + Y$, $X - Y$, $Z + W$, W^2 and XZ.
2. (a) Are any two of the random variables in Problem 1 independent?
   (b) Write down the partitions induced by the random variables $Z + W$ and XZ, defined in Problem 1.
   (c) Find the probability distributions of X, Y, Z, W, $Z + W$ and XZ.

*3. Let X be a random variable with a probability distribution given by

X	-2	-1	0	1	2	3	7	12
$f(x)$	.1	.15	.2	.2	.15	.1	.05	.05

What is the probability of the event that
(a) X is even?
(b) X is negative?
(c) X is less than -1 and odd?
(d) X takes a value between 1 and 8 inclusive?
(e) X takes a value between 1 and 7 inclusive?

4. A hunter finds a rock ptarmigan sitting high in a tree. (It is well known that this bird will not fly away even if it is being shot at.) He has only three shells in his 1100 Remington autoloader (and no other ammunition). If the probability of hitting the bird by each shot is .4, find the distribution of the number of shots. (Assume that the hunter keeps shooting till he gets the bird or has no shells left.)

*5. In a certain camp there are 15 junior and 25 senior campers. Each day seven are chosen by lot for kitchen duty. If Y denotes the number of juniors in this group, what is the probability that
(a) $Y = 0$?
(b) $Y = 2$?
(c) $6 \leq Y \leq 7$?

6. Let X and Y be two random variables with a joint probability distribution given by

$X \backslash Y$	-1	1	2	5
-1	$\frac{1}{27}$	$\frac{1}{9}$	$\frac{1}{9}$	$\frac{1}{27}$
1	$\frac{1}{9}$	$\frac{2}{9}$	$\frac{1}{9}$	0
5	$\frac{4}{27}$	$\frac{1}{9}$	0	0

What is the probability of the event that
(a) Y is even?
(b) Y is even and X^2 is less than 2?
(c) XY is odd?
(d) X is positive but less than 5?
(e) Y is positive but less than 5?

7. An urn contains eight balls, of which three are red and five are black. Four balls are drawn from the urn without replacement in such a way that each combination of balls is equally likely. What is the probability distribution of X, the number of red balls drawn? What is the joint probability distribution of X and the number Y of black balls drawn?

*8. Three balls are placed successively in one of three boxes numbered 1, 2 and 3, and all possible $3^3 = 27$ placements are regarded as equally likely. Find the probability distribution of the number X_1 of balls in box number 1 and of the number X_2 of balls in box number 2. Find the joint probability distribution of X_1 and X_2.

9. A special deck of cards contains only tens, jacks, queens, kings and aces of all four suits. A hand of three cards is dealt. Let X denote the number of hearts, and Y denote the number of kings. Find the joint distribution of X and Y, and its marginals.

*10. A box contains five red and five blue marbles. We know that three of the red marbles and four of the blue ones are wooden; the remaining marbles are plastic. A sample of two marbles is drawn at random. X is defined as the number of red marbles and Y as the number of plastic marbles in the sample. Find the joint probability distribution of X and Y, and its marginals. Are X and Y independent?

11. Your friend has two dice. The first one is fair and the second one is loaded as shown in the table:

outcome	1	2	3	4	5	6
probability	$\frac{1}{4}$	$\frac{1}{4}$	$\frac{1}{4}$	$\frac{1}{12}$	$\frac{1}{12}$	$\frac{1}{12}$

You do not know which is which, so you pick up one at random and roll it twice. Let X be the minimum in two rolls, and Y be the absolute value of the difference. Find the joint probability distribution of X and Y, and its marginals.

*12. Let X and Y be two random variables with the following distributions:

$X \backslash Y$	0	1	2	
1	0			.2
2		.3		.4
3	0		.1	
	.1		.2	

Are X and Y independent? Calculate the missing values.

13. Let X, Y be two independent random variables with a joint probability distribution partly given as follows:

$X \backslash Y$	0	1	
1	$\frac{1}{6}$		
2			$\frac{2}{3}$
3			
		$\frac{1}{2}$	

Fill the gaps.

14. X and Y are independent r.v.'s. Find their distributions.

$X \backslash Y$	1	2	3	
1				$\frac{1}{8}$
2	$\frac{1}{16}$			
3		$\frac{1}{6}$		
			$\frac{1}{2}$	

15. X and Y are independent r.v.'s. Find their distributions.

$X \backslash Y$	2	4	6	
3	.06			
5				.2
7			.3	
		.2		

16. Bob and Ray each toss a coin twice. Bob's coin is fair but Ray's coin is biased, giving in the long run twice as many heads as tails. Find the joint probability distribution of the numbers of heads obtained by each boy (assume that their results are independent).

17. A little experimental pool contains 25 fish, of which 10 are cutthroat trout and the rest are rainbow trout. There are 15 female fish in the pool, of which 6 are cutthroat trout. If a random sample of size 10 fish is chosen, find the probability that
*(a) There are more rainbow than cutthroat trout in the sample.
*(b) There are exactly four females in the sample.
*(c) The majority of the fish in the sample are females.
(d) There are exactly four males in the sample given that there are three rainbow trout there.

FUNDAMENTAL PROBABILITY DISTRIBUTIONS

In Chapters VI through X, we encountered several groups of problems which, regardless of their seemingly rather distinct formulations, actually had identical logical structure. In this chapter, we shall disregard all particular and special features of these problems and generalize their common characteristic properties. In this way we shall derive some fundamental probability distributions; namely the binomial, the multinomial, the hypergeometric and the Polya distributions, as well as the distribution of runs.

BINOMIAL DISTRIBUTION

EXAMPLE 1

Let us consider a student who is going to take a multiple choice examination. Each question has four possible answers printed on the exam sheet, and he is to select and circle exactly one of them. The student had been skipping lectures and did not study for the exam, therefore he is completely ignorant of the subject matter. For that reason he decides to circle answers at random. What is the probability that he gets exactly two correct answers if the exam paper contains (a) three questions? (b) twenty questions?

SOLUTION. The student's answer to each question can be either correct—let us call this outcome a success and denote it by S, or false—let us call this outcome a failure and denote it by F. If the student were to answer only one question, then a suitable sample space to describe this experiment would be $\Omega = \{S,F\}$. Let us denote the probability of a success by p, hence $p = \frac{1}{4}$, and the probability of failure by $q = 1 - p$, hence $q = \frac{3}{4}$ because the student is choosing answers at random.

In part (a), we have to investigate all possible outcomes provided the student is to answer three questions. This "new experiment" consists actually of *independent repetitions* of answering a single question. A new sample space, or in other words, a representation of the set of all possible outcomes, consists of all ordered triplets of letters S and F where the i-th letter means whether the i-th question has a correct (S) or a false (F) answer and it will be denoted by Ω^3.

Hence $\qquad \Omega^3 = \{SSS, SSF, SFS, FSS, FFS, FSF, SFF, FFF\}.$

It is clear that the new sample space is a threefold Cartesian product of Ω,

$$\Omega^3 = \Omega \times \Omega \times \Omega$$

which consists of all ordered sequences of length 3 formed from the letters S and F. Let us find the probabilities of all the sample points. Because of independence, we have:

Point of Ω^3	Its Probability
SSS	$\frac{1}{4}\frac{1}{4}\frac{1}{4} = p^3$
SSF	$\frac{1}{4}\frac{1}{4}\frac{3}{4} = p^2q$
SFS	$\frac{1}{4}\frac{3}{4}\frac{1}{4} = p^2q$
FSS	$\frac{3}{4}\frac{1}{4}\frac{1}{4} = p^2q$
FFS	$\frac{3}{4}\frac{3}{4}\frac{1}{4} = pq^2$
FSF	$\frac{3}{4}\frac{1}{4}\frac{3}{4} = pq^2$
SFF	$\frac{1}{4}\frac{3}{4}\frac{3}{4} = pq^2$
FFF	$\frac{3}{4}\frac{3}{4}\frac{3}{4} = q^3$

There are three ways to get exactly two correct answers out of three, namely, SSF, SFS, FSS, and each of them has the same probability $\left(\frac{1}{4}\right)^2 \frac{3}{4} = p^2q$. Hence

$$P(\text{exactly two correct answers}) = 3\left(\frac{1}{4}\right)^2 \frac{3}{4} = \binom{3}{2} p^2 q.$$

This investigation was a bit tedious, but we could easily manage it. Part (b) is much worse. In this case, we would have to investigate sequences of the letters S,F of the length 20, and find those containing exactly two letters S. However, there are exactly $\binom{20}{2}$ such sequences, and the probability of each is $\left(\frac{1}{4}\right)^2 \left(\frac{3}{4}\right)^{18} = p^2q^{18}$ because each sequence consists of exactly two letters S (each has probability p) and 18 letters F (each has probability q). Hence

$$P(\text{exactly two correct answers}) = \binom{20}{2}\left(\frac{1}{4}\right)^2 \left(\frac{3}{4}\right)^{18} = \binom{20}{2} p^2 q^{18}.$$

Let us investigate a general case. We have an experiment with two outcomes: success S and failure F, $\Omega = \{S,F\}$. Further, we denote the probability of a success by p and the probability of a failure by $q = 1 - p$. The new experiment to be considered consists of *n independent repetitions* of the original experiment. Its sample space Ω^n is, therefore, the set of all possible sequences of the letters S,F of the length n:

$$\Omega^n = \underbrace{\Omega \times \Omega \times \ldots \times \Omega}_{n \text{ times}}$$

Let us consider any sequence that contains the letter S exactly k times and the letter F exactly $(n - k)$ times (we assume $0 \leq k \leq n$). The probability of such a sequence is $p^k q^{n-k}$. In order to specify such a sequence, we have to specify k places (out of n) for the letter S (the remaining places are filled by the letter F). Therefore, there are exactly $\binom{n}{k}$ such sequences. The probability of getting exactly k successes in n independent trials is hence $\binom{n}{k} p^k q^{n-k}$. This proves the following:

THEOREM 1: If X is the number of successes in n independent repetitions of an experiment with two outcomes: a success S and a failure F, and if the probability of a success S is p and the probability of a failure F is $q = 1 - p$ in each repetition, then

$$P(\{X = k\}) = \binom{n}{k} p^k q^{n-k}. \qquad 0 \leq k \leq n.$$

DEFINITION 1: The r.v. X in Theorem 1 is called a binomial random variable, and its distribution is called the *binomial distribution.*

Corresponding to Theorem 1 of Chapter X is the following theorem.

THEOREM 2: Let $0 < p < 1$, $q = 1 - p$ and n be a natural number. Then

$$\sum_{k=0}^{k} \binom{n}{k} p^k q^{n-k} = 1.$$

Proof: Using the binomial theorem, we get

$$\sum_{k=0}^{n} \binom{n}{k} p^k q^{n-k} = (p + q)^n = 1^n = 1. \qquad \text{q.e.d.}$$

EXAMPLE 2

Toss a coin and call a success if a head appears. If you toss the coin six times, what is the probability of getting (a) exactly two heads? (b) at least four heads?

SOLUTION. In this case $n = 6$, $p = \frac{1}{2}$, $q = \frac{1}{2}$.

(a) $P(\text{exactly two heads}) = \binom{6}{2}\left(\frac{1}{2}\right)^2\left(\frac{1}{2}\right)^4 = \frac{15}{64}$.

(b) $P(\text{at least four heads}) = P(\text{exactly four heads}) + P(\text{exactly five heads}) + P(\text{exactly six heads})$

$$= \binom{6}{4}\left(\frac{1}{2}\right)^4\left(\frac{1}{2}\right)^2 + \binom{6}{5}\left(\frac{1}{2}\right)^5\left(\frac{1}{2}\right) + \binom{6}{6}\left(\frac{1}{2}\right)^6\left(\frac{1}{2}\right)^0 = \frac{11}{32}$$

EXAMPLE 3

You decide to throw a grandiose New Year's Day party and, therefore, you invite 500 people to attend. In order to impress your guests, you decide to give a bottle of champagne to everybody who has a birthday on that day. If all of the invited people attend the party, what is the probability $P(k)$ that you will have to give away exactly k bottles of wine (exactly k people will have a birthday on that day)?

SOLUTION. We have to assume that birthdays are independent (that is, that you have not invited any quintuplets), and that they have a constant rate during a year (that is, that any person can be born equally likely on any day of the year). (This assumption is rather irrational because it is a well known fact that some important events—like shortages of electricity and light in England, or public transportation strikes in the United States—are followed by a surge of births nine months later.) Under these assumptions, we have $n = 500$ independent repetitions with a probability of a success (getting a bottle of champagne) $p = \frac{1}{365}$, therefore

$$P(k) = \binom{500}{k}\left(\frac{1}{365}\right)^k\left(\frac{364}{365}\right)^{500-k}.$$

Following are the numerical values of $P(k)$:

k	0	1	2	3	4	5	6	7	8
$P(k)$	.2537	.3484	.2388	.1089	.0372	.0101	.0023	.0004	.0001

COMMENT. Notice that the probability $P(k)$, depending on k, first increases and then decreases. This behavior is explained by the following theorem.

→ **THEOREM 3:** Let us define $P(k)$ for $k = 0,1,2, \ldots, n$ as follows: $P(k) = \binom{n}{k}p^kq^{n-k}$ $(p + q = 1)$. Then the function $P(k)$ is monotonically nondecreasing for $k \leq (n + 1)p$ and monotonically decreasing for the remaining values of k. If $K_0 = (n + 1)p$ is an integer, then

$$P(k_0) = P(k_0 - 1).$$

Proof: $P(k) \geq P(k - 1)$ if and only if

$$\frac{P(k)}{P(k-1)} = \frac{\binom{n}{k}p^kq^{n-k}}{\binom{n}{k-1}p^{k-1}q^{n-(k-1)}} = \frac{(n-k+1)p}{kq} \geq 1, \quad \text{for} \quad P(k-1) \neq 0.$$

This is equivalent to the inequality

$$(n - k + 1)p \geq kq$$

or

$$(n + 1)p \geq k(p + q) = k.$$

Hence

$$P(k - 1) \leq P(k) \text{ if and only if } k \leq (n + 1)p.$$

Obviously, if $k_0 = (n + 1)p$ is an integer, then

$$P(k_0 - 1) = P(k_0).$$

q.e.d.

It is interesting to depict the binomial probabilities $P(k) = \binom{n}{k}p^kq^{n-k}$ in the form of a graph. Then we can easily see for which value k is the probability $P(k)$ largest. For $n = 4$ and $p = \frac{1}{2}$, we get the following graph of the probabilities $P(k)$:

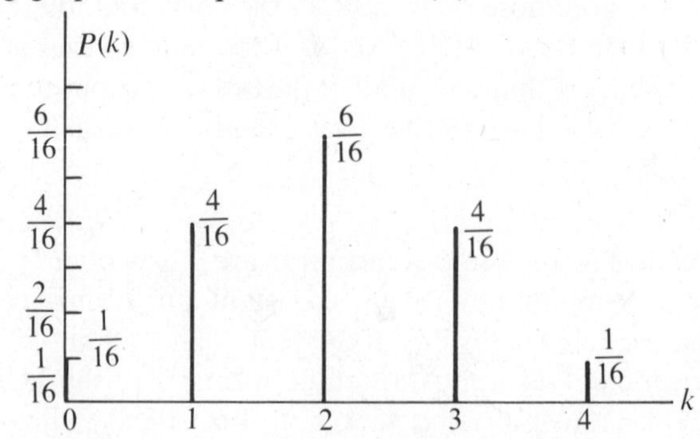

EXAMPLE 4

A coin is biased so that the odds are 4:3 in favor of heads. If you flip such a coin 76 times, what is the most likely number of heads you may get? What is the probability of getting this number of heads?

SOLUTION. We have a binomial experiment with $n = 76$, $p = P(H) = \frac{4}{7}$. The greatest probability of occurrence is obtained (by Theorem 3) for the following number of heads k_0:

$$k_0 = (n + 1)p = 77 \cdot \frac{4}{7} = 44.$$

Therefore, 43 and 44 are the most likely outcomes (by Theorem 3) and

$$P(43) = .09156362. \qquad P(44) = .09156362.$$

We can easily find that

$$P(40) = .06676654 \qquad P(41) = .07816571$$

$$P(42) = .08685079 \qquad P(45) = .08681588$$

$$P(46) = .07800847 \qquad P(47) = .06639019.$$

If we compare the numerical values of the probabilities $P(k)$, $k = 40, 41, \ldots, 47$, then we can readily see that the solution obtained from Theorem 3 is correct.

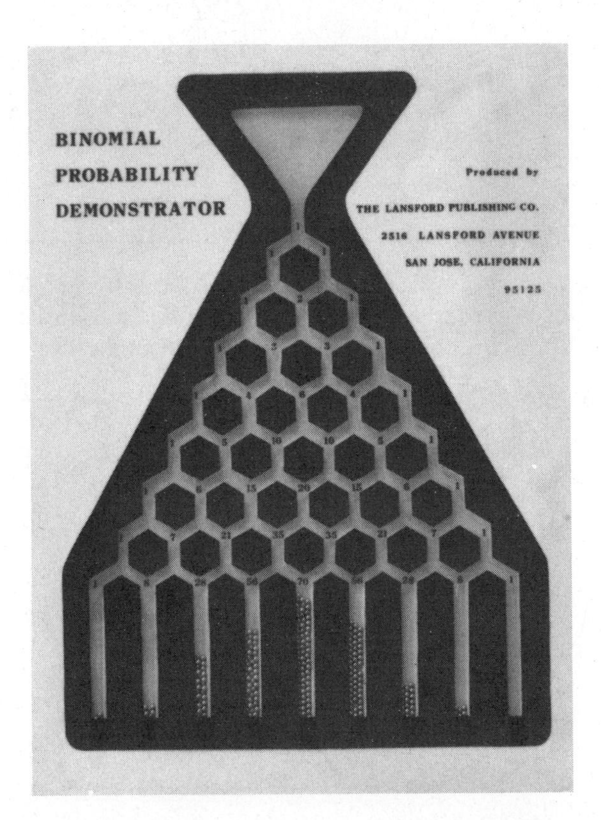

Binomial probability demonstration. 256 balls are dropped and the probability of any ball going left or right at any branching point is $\frac{1}{2}$. The distribution of the balls at the bottom corresponds to the binomial distribution with $n = 8$, $p = \frac{1}{2}$.

EXAMPLE 5

A person playing a game of cards can win \$5 with probability $\frac{1}{3}$, and can lose \$1 with probability $\frac{2}{3}$. If he plays 20 games, what is the probability that he wins at least \$15?

SOLUTION. Let x be the least number of games out of 20 he must win to secure at least \$15. Then

$$5x - (20 - x) \geq 15$$

or

$$x \geq \frac{35}{6}.$$

He has to win at least six games.

$$\begin{aligned} P(\text{winning at least \$15}) &= P(\text{winning six games or more}) \\ &= \sum_{i=6}^{20} \binom{20}{i} \left(\frac{2}{3}\right)^{20-i} \left(\frac{1}{3}\right)^{i} \\ &= 1 - \sum_{i=0}^{5} \binom{20}{i} \left(\frac{2}{3}\right)^{20-i} \left(\frac{1}{3}\right)^{i} = .7028. \end{aligned}$$

It is more convenient to use the second sum, which has only 6 terms, while the first sum has 15 terms.

EXAMPLE 6

Banach Matchbox Problem: This example was solved on page 69, without using the binomial distribution. However, we can understand a mathematician's decision as a sequence of independent trials if the selection of one box is understood as a success and selection of the other one as a failure. The probability of an individual success is $p = \frac{1}{2}$. The mathematician can discover an empty box in the $(2n - r + 1)^{st}$ trial in two ways: either the first $(2n - r)$ trials result in exactly n successes and the $(2n - r + 1)^{st}$ trial is a success (the probability of which is $\binom{2n - r}{n} 2^{-2n+r} \cdot \left(\frac{1}{2}\right)$, or the first $(2n - r)$ trials result in exactly n failures and the $(2n - r + 1)^{st}$ trial is a failure $\left(\text{the probability of which is also } \binom{2n - r}{n} 2^{-2n+r} \cdot \left(\frac{1}{2}\right)\right)$. The desired probability is hence

$$P(r) = \binom{2n - r}{n} 2^{-2n+r}.$$

If $n = 50$, we get the following values:

r	0	1	2	10	20	30 or more
$P(r)$	.080	.080	.079	.048	.007	0

EXAMPLE 7

Samuel Pepys (1633–1703) was a peculiar Englishman who is known today mainly for his fascinating diary. The book was written during the period 1660–1669, and he recorded in it his personal impressions of contemporary London life, observations of life at the Court and the administration of the navy. His successful career culminated in the position of secretary of the admiralty. Pepys was a man of insatiable curiosity, and in 1693 he asked *Isaac Newton* the following question: "Three people play dice. The first rolls a die six times, and considers getting at least one 6 a success; the second rolls a die 12 times and his success is getting at least two 6's; finally, the third rolls a die 18 times and his success is getting at least three 6's. Are their chances for success equal?"

SOLUTION. At first glance it may seem that their chances are equal. However, the calculations show that this is not true. If we denote the required probabilities by p_1, p_2 and p_3 then

$$p_1 = \sum_{i=1}^{6} \binom{6}{i} \left(\frac{1}{6}\right)^i \left(\frac{5}{6}\right)^{6-i} = 1 - \binom{6}{0}\left(\frac{1}{6}\right)^0\left(\frac{5}{6}\right)^6.$$

$$p_2 = \sum_{i=2}^{12} \binom{12}{i} \left(\frac{1}{6}\right)^i \left(\frac{5}{6}\right)^{12-i} = 1 - \binom{12}{0}\left(\frac{1}{6}\right)^0\left(\frac{5}{6}\right)^{12} - \binom{12}{1}\left(\frac{1}{6}\right)^1\left(\frac{5}{6}\right)^{11}.$$

$$p_3 = \sum_{i=3}^{18} \binom{18}{i} \left(\frac{1}{6}\right)^i \left(\frac{5}{6}\right)^{18-i} = 1 - \binom{18}{0}\left(\frac{1}{6}\right)^0\left(\frac{5}{6}\right)^{18} - \binom{18}{1}\left(\frac{1}{6}\right)^1\left(\frac{5}{6}\right)^{17} - \binom{18}{2}\left(\frac{1}{6}\right)^2\left(\frac{5}{6}\right)^{16}.$$

After some calculations, we get

$$p_1 = .6651, \qquad p_2 = .6187, \qquad p_3 = .5973.$$

Therefore, the chances are decreasing with the number of 6's that is required. Newton answered Pepys that "an easy computation" gives $p_1 > p_2 > p_3$. However, Pepys did not believe him, and was not convinced even when Newton later delivered detailed calculations.

Incidentally, this episode shows the extent of scientific progress. A college student nowadays can easily cope with problems which not so long ago required the efforts of a genius.

MULTINOMIAL DISTRIBUTION

In the case of a binomial distribution, we considered independent repetitions of an experiment with only two different outcomes. However, in many practical situations, each independent trial can have more than two outcomes. Let us consider the following example.

EXAMPLE 8

Students in a class are asked to fill out a questionnaire evaluating a course in probability. The question rating the professor has three possible answers: excellent, average, poor. It is known that about 40% of the students always rate the professor as average, while twice as many give him an excellent rather than poor rating. If two questionnaires are picked at random, what is the probability that the professor gets one excellent and one average rating?

SOLUTION. Denote the possible answers as follows:

excellent–F_1 average–F_2 poor–F_3

Then obviously,

$$P(F_1) = p_1 = .4,$$
$$P(F_2) = p_2 = .4,$$
$$P(F_3) = p_3 = .2$$

and $p_1 + p_2 + p_3 = 1$. The following probability tree shows all the possibilities for two questionnaires.

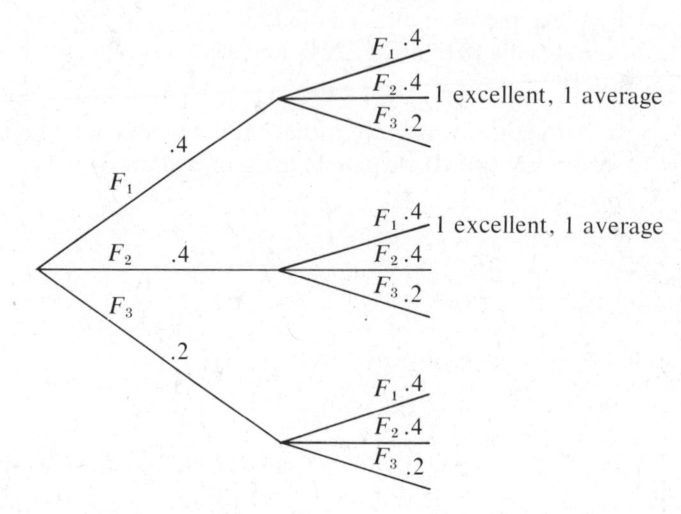

Therefore, the required probability is

$$2 \times (.4) \times (.4) = \binom{2}{1,1,0} p_1^1 p_2^1 p_3^0 = .32.$$

(See multinomial coefficients, page 61.)

EXAMPLE 9

Team A is going to play a series of five games with team B. It is known that team A wins any match with team B with the probability .5, while team B wins with probability .3 (with the remaining games being tied). What is the probability that team A will win two games and lose one? (Consider successive games independent.)

SOLUTION. Each game has three possible outcomes with the following probabilities:

Team A wins—F_1	$P(F_1) = p_1 = .5$
Team A ties—F_2	$P(F_2) = p_2 = .2$
Team A loses—F_3	$P(F_3) = p_3 = .3$

We could solve this problem using a probability tree but it would be quite awkward: the tree would have 243 branches. It is better to consider the following approach. An outcome of a series can be expressed by an ordered quintuple like (F_1,F_1,F_3,F_2,F_2) where i-th place denotes the outcome of the i-th game—in our case A wins the first two games, B wins the third game, and the last two games are tied. We are interested in all quintuples which contain two letters F_1, one letter F_2 and two letters F_3. Because of the independence of the successive games, the probability of each quintuple of this kind is

$$(.5)^2 \times (.2)^2 \times (.3)^1 = p_1{}^2 p_2{}^2 p_3{}^1.$$

According to Theorem 3 from Chapter VI, there are $\binom{5}{2,2,1}$ such quintuples and therefore the required probability is

$$\binom{5}{2,2,1}p_1{}^2p_2{}^2p_3{}^1 = \binom{5}{2,2,1}(.5)^2(.2)^2(.3) = .09.$$

We can now generalize this method to n independent repetitions of an experiment with k distinct outcomes.

Let us investigate a sequence of n independent repetitions of an experiment with k possible outcomes $\Omega = \{F_1,F_2, \ldots ,F_k\}$. Let the probability of the i-th outcome F_i be p_i, where $p_1 + p_2 + \ldots + p_k = 1$. The sample space Ω^n of the new experiment will be the set of all sequences of the length n consisting of the letters $F_1,F_2, \ldots ,F_k$,

$$\Omega^n = \underbrace{\Omega \times \Omega \times \ldots \times \Omega.}_{n\text{-times}}$$

Let us consider a sequence containing the letter F_1 n_1 times, the letter F_2 n_2 times, . . . and the letter F_k n_k times, where $n_1 + n_2 + \ldots + n_k = n$. Due to the independence, the probability of such a sequence is $p_1{}^{n_1}p_2{}^{n_2} \ldots p_k{}^{n_k}$, and there are exactly $\binom{n}{n_1,n_2, \ldots ,n_k}$ such sequences (see Chapter VI, Theorem 3). This proves the following theorem:

THEOREM 4: Let X_i be the number of occurrences of the outcome F_i in n independent repetitions of an experiment with k possible outcomes $\{F_1,F_2, \ldots ,F_k\}$. Let the probability of F_i be $p_i(p_1 + p_2 + \ldots + p_k = 1)$. Then for any k natural numbers $n_1,n_2, \ldots ,n_k$ such that $n_1 + n_2 + \ldots + n_k = n$, we have

$$P(\{X_1 = n_1, \ X_2 = n_2, \ldots X_k = n_k\}) = \binom{n}{n_1,n_2 \ldots n_k}p_1{}^{n_1}p_2{}^{n_2} \ldots p_k{}^{n_k}$$

COMMENT. The joint distribution of the random variables $X_1,X_2, \ldots ,X_k$ given in Theorem 4 is called the *multinomial distribution.*

EXAMPLE 10

A fair die is tossed eight times. Find the probability that 5 appears twice, 6 appears twice and all the other numbers appear once.

$$P = \frac{8!}{1!1!1!1!2!2!}\left(\frac{1}{6}\right)^1\left(\frac{1}{6}\right)^1\left(\frac{1}{6}\right)^1\left(\frac{1}{6}\right)^1\left(\frac{1}{6}\right)^2\left(\frac{1}{6}\right)^2 = \frac{35}{5832}$$

EXAMPLE 11

Of the painted light bulbs produced by a company, 50% are red, 30% blue and 20% green. Find the probability that a sample of five bulbs taken at random will contain two red, one green and two blue bulbs. (Sampling is done in such a way that a bulb is selected, checked, replaced and then another bulb is selected, etc.)

$$P = \frac{5!}{2!2!1!}(.5)^2(.3)^2(.2) = .135.$$

EXAMPLE 12

Assume that the USSR national hockey team can beat the Canadian national hockey team with the probability .5, while Canada will defeat Russia with the probability .3 and there is some chance of ties. What is the probability that Russia will win a play-off series consisting of five games? (Consider the successive games to be independent.)

SOLUTION. Denote the event as follows

$$\text{USSR wins} - W \qquad \text{a tie} - T \qquad \text{USSR loses} - L$$

Then obviously, $P(W) = .5$, $P(T) = .2$ and $P(L) = .3$, and we have $n = 5$ independent repetitions. Therefore,

$$\begin{aligned} P(\text{USSR wins the series}) = {} & P(5W{,}0T{,}0L) + P(4W{,}1T{,}0L) + P(4W{,}0T{,}1L) \\ & + P(3W{,}2T{,}0L) + P(3W{,}1T{,}1L) \\ & + P(3W{,}0T{,}2L) + P(2W{,}3T{,}0L) \\ & + P(2W{,}2T{,}1L) + P(1W{,}4T{,}0L). \end{aligned}$$

We can evaluate each of these probabilities using the multinomial distribution formula. For example,

$$P(2W{,}2T{,}1L) = \binom{5}{2{,}2{,}1}(.5)^2(.2)^2(.3)^1 = .09. \qquad \text{(See Example 8.)}$$

$$P(2W{,}3T{,}0L) = \binom{5}{2{,}3{,}0}(.5)^2(.2)^3(.3)^0 = .02.$$

$$P(1W{,}4T{,}0L) = \binom{5}{1{,}4{,}0}(.5)^1(.2)^4(.3)^0 = .004.$$

Similarly, we can calculate all the remaining probabilities and after adding them, we get

$$P(\text{USSR wins the series}) = .614.$$

However, this method is rather clumsy. We can get a solution much faster using *both* binomial and multinomial distributions. Let us denote the event that the USSR does not win a game by N. Then obviously,

$$P(W) = .5 \qquad P(N) = .5$$

$$\begin{aligned} P(\text{USSR wins the series}) = {} & P(5W{,}0N) + P(4W{,}1N) + P(3W{,}2N) + P(2W{,}3T{,}0L) \\ & + P(2W{,}2T{,}1L) + P(1W{,}4T{,}0L). \end{aligned}$$

We can calculate the first three probabilities using the binomial distribution.

$$P(5W{,}0N) = \binom{5}{5}(.5)^5(.5)^0 = .03125.$$

$$P(4W{,}1N) = \binom{5}{4}(.5)^4(.5)^1 = .15625.$$

$$P(3W{,}2N) = \binom{5}{3}(.5)^3(.5)^2 = .3125.$$

The remaining three probabilities we calculated above, using the multinomial distribution. Adding all the probabilities together gives

$$P(\text{USSR wins the series}) = .614.$$

EXAMPLE 13

Sampling with Replacement. Let us have a box containing c_1 balls of color 1, c_2 balls of color 2, . . . , c_k balls of color k. A ball is drawn at random, its color recorded, and the ball is then returned. Let us assume that this experiment is repeated independently n times. What is the probability that our record will show n_1 balls of the color 1, n_2 balls of the color 2, . . . , n_k balls of the color k?

SOLUTION. In this case, the probability of drawing a ball of the color i is

$$p_i = \frac{c_i}{\sum_{j=1}^{k} c_j}$$

and therefore the required probability is

$$\binom{n}{n_1, n_2, \ldots, n_k} \left(\frac{c_1}{\sum_{j=1}^{k} c_j}\right)^{n_1} \left(\frac{c_2}{\sum_{j=1}^{k} c_j}\right)^{n_2} \cdots \left(\frac{c_k}{\sum_{j=1}^{k} c_j}\right)^{n_k}.$$

THEOREM 5: The sum of all the probabilities of the multinomial distribution is equal to 1:

$$\sum \binom{n}{n_1, n_2, \ldots n_k} p_1^{n_1} p_2^{n_2} \ldots p_k^{n_k} = 1$$

where the sum is taken over all possible $n_1, n_2, \ldots, n_k$ such that $n_1 + n_2 + \ldots + n_k = 1$.

Proof: Using the multinomial theorem (page 61), we get

$$\sum_{n_1+n_2+\ldots+n_k} = \binom{n}{n_1, n_2, \ldots n_k} p_1^{n_1} p_2^{n_2} \ldots p_k^{p_k} = (p_1 + p_2 + \ldots + p_k)^n = 1^n = 1. \qquad \text{q.e.d.}$$

Interesting examples of multinomial distributions can be found in Chapter XV dealing with population genetics and cases of disputed paternity.

HYPERGEOMETRIC DISTRIBUTION

In previous chapters, we encountered many examples which had a common nature: getting aces from a deck of cards (Examples 1 and 2, Chapter VII), distribution of trumps in bridge (Example 5, Chapter VII), getting a 6 on a domino (Example 6, Chapter VII), renting new ski boots (Example 11, Chapter X), defective items in a sample (Example 10, Chapter X) and many others. All these problems represent so-called *sampling without replacement* and can be described by one model as follows.

Imagine an urn containing k black balls and $N - k$ white balls ($k < N$). Let us draw $n \leq N$ balls at random. We can imagine that we successively draw one ball after another *without replacement.* The probability of getting a black ball on the second draw obviously depends on the outcome of the first draw; the probability of a white ball on the third draw obviously depends on the outcomes of the first and second draws. Therefore, we have a sequence of *dependent trials.* Any final outcome will contain n balls, of which not more than k are black and not more than $(N - k)$ are white, and can be represented by a sequence of letters B (for black) and W (for white) of the length n which contains not more than k B's and not more than $(N - k)$ W's. Let us denote the set of such sequences Ω. The set Ω can obviously be used as a sample space for drawing n balls from the described urn. Any two sequences of n balls that have a different *order* of black and white balls are different, and therefore the space Ω has $N \cdot (N - 1) \ldots (N - n + 1) = \dfrac{N!}{(N - n)!}$ equally likely points.

For any outcome $\omega \in \Omega$ (that is, for any such sequence of letters B and W), let $X(\omega)$ mean the number of letters B. Obviously $X(\omega)$ is a random variable, and can attain the values r such that $0 \leq r \leq k, r \leq n, r \geq n + k - N$. The last condition means that among n balls we cannot have more than $(N - k)$ white balls. Let us find the distribution of $X(\omega)$, that is, the probabilities $f(r) = P(\{\omega: X(\omega) = r\})$. We can draw n balls out of N balls in $\binom{N}{n}$ different ways (regardless of order and equally likely). Out of these n balls we can get r black balls in $\binom{k}{r}$ different ways, and the remaining $(n - r)$ white balls in $\binom{N-k}{n-r}$ different ways. (All these selections are regardless of order.)

→ **THEOREM 6:** The distribution of the number of black balls is given as follows:

$$f(r) = P(\{\omega: X(\omega) = r\}) = \frac{\binom{k}{r}\binom{N-k}{n-r}}{\binom{N}{n}} \quad \text{where} \quad 0 \leq r \leq k, \quad r \leq n, \quad r \geq n + k - N$$

This distribution is called the *hypergeometric distribution.*

EXAMPLE 14

Each of 50 states has 2 senators. A committee of 50 senators is chosen at random. What is the probability that a given state is represented?

SOLUTION. Consider the two senators of the given state as "blue" and all other senators as "green." Then $N = 100$, $k = 2$, $n = 50$.

$$P(0) = P(\text{no "blue senator" selected}) = \frac{\binom{2}{0}\binom{98}{50}}{\binom{100}{50}} = \frac{50.49}{100.99} = .24747.$$

$$P(1) = P(\text{one "blue senator" selected}) = \frac{\binom{2}{1}\binom{98}{49}}{\binom{100}{50}} = \frac{2.50.50}{100.99} = .50505.$$

$$P(2) = P(\text{two "blue senators" selected}) = \frac{\binom{2}{2}\binom{98}{48}}{\binom{100}{50}} = \frac{50.49}{99.100} = .24747.$$

TOTAL .99999

The total should be 1.00000. However, we calculated the probabilities only to five decimal places, and thus committed a rounding error. The probability that a given state is represented is $P(1) + P(2) = 1 - P(0) = .75252$.

EXAMPLE 15

Estimation of the Size of Animal Population. Assume that 100 pheasants were caught alive in a certain area, marked by legbands and released. After several weeks, another sample of 150 pheasants was caught. It was found that of these birds, 29 were marked. Let us assume that the birds do not learn from experience, and therefore one is as likely to catch a particular bird the first time as to catch it after banding. We can use the obtained results for an estimation of the number of pheasants in the area. Let us denote N, n, k and r as follows:

N—the total number of pheasants in the area

k—the number of pheasants in the first catch

n—the number of pheasants in the second catch

r—the number of marked birds in the second catch

The probability $P(n,r)$ of getting r marked birds in the second catch, if we have released k banded birds, is

$$P(N,r) = \frac{\binom{k}{r}\binom{N-k}{n-r}}{\binom{N}{n}}$$

The values $k(= 100)$, $n(= 150)$ and $r(= 29)$ are known; the only unknown value is the total number of pheasants N. It does not seem unreasonable to take as an estimate of N the value which makes $P(N,r)$ maximal. This method is called the *maximum likelihood* method of estimation. In order to find this N, consider the ratio

$$\frac{P(N,r)}{P(N-1,r)} = \frac{\binom{k}{r}\binom{N-k}{n-r}}{\binom{N}{n}} \frac{\binom{N-1}{n}}{\binom{k}{r}\binom{N-1-k}{n-r}} = \frac{(N-n)(N-k)}{N(N-k-n+r)}.$$

$$\frac{P(N,r)}{P(N-1,r)} \geq \text{ if and only if } (N-n)(N-k) \geq N(N-k-n+r)$$

or
$$N^2 - Nk - Nn + nk \geq N^2 - Nk - Nn + Nr$$

or equivalently
$$N \leq \frac{nk}{r}.$$

Hence $P(N,r)$ is increasing for $N \leq \frac{nk}{r}$ and decreasing for $N > \frac{nk}{r}$, reaching its maximum for N equal to the largest integer, which is $\leq \frac{nk}{r}$. In our case $\frac{nk}{r} = \frac{100 \times 150}{29} = 517.24$, hence $N = 517$ is the population of pheasants most likely yielding catches as indicated. Our result is in agreement with common sense because one would expect that $k : N$ should be close to $r : n$ if the samples (catches) are representative.

EXAMPLE 16

Lady and Tea (from R. A. Fisher). A lady claims that she can tell whether milk or tea has been put into a cup first just by tasting it. To test her claim the following experiment is done: eight cups of tea with milk, four of each kind, are presented to the lady randomly. It is considered a success if she can tell them apart correctly. If this trial is repeated 10 times and the lady is successful at least 2 times, does it say anything about a significant finesse of her taste?

SOLUTION. Let us find first the probability p that she will be successful in telling eight cups apart correctly if she guesses at random. There are four cups of each kind, therefore

$$p = \frac{\binom{4}{4}\binom{4}{4}}{\binom{8}{4}} = \frac{1}{70} = .01429.$$

Now using the binomial distribution with this p, we get

$$P(\text{at least 2 successes in 10 trials}) = 1 - \binom{10}{0}p^0(1-p)^{10} - \binom{10}{1}p^1(1-p)^9 = .0085.$$

Therefore, if the lady guesses at random, she has a negligible chance of at least 2 successes in 10 trials. However, if she can do it she is unlikely to be guessing at random, and probably has definite tasting prowess.

→ **THEOREM 7:** The sum of all the probabilities of the hypergeometric distribution is equal to 1.

$$\sum_r \frac{\binom{k}{r}\binom{N-k}{n-r}}{\binom{N}{n}} = 1.$$

Proof 1: Using Theorem 12 of Chapter VI, we can see that

$$\sum_r \binom{k}{r}\binom{N-k}{n-r} = \binom{N}{n}. \qquad \text{q.e.d.}$$

Proof 2: Let us denote the event of getting exactly r black balls if we choose n balls out of the urn by A_r. Then obviously A_r is a partition of Ω,

$$\Omega = \bigcup_r A_r$$

hence

$$1 = P(\Omega) = P(\bigcup_r A_r) = \sum_r P(A_r) = \sum_r \frac{\binom{k}{r}\binom{N-k}{n-r}}{\binom{N}{n}}. \qquad \text{q.e.d.}$$

POLYA DISTRIBUTION

In Example 14, Chapter IX, we investigated Polya's urn scheme as a model for the spreading of contagious diseases. This model is actually a generalization of the hypergeometric distribution. An urn contains m black and n white balls. Balls are drawn randomly and replaced after each drawing. At the same time, a number k of balls of the same color as the ball drawn are added to the urn. Let the r.v. X denote the number of black balls in N drawings. Obviously, for $k = -1$, we are getting the hypergeometric r.v. In Chapter IX, we derived the following:

THEOREM 8: The distribution of the number of black balls is given by

$$f(i) = P(\{X = i\}) = \frac{\binom{\frac{m}{k} + i - 1}{i}\binom{\frac{n}{k} + N - i - 1}{N - i}}{\binom{\frac{m+n}{k} + N - 1}{N}}.$$

This distribution is called the *Polya distribution.*

EXAMPLE 17

The urn contains $m = 3$ black and $n = 2$ white balls, and on each draw you add $k = 1$ more balls of the color drawn. Find the probability distribution of the number of black balls in $N = 4$ draws.

SOLUTION. In this case,

$$f(i) = \frac{\binom{2+i}{i}\binom{5-i}{4-i}}{\binom{8}{4}}, \qquad i = 0,1,2,3,4.$$

i	0	1	2	3	4
$f(i)$	.0714	.1714	.2572	.2857	.2143

EXAMPLE 18

Among other attractions, the Annual Outdoors Show offers a little pool with fish for kids. Eight trout and six whitefish are released on the first day, and any child is allowed to try her or his angling luck. However, each father has to pledge that in the event of a success he will buy three fish of the same kind at the hatchery stand and release them in the pool while his offspring may keep the catch. Assuming that the fish are biting at random, what is the probability distribution of the number of trout among the first five catches?

SOLUTION. In this case, we have $m = 8$, $n = 6$, $k = 2$ (one fish being a replacement of the catch), $N = 5$.

$$f(i) = \frac{\binom{3+i}{i}\binom{7-i}{5-i}}{\binom{11}{5}} \qquad i = 0,1,2,3,4,5.$$

The table of the probability distribution is as follows:

i	0	1	2	3	4	5
$f(i)$	.0454	.1299	.2165	.2597	.2273	.1212

DISTRIBUTION OF RUNS

One of the most important elementary distributions from the point of view of practical applications is certainly the *distribution of runs.* Imagine a sequence of symbols of two different types, say 0's and 1's. A *run* is defined as a succession of identical symbols, which is limited on both ends by different symbols or by a blank space. For example, the sequence 001010001110 contains seven runs, namely 00; 1; 0; 1; 000; 111; 0. Applications of the theory of runs are wide ranging. Let us mention just a few.

EXAMPLE 19

Communication Theory. Imagine that an astronomer observes a sequence of signals of two different types (physically represented, perhaps, by a high and a low intensity of the magnetic field of a pulsating star). Is it possible that these signals have been sent at random, or do they contain some information about a recent state of the star or perhaps about other intelligences? If the sequence is of a random origin, then the number of runs should be neither too high nor too low, and should not show any significant changes during a period of time.

EXAMPLE 20

Neurocybernetics. Nerves in the human body can be considered as sequences of cells, each of which can be in one of two different states: either excited (1) or not excited (0). Any stimulus transferred through the nerve is considered to excite or dampen some of the cells according to physiological laws. At any time, a nerve can be considered as a sequence of 0's and 1's and the number of runs is of importance in an evaluation of the functioning of the nerve.

EXAMPLE 21

Quality Control. A machine produces a certain number of good items (labeled by 1's) and a certain number of defectives (labeled by 0's) each day. It is impossible to avoid defectives completely, but if the machine is set up correctly then they will appear randomly. However, as soon as they start appearing periodically, causing runs of defectives, it might be a signal that the machine is out of proper alignment.

In all of the above examples it is important to know the probability of a certain number of runs, if we assume that both symbols are distributed randomly in the sequence.

→ **THEOREM 9:** Let m 0's and n 1's be generated randomly so that all distinguishable sequences are equally likely. Then the probability of exactly $2k$ runs is equal to

$$P_{2k} = \frac{2\binom{m-1}{k-1}\binom{n-1}{k-1}}{\binom{m+n}{m}} \quad \text{for } 1 \le k \le \min(m,n)$$

and the probability of exactly $(2k + 1)$ runs is equal to

$$P_{2k+1} = \frac{\binom{m-1}{k}\binom{n-1}{k-1} + \binom{m-1}{k-1}\binom{n-1}{k}}{\binom{m+n}{m}} \quad \text{for } 1 \leq k \leq \min(m-1, n-1).$$

Proof: Having m 0's and n 1's, we can write $\binom{m+n}{m}$ distinguishable sequences of these symbols. Let us investigate the sequences that will have an even number of runs, say $2k$. We can imagine $2k$ cells in a row, some of them containing only 0's, and others containing only 1's such that cells containing 0's and 1's alternate and no cell remains empty. By Theorem 9 of Chapter VI, we can place m 0's into the odd-numbered boxes in $\binom{m-1}{k-1}$ distinct ways so that none of the k boxes remains empty. Further, we can place n 1's in k even-numbered boxes in $\binom{n-1}{k-1}$ so that none remains empty. The total number of ways yielding $2k$ runs starting with 0's is therefore $\binom{m-1}{k-1}\binom{n-1}{k-1}$. However, there are an equal number of arrangements giving $2k$ runs starting with 1's. Therefore,

$$P_{2k} = \frac{2\binom{m-1}{k-1}\binom{n-1}{k-1}}{\binom{m+n}{m}}.$$

Consider now an odd number of runs, say $(2k + 1)$. In this case, there are k even-numbered and $(k + 1)$ odd-numbered boxes. We can put 0's in the even boxes and 1's in the odd boxes in $\binom{m-1}{k-1}\binom{n-1}{k}$ distinct ways. Further, we can put the 1's in the odd boxes and the 0's in the even boxes in $\binom{m-1}{k}\binom{n-1}{k-1}$ distinct ways.

Therefore

$$P_{2k+1} = \frac{\binom{m-1}{k}\binom{n-1}{k-1} + \binom{m-1}{k-1}\binom{n-1}{k}}{\binom{m+n}{m}}.$$

q.e.d.

EXAMPLE 22

Let four 0's and four 1's be randomly generated, and let the r.v. X denote the number of runs. Find numerically the probability distribution of X.

SOLUTION. X can attain any natural number between 2 and 8 with the following probabilities:

X	2	3	4	5	6	7	8
$f(x)$	.0286	.0857	.2571	.2572	.2571	.0858	.0286

EXAMPLE 23

Tests of Randomness. An operator has received a strange message on his short-wave receiver. It consists of 17 dots and 20 dashes, and resembles a Morse code sequence. However, he can not decode it and he feels that it might be just one of those "random noises," without information content. He would like to consider it as a random sequence of dots and dashes, but he is worried lest he overlook some important message, since he feels that the number of runs, which is equal to 27, is rather high and it might be some sort of communication. Assuming that the signals really are a random noise, what is the probability of getting a sequence with more than 26 runs?

SOLUTION. We have to add all the probabilities $\sum_{i=27}^{34} P_i$ for $m = 17$, $n = 20$ because there could be up to 34 runs in a sequence of 17 dots and 20 dashes. It takes a while to carry out the calculations, but it is quite easy to find that

$$\sum_{i=27}^{34} P_i = .0076.$$

Therefore, the event that a random noise would yield more than 26 runs is practically impossible; its probability is less than 1%. The message could be considered as some sort of a communication.

EXAMPLE 24

Randomness in Quality Control. A machine is set to pump 22.5 cubic centimeters of liquid into each container on an assembly line. Some random fluctuations are unavoidable, but the actual amounts should not deviate too much from the specified volume. Following is a sequence of 17 actual successive measurements: 22.4, 22.3, 22.1, 21.9, 22.4, 22.6, 22.7, 22.8, 22.9, 23.0, 22.6, 22.4, 22.1, 22.6, 22.7, 22.8, 22.9. Is there any sign which might show that the process is not under control?

SOLUTION. Let us denote any deviations above 22.5 cubic centimeters by + and deviations below 22.5 by −. Then the observations can be expressed as

$$- - - - - + + + + + + - - + + + +.$$

There are $m = 7$ signs −, $n = 10$ signs + and four runs altogether. The probability of getting fewer than five runs if the fluctuations in production are really random is equal to

$$\sum_{i=2}^{4} P_i = .00643.$$

It is practically impossible to obtain so few runs if the fluctuations were at random, therefore we have to conclude that there are probably some periodical changes in the machine, and that the process is out of control.

The decision to denote the measurements above 22.5 by + and below 22.5 by − is quite arbitrary. As a matter of fact, we can make a different assignment of symbols as follows: If a measurement is bigger than the previous one, denote this fact by +, otherwise use −. In this case, we have to skip the first observation, and we get the following sequence:

$$- - - + + + + + + - - - + + + +$$

There are $m = 6$ signs −, $n = 10$ signs + and four runs. The probability of getting fewer than five runs if the fluctuations are really random is equal to

$$\sum_{i=2}^{4} P_i = .0132$$

and we have to make the same conclusion as before.

EXAMPLE 25

A group of 14 teenagers, of whom 6 are boys and the rest are girls, arrived at a movie just after the beginning. They crowded noisily at random into one row of seats. Let X be the number of situations in which a boy sits next to a girl. Find the probability distribution of X. (Assume that no seats among them are empty.)

SOLUTION. Consider the number of runs Y of the teenagers (imagine boys as 0's and girls as 1's). Then obviously $X = Y - 1$. For example, if the order is as follows

bggbbgggbgbbgg

then there are $Y = 8$ runs and $X = 7$ cases in which a boy sits next to a girl. (If a boy sits between two girls, we have to count such a situation twice.) Therefore, there can be $i = 1,2,3, \ldots, 12$ such cases and

$$f(i) = P(\{X = i\}) = P(\{Y = i + 1\}).$$

We can calculate these probabilities from the distribution of runs. Following is the distribution of X:

i	1	2	3	4	5	6
$f(i)$	.0007	.0040	.0233	.0582	.1399	.1865

i	7	8	9	10	11	12
$f(i)$	.2331	.1748	.1166	.0466	.0140	.0023

EXERCISES

(Answers in Appendix to exercises marked)*

*1. A student has a probability .7 of solving any problem assigned to him. If an examination consists of eight problems and he must solve a minimum of five problems to pass, what is his chance of passing?

2. The probability that a student will not get married during his undergraduate years at a university is $\frac{2}{3}$. In the graduating class of 15 students (of whom nobody was married before entering the university), what is the
(a) most likely number of married students?
(b) probability that a majority will be married?
(c) probability that at most three students will be married?

*3. A person playing a game of cards can win $4 with probability $\frac{1}{4}$ or lose $1 with probability $\frac{3}{4}$. If he plays 10 such games, what is the probability that he ends up winning between $5 and $10, inclusively?

4. The probability of hitting a grizzly bear on each shot as he runs towards you is $\frac{1}{5}$. You have time for four shots only, and you must hit him at least twice to stop him (otherwise he will devour you). What is the probability that you end up as grizzly droppings? A distant observer noticed that at least one of your shots did hit the bear before the bruin disappeared from sight in the bush. What is the probability of stopping a grizzly, given this fact?

*5. The majority of scuba divers have their own regulators. Only about 30% of those who rent an air tank also ask for a regulator. A manager of an equipment shop therefore stocks 3 regulators and 10 tanks for rent. Find the probability that on a certain day when he has all 10 tanks booked, he will not have enough rental regs (nobody rents a reg without a tank), and will have to rent out a new one.

6. You get your morning tea too hot to be able to

drink 40% of the time. However, being in a hurry every day you always attempt to drink it at once. What is the probability that during one week you get burned
(a) exactly twice?
(b) not more than three times?

7. It is estimated that about 40% of people suffer from a common cold during a winter. An advertisement claims that a certain patented medicine is very powerful in preventing this kind of trouble. Among six people taking the medicine regularly, only two caught cold. Does this fact support the manufacturer's claim? Calculate the probability that among six people who do not take the medicine
(a) exactly two will suffer the cold.
(b) two or less will catch the cold.

*8. It was observed that about 60% of rats exposed to stress in an overcrowded cage will develop definite signs of aggressiveness within 48 hours. In order to test a new type of sedative, 15 rats were given this drug in their food, and left in one small cage for two days. After this period of time, four rats showed definite signs of aggression. Can anything be concluded about the effectiveness of the drug? Calculate the probability that, if no drug is used and they are otherwise given the same treatment,
(a) exactly four rats will be aggressive.
(b) at most four rats will be aggressive.

9. A motel manager has 10 rooms for rent. He has four TV sets which he installs for an extra charge. Find the probability that on a night when all the rooms are rented he will not be able to satisfy the demand for a TV. Assume that there is a fifty-fifty chance that a party renting a room will ask for a TV set.

10. Toss three fair coins seven times. What is the probability that
(a) three heads will appear in at least one toss?
(b) at least one head on the three coins will appear at most twice?

*11. A coin is biased so that the odds in favor of heads are 6:4. Two people flip such a coin six times. Find the probability that
(a) they score the same number of heads.
(b) they score together a total of nine heads.

12. The probability that a doctor can diagnose a certain fatal disease is .6. If the disease is recognized, the patient has probability .9 of being cured. If the doctor's diagnosis is wrong, the patient still has a 40% chance of being cured by the natural development of his body's protective systems. Ten patients having the disease (but not knowing it) consulted a doctor recently. Find the probability that
(a) at most three will die because of the disease.
(b) exactly two patients will be diagnosed incorrectly but cured.
(c) exactly seven patients will be diagnosed correctly and cured.

*13. In 10 independent tosses of a perfect die, what is the probability of getting
(a) no 1 and each other number twice?
(b) each number except 6 once, and the remainder 6's?
(c) more 6's than other numbers together?

14. A man has to decide each morning whether he will use his car, take a bus or take the subway to get to his office. His wife observed that the probabilities of decisions for these means of transportation are .5, .3 and .2, respectively. Assume that the man's choice is independent of what he took the previous days. Find the probability that in one week (five working days), he will
(a) take the car exactly three times, a bus and the subway each once.
(b) take the car more often than the other means together.
(c) will not take the subway.
(d) will take the car at most twice.
(e) will take the subway exactly once, but use the car more often than a bus.
(f) use the car and a bus the same number of times.

15. When a big-petalled yellow chrysanthemum is crossbred with a small-petalled white chrysanthemum, the resulting plant may be big-petalled yellow, small-petalled yellow, big-petalled white or small-petalled white. According to Mendel's laws of genetics, the respective proportions in which these four kinds will occur are 9:3:3:1. Find the probability that of 12 plants produced by such a crossbreeding, the numbers of plants in these four categories are 6, 3, 2 and 1, respectively. (Assume that the plants are crossbred independently.)

*16. A father has three children: Jan, Johanna and Bjorne. Each Saturday he gives a piece of chocolate to one of the children according to the following rules: he calls the children together, then he thinks (at random) of one of the numbers 1,2, or 3 and Jan has to guess which one it is. If Jan is right, he gets the candy, otherwise it is Johanna's turn. If she guesses correctly she gets it, otherwise small Bjornie will enjoy the sweet delight. However, "little big man" Bjorne protests vehemently because he is always last, has no choice and feels cheated. Is the father really unfair to him? What is the probability that over the summer holidays (eight weeks) Jan and Johanna get three bars each, and Bjorne gets only two? (Hint: calculate first the probability of each child getting the chocolate on Saturday given that they hear previous guesses.)

17. Assume that team A can beat team B with

probability .4 while team B will defeat team A with probability .3, and there is also some chance of a tie. What is the probability that team A will win a play-off series consisting of four games?

18. A receiver contains six transistors, of which two are defective. Three transistors are selected at random and inspected. Let X be the number of defectives observed. Find the probability distribution of X.

*19. There are 3 cartons of 12 apples each. Two of these cartons contain four rotten apples and the remaining one contains three rotten apples. You buy a carton at random and choose three apples from it. Let X be the number of rotten apples chosen. Find the probability distribution of X.

20. A bag contains 15 nuts, of which 8 are walnuts and the rest are Brazil nuts. The mother picks six nuts at random and gives them to Jan and Johanna to split equally. However, both children prefer walnuts because they are easier to crack. What is the probability that dividing the nuts will not result in a fight (that is, that the number of walnuts will be even)?

*21. A manufacturer of a magnetic recording tape introduces a special deal: he offers packages containing 15 reels of a high output, low noise tape and 5 reels of ordinary tape at a discount price. A buyer for an electronic retail sales chain decides to test before purchasing a large amount of tape. He selects 10 packages; from each he chooses 4 reels at random and has them checked in a laboratory. He feels that there should be $\frac{1}{4}$ of ordinary quality tapes only, and therefore he considers a package as unsatisfactory if the lab finds more than one ordinary tape among four selected. Furthermore, he decides to refuse the whole shipment if among 10 packages more than 3 are unsatisfactory. Is the buyer reasonable in his requirements? Find the probability that he will refuse the shipment even if the manufacturer's claim is true.

22. Jan has four red and five yellow marbles in his bag while Johanna has five red and six yellow marbles. Each of them chooses (at random) three marbles to play with. All marbles are tossed on the ground. Find the probability that there are among them
(a) exactly four red marbles.
(b) at least five yellow marbles.

23. It is rather surprising, but one can often find mixed flocks of spruce grouse and ruffed grouse wintering together in the foothills of the Canadian Rocky Mountains. These birds have the peculiar habit of walking in each other's tracks in a long line when a deep snow comes. Assume that the birds mix at random. Find the probability that in a flock of 15 birds, of which 8 are spruce grouse, X birds walk behind a bird of the other kind. (That is, find the probability distribution of X.)

24. A professional poker player claims that he is going to deal cards from a well shuffled deck of cards. The cards are dealt in the following order (J = jack, Q = queen, K = king, A = ace, H = heart, S = spade, C = club, D = diamonds):

2D, 4H, 7D, QD, KC, 10S, 5S,
4C, 7H, 10D, JH, AD, KS, JC

Can you believe that he is not cheating? (Hint: If the deck is well shuffled, then cards have to appear at random.) Denote cards as black = b and red = r, find the number of runs in r and b and the probability that there will be at most this number of runs if the deck has been well shuffled. A very low probability would indicate that the deck might have been prepared to give a good chance of a flush. Further denote + if the next card has a higher face value than the preceding one, and − if it is vice versa. Find the number of runs in + and −, and the probbility that there are at most as many runs as counted. A very low probability of this type might indicate that the deck has been prepared to favor a straight.

*25. Consider the behavior of a peculiar collector of LP records. He likes only classical and rock music; he always selects a record to play at random from a big heap representing his entire collection. Whenever he listens to a record he is so infatuated by it that the very same day he buys two more records of the same genre, and dumps them on the heap without even playing them. Imagine that he begins his collection with six rock and two classical records, and never plays more than one record a day. Investigate his claims of musical versatility after he has played four records. Find the probability that he has listened to
(a) more classical than rock records.
(b) equal numbers of classical and rock records.
(c) more rock than classical records.

EXPECTATIONS

Dew successe shoulde not chaunce according unto
theyr hope & expectation.

Richard Eden, 1553

In the previous chapters, we investigated a random variable X defined on a probability measure space (Ω,P), and its probability distribution $f(x_i)$ $i = 1,2, \ldots ,k$. The set of all possible values $x_1,x_2, \ldots ,x_k$ can sometimes be very extensive and the probability distribution therefore rather confusing and perplexing even if it is given in the form of a table, as in the examples of Chapter XI. For this reason, we are often interested in one or two characteristics that will *represent* the whole distribution in some way or another.

EXPECTED VALUE

Imagine that we represent the distribution $f(x_i)$ on the real axis in such a way that we place a metal weight of the size $f(x_i)$ at the point x_i, $i = 1,2 \ldots ,k$. Further, we assume that the axis itself has no weight but is rigid enough to be used as the bar of this sophisticated dumbbell. Where should a one-handed weightlifter grab the axis so that the whole dumbbell is perfectly balanced?

"Mr. Probability Universe"

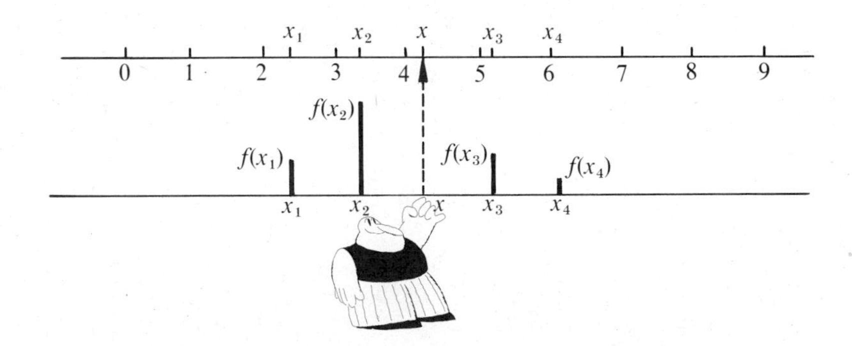

Readers familiar with the physical laws of mechanics can remember that the distance x of this ideal point (called the center of gravity) from the origin can be calculated as

$$x = \sum_{i=1}^{k} x_i f(x_i).$$

Every reasonable weightlifter is hence "expected" to grab the dumbbell at the point x. This point in a sense represents the whole distribution of weights-probabilities, and is some sort of a center or "measure of location" of the distribution on the axis. We shall see that this point is of high importance also in probability theory.

For another motivation of the abstract notion of expected value of a random variable, we shall return to the abandoned frequency definition of probability. Our reasoning will be, therefore, perhaps slightly nebulous. Nevertheless, we shall get clear ideas about the *meaning of the expectation in practical applications.* Let us make a long sequence of n independent repetitions of an experiment, and always observe the value which the random variable X attains. Let us assume that outcome x_1 occurs m_1 times, x_2 occurs m_2 times, . . . , etc., and x_k occurs m_k times (out of n observations). Then the common sense intuitive statistical *average* or the *mean* of these observations is

$$M_n = \text{mean of } n \text{ observations} = \frac{1}{n}\sum_{i=1}^{k} m_i x_i = \sum_{i=1}^{k} \frac{m_i}{n} x_i.$$

However, from the frequency definition of probability, it follows that (almost always) the frequencies $\frac{m_i}{n}$ tend with increasing n to the corresponding probabilities $f(x_i) = P(\{X = x_i\})$. It may, therefore, be reasonably expected that the mean of n observations M_n will (almost always) tend to the value $\sum_{i=1}^{k} x_i f(x_i)$, with n increasing above all limits:

$$M_n \longrightarrow \sum_{i=1}^{k} x_i f(x_i).$$

(We shall discuss this fact in more detail in Chapter XIII dealing with the Law of Large Numbers.) In both examples we got the same expression $\sum_{i=1}^{k} x_i f(x_i)$, which appears to have some practical meaning. This suggests the following definition.

DEFINITION 1: Let X be a r.v. with the probability distribution $f(x_i)$, $i = 1,2, \ldots, k$. Then the expected value of X, denoted by $E(X)$, is defined as

$$E(X) = \sum_{i=1}^{k} x_i\, f(x_i).$$

Girl croupiers at the craps table bring the Las Vegas style to London. (© The Press Association Ltd., London. Used with permission.) For the analysis of the actual casino craps, see page 321.

EXAMPLE 1

Simplified Craps. Consider the following simplified version of the game of craps. A bet of \$1 is taken as a charge of admission to play the game. Then two fair dice are rolled, and the total score is recorded. If it is 12, you win \$5 (and you can keep your \$1 bet). If 2 appears you win \$3, if 10 or 11 appears you win \$2, if 3 or 4 appears you win \$1 and you always keep your \$1 bet. If any other sum appears you lose the bet. Should you play the game?

SOLUTION. At first glance, the game looks quite generous. From 11 possible outcomes (the sum can be 2,3, . . . ,12), the majority are winning numbers; some of them pay up to 5 to 1. However, the situation is more complicated. Let the r.v. X denote the actual payoff. Then X attains the values -1, 1, 2, 3, 5 with the following probabilities:

$$f(5) = P(\{X = 5\}) = P(\{(6,6)\}) = \frac{1}{36}$$

$$f(3) = P(\{X = 3\}) = P(\{(1,1)\}) = \frac{1}{36}$$

$$f(2) = P(\{X = 2\}) = P(\{(4,6),(5,5),(6,4),(6,5),(5,6)\}) = \frac{5}{36}$$

$$f(1) = P(\{X = 1\}) = P(\{(1,3),(2,2),(3,1),(1,2),(2,1)\}) = \frac{5}{36}$$

$$\begin{aligned} f(-1) = P(\{X = -1\}) = P(\{&(1,4),(2,3),(3,2),(4,1),(1,5),(2,4),(3,3),(4,2),(5,1),(1,6),(2,5),(3,4),\\ &(4,3),(5,2),(6,1),(2,6),(3,5),(4,4),(5,3),(6,2),(3,6),(4,5),(5,4),\\ &(6,3)\}) = \frac{24}{36}. \end{aligned}$$

Hence the expected payoff in the game is

$$E(X) = 5 \cdot \frac{1}{36} + 3 \cdot \frac{1}{36} + 2 \cdot \frac{5}{36} + 1 \cdot \frac{5}{36} + (-1) \cdot \frac{24}{36} = \frac{-1}{36}\$ \approx -3¢.$$

If we take into consideration the above mentioned frequency interpretation of the expected value $E(X)$, then we can see that in a *very long run* one may reasonably *expect* to lose an average of 3¢ per game. Therefore this game is unfavorable—the player will lose money, but it is not so bad to pay 3¢ per game for the fun of playing! However, if one plays only a few games, it is quite possible to lose on the average much more than 3¢ per game. In this case the expected value $E(X)$ does not have so clear a practical meaning as for a very long run of games.

EXAMPLE 2

Roulette. A roulette wheel usually has the numbers 00, 0, 1, 2, . . . , 36 on its perimeter. Both 0 and 00 are green, one half of the remaining numbers are red and the others are black. The numbers in the roulette table layout are divided into three columns, and also into three groups according to their magnitude. A little ball is spun on the wheel, and stops at one of the numbers. Let us calculate the expected payoffs for some standard bets.

(a) Betting on any single number 1 to 36, including 0 and 00, pays "35 to 1." This means that if a player puts his wager on a particular number on the table layout and this number comes up on the wheel, he wins 35 times the multiple of his bet (and keeps his bet too). Imagine

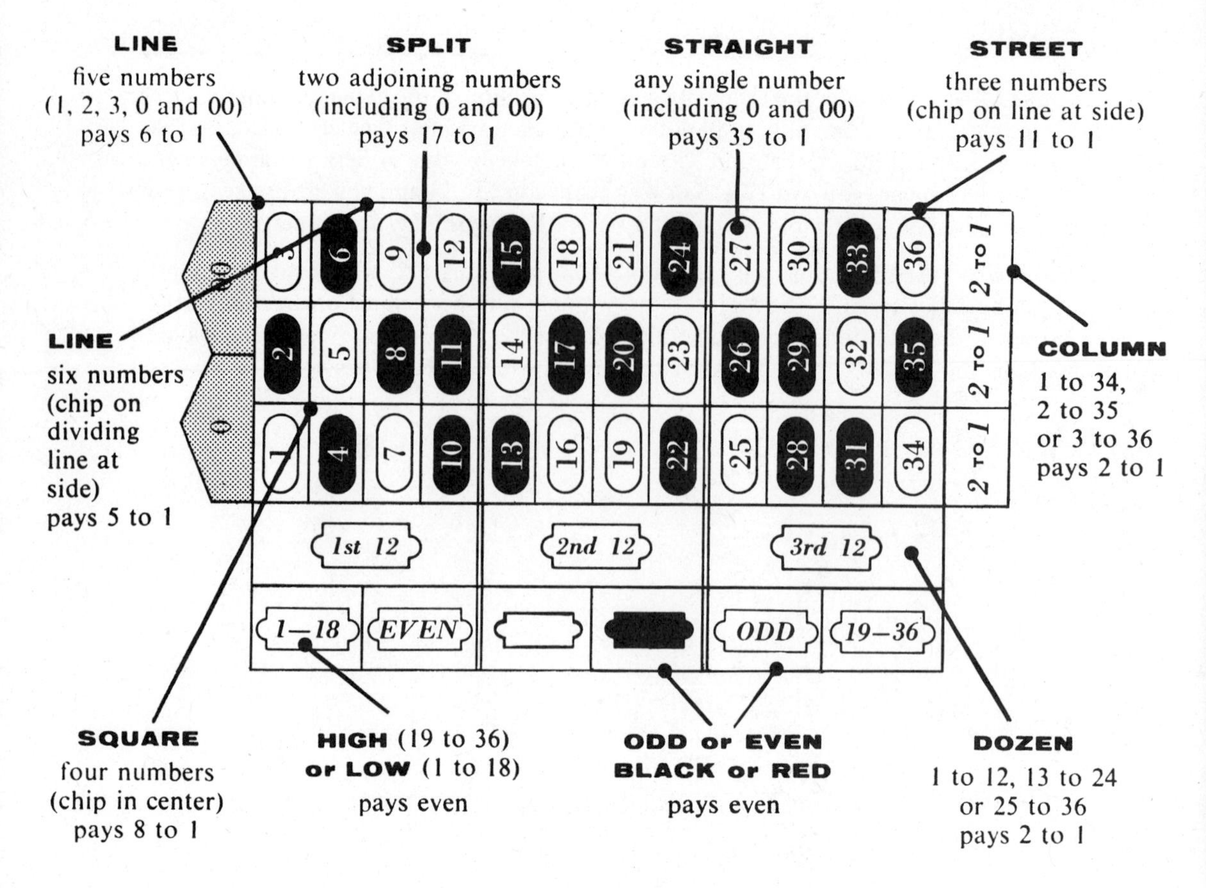

Roulette layout

that he puts a one-dollar chip on a number. If the number comes up, he wins \$35; otherwise he loses \$1. Let X be the payoff. If the wheel is well balanced and the table is not tilted, then the probability distribution of X is as follows:

x_i	35	-1
$f(x_i)$	$\frac{1}{38}$	$\frac{37}{38}$

The expected win of the player is, therefore,

$$E(X) = 35 \cdot \frac{1}{38} + (-1) \cdot \frac{37}{38} = \frac{-2}{38} = -5.26¢.$$

Hence, if the player bets \$1 on "singles," in a long run he will lose 5.26¢ per game. The loss is 5.26% of his wagers, and this percentage is called the *house advantage.*

(b) Betting on any two numbers at a time pays "17 to 1." In this case, a player puts a wager on any two numbers, and wins if one of them comes up. The distribution of payoffs is:

x_i	17	-1
$f(x_i)$	$\frac{2}{38}$	$\frac{36}{38}$

The expected payoff is

$$E(X) = 17 \cdot \frac{2}{38} + (-1) \cdot \frac{36}{38} = \frac{-2}{38} = -5.26¢.$$

Surprisingly (or not), the house advantage is again 5.26%.

(c) Betting on triples pays "11 to 1." The distribution of payoffs and its expectation are as follows:

x_i	11	−1
$f(x_i)$	$\frac{3}{38}$	$\frac{35}{38}$

$$E(X) = 11 \cdot \frac{3}{38} + (-1) \cdot \frac{35}{38} = \frac{-2}{38} = -5.26¢.$$

The house advantage is again 5.26%.

(d) Betting a column of numbers against the two remaining columns or betting one group of 12 numbers (1–12, 13–24, 25–36) against the two remaining groups pays off "2 to 1." If 0 or 00 comes up, the player loses. The distribution of payoffs and its expectations are as follows:

x_i	2	−1
$f(x_i)$	$\frac{12}{38}$	$\frac{26}{38}$

$$E(X) = 2 \cdot \frac{12}{38} + (-1) \cdot \frac{26}{38} = \frac{-2}{38} = -5.26¢.$$

The house advantage is again 5.26%.

(e) There are several so-called "even-money bets" in roulette. One can bet on red against black, on even against odd or on high against low numbers at even payoffs. The 0's are always in favor of the casino. The distribution of payoffs and the expectation are as follows:

x_i	1	−1
$f(x_i)$	$\frac{18}{38}$	$\frac{20}{38}$

$$E(X) = 1 \cdot \frac{18}{38} + (-1) \cdot \frac{20}{38} = \frac{-2}{38} = -5.26¢.$$

We can see that the house advantage is always uniformly 5.26% and that there is no way to beat the house unless some manipulations with the table are considered. In some cases (as in Monte Carlo), the roulette wheel does not have the 00 and the house advantage is therefore smaller. The game described above is sometimes referred to as Las Vegas roulette.

Roulette in Swaziland. Photograph by Jan Kopec, Camera Press, London.

EXAMPLE 3

Keno. One of the boldest and worst rip-offs must be keno as played by thousands of unsuspecting visitors to Las Vegas. Keno is a very old American game of chance with many variations and versions. Let us analyze so-called "$25,000 Keno" which is actually played in some of the casinos. The main attraction of the game is that for a 60¢ bet, one has a positive chance of winning $25,000 with no particular knowledge or skill.

A player gets a ticket containing 80 numbers 1,2,3, . . . ,80, called spots. He can mark any number n of spots, $n = 1,2, \ldots ,15$.

1	2	3	4	5	6	7	8	9	10
11	12	13	14	~~15~~	~~16~~	17	18	19	20
21	22	23	24	25	26	~~27~~	28	29	30
31	32	33	34	35	36	37	38	39	40
41	42	43	44	~~45~~	46	47	48	49	50
51	52	53	54	55	~~56~~	57	58	59	60
61	62	63	64	65	66	~~67~~	68	69	70
71	72	73	74	75	76	77	~~78~~	79	80

REGULAR TICKET COSTS AND PAYOFFS

MARK 1 SPOT

Catch	Play 60¢	Play 1.20	Play 3.00
1 Pays	1.80	3.60	9.00

MARK 2 SPOTS

Catch	Play 60¢	Play 1.20	Play 3.00
2 Pays	7.50	15.00	37.50

MARK 3 SPOTS

Catch	Play 60¢	Play 1.20	Play 3.00
2 Pays	.60	1.20	3.00
3 Pays	26.00	52.00	130.00

MARK 4 SPOTS

Catch	Play 60¢	Play 1.20	Play 3.00
2 Pays	.60	1.20	3.00
3 Pays	2.50	5.00	12.50
4 Pays	70.00	140.00	350.00

MARK 5 SPOTS

Catch	Play 60¢	Play 1.20	Play 3.00
3 Pays	1.00	2.00	5.00
4 Pays	14.00	28.00	70.00
5 Pays	300.00	600.00	1,500.00

MARK 6 SPOTS

Catch	Play 60¢	Play 1.20	Play 15.00
3 Pays	.50	1.00	12.50
4 Pays	3.00	6.00	75.00
5 Pays	55.00	110.00	1,375.00
6 Pays	1,000.00	2,000.00	25,000.00

MARK 7 SPOTS

Catch	Play 60¢	Play 1.20	Play 6.00
3 Pays	.30	.60	3.00
4 Pays	1.00	2.00	10.00
5 Pays	14.00	28.00	140.00
6 Pays	200.00	400.00	2,000.00
7 Pays	3,000.00	6,000.00	25,000.00

MARK 8 SPOTS

Catch	Play 60¢	Play 1.20	Play 3.00
5 Pays	5.00	10.00	25.00
6 Pays	50.00	100.00	250.00
7 Pays	1,100.00	2,200.00	5,500.00
8 Pays	12,500.00	25,000.00	25,000.00

MARK 9 SPOTS

Catch	Play 60¢	Play 1.20	Play 3.00
4 Pays	.20	.40	1.00
5 Pays	2.00	4.00	10.00
6 Pays	28.00	56.00	140.00
7 Pays	180.00	360.00	900.00
8 Pays	2,500.00	5,000.00	12,500.00
9 Pays	12,500.00	25,000.00	25,000.00

MARK 10 SPOTS

Catch	Play 60¢	Play 1.20	Play 3.00
5 Pays	1.20	2.40	6.00
6 Pays	12.00	24.00	60.00
7 Pays	90.00	180.00	450.00
8 Pays	660.00	1,320.00	3,300.00
9 Pays	2,400.00	4,800.00	12,000.00
10 Pays	12,500.00	25,000.00	25,000.00

MARK 11 SPOTS

Catch	Play 60¢	Play 1.20	Play 3.00
6 Pays	6.00	12.00	30.00
7 Pays	50.00	100.00	250.00
8 Pays	250.00	500.00	1,250.00
9 Pays	1,200.00	2,400.00	6,000.00
10 Pays	7,500.00	15,000.00	25,000.00
11 Pays	12,500.00	25,000.00	25,000.00

MARK 12 SPOTS

Catch	Play 60¢	Play 1.20	Play 3.00
5 Pays	.50	1.00	2.50
6 Pays	3.00	6.00	15.00
7 Pays	18.00	36.00	90.00
8 Pays	130.00	260.00	650.00
9 Pays	375.00	750.00	1,875.00
10 Pays	900.00	1,800.00	4,500.00
11 Pays	5,000.00	10,000.00	25,000.00
12 Pays	25,000.00	25,000.00	25,000.00

MARK 13 SPOTS

Catch	Play 60¢	Play 1.20	Play 3.00
6 Pays	1.00	2.00	5.00
7 Pays	10.00	20.00	50.00
8 Pays	50.00	100.00	250.00
9 Pays	450.00	900.00	2,250.00
10 Pays	2,400.00	4,800.00	12,000.00
11 Pays	4,500.00	9,000.00	22,500.00
12 Pays	10,000.00	20,000.00	25,000.00
13 Pays	25,000.00	25,000.00	25,000.00

MARK 14 SPOTS

Catch	Play 60¢	Play 1.20	Play 3.00
6 Pays	2.00	4.00	10.00
7 Pays	5.00	10.00	25.00
8 Pays	20.00	40.00	100.00
9 Pays	160.00	320.00	800.00
10 Pays	500.00	1,000.00	2,500.00
11 Pays	1,500.00	3,000.00	7,500.00
12 Pays	5,000.00	10,000.00	25,000.00
13 Pays	15,000.00	25,000.00	25,000.00
14 Pays	25,000.00	25,000.00	25,000.00

MARK 15 SPOTS

Catch	Play 60¢	$1.20	Play 3.00
6 Pays	1.00	2.00	5.00
7 Pays	5.00	10.00	25.00
8 Pays	14.00	28.00	70.00
9 Pays	50.00	100.00	250.00
10 Pays	150.00	300.00	750.00
11 Pays	1,500.00	3,000.00	7,500.00
12 Pays	5,000.00	10,000.00	25,000.00
13 Pays	15,000.00	25,000.00	25,000.00
14 Pays	20,000.00	25,000.00	25,000.00
15 Pays	25,000.00	25,000.00	25,000.00

EXAMPLE TICKETS

3.00

6.00

Keno payoff tables

The player can bet any multiple of 60¢ on his ticket, and register his wager with a runner or "keno girl." After the registration of all bets, 20 numbers out of 80 are chosen at random (??) as winning spots. If the player has "caught" more than a certain number of winning spots among his marked spots, he wins according to payoff tables.

All information about the possible bets and the corresponding payoffs is freely available to the player. Let us calculate the probability that if he marks n spots ($n = 1,2, \ldots ,15$) he will catch i winning spots, $i = 0,1,2, \ldots ,n$. The corresponding formula for the hypergeometric distribution gives:

$$P(i) = \frac{\binom{20}{i}\binom{60}{n-i}}{\binom{80}{n}} \qquad i = 0,1,2, \ldots ,n.$$

If a player marks one spot, then the probability distribution for the payoffs is

$$P(0) = \frac{\binom{20}{0}\binom{60}{1}}{\binom{80}{1}} = .75, \qquad \text{lose } 60¢.$$

$$P(1) = \frac{\binom{20}{0}\binom{60}{0}}{\binom{80}{1}} = .25, \qquad \text{win } (1.80 - .60) = \$1.20.$$

The corresponding r.v. X, which means the payoff, is

x_i	$-.6$	1.2
$f(x_i)$	.75	.25

$$E(X) = (-.6) \cdot (.75) + (1.2) \cdot (.25) = -.15$$

The player will therefore lose, on average, 15¢ per game if he bets 60¢ each game and keeps playing a long time. The house advantage is exactly 25%. If a player marks four spots, then the probability distribution of the profit X is as follows (subtract 60¢ from the pay tables as the charge per game).

x_i	$-.6$	0	1.9	69.4
$f(x_i)$	.7411	.2126	.0432	.0031

The expected value is

$$E(X) = -.1499$$

and the house advantage is 24.98%. Some players feel that marking a higher number of spots will give them a better chance of winning. Let us calculate the distribution of profits if the player marks 12 spots. This gives him a "real chance" to win $25,000 from a 60¢ bet.

x_i	$-.6$	$-.1$	2.4	17.4	129.4	374.4
$f(x_i)$	.86025	.09939	.03221	.00703	$.10195 \times 10^{-2}$	$.9540 \times 10^{-4}$

899.4	4999.4	24999.4
$.5428 \times 10^{-5}$	$.1673 \times 10^{-6}$	$.21 \times 10^{-8}$

His expected profit is $E(X) = -.153089$, or in other words, he will again lose, on average, 15¢ per game if he bets 60¢ and keeps on playing long enough. Note the accompanying table of expected payoffs and house advantages for different numbers of marked spots, if the player bets 60¢ per ticket:

Marked spots	Expected payoff	House advantage, %
1	$-.15$	25
2	$-.149050$	24.84
3	$-.155988$	26.00
4	$-.149861$	24.98
5	$-.153295$	25.55
10	$-.150062$	25.01
12	$-.153089$	25.51
15	$-.153316$	25.55

Whichever way the player bets, he will always lose in a long run about 25% of his money per game, which is really an atrociously high house advantage. But there is a SPECIAL BONUS deal on marking four spots, which pays off as follows: Catch 3 pays 1.70; Catch 4 pays 1.45. The price of this "treat" is 70¢, only 10¢ more than an ordinary 4 spot ticket and one can win more than twice as much. However, let us have a look at this great deal. The probability distribution of the payoffs is as follows:

x_i	$-.7$	$+1$	144.3
$f(x_i)$	.953689	.043248	.003063

The expected payoff is $E(X) = -.182287$ and therefore the house advantage is 26.04%, well above any ordinary bet. This is apparently the reason for the name.

Why do people play keno? It is not easy to calculate the house advantage, even if they know enough about probability theory, because the numerical obstacles are great (the above distribution and house advantages had to be calculated on a computer). But the main attraction is purely psychological: who would not like to win $25,000 with a 60¢ bet? However, very few people can find the probability that it will really happen. The best chance is to mark 12 spots, and the above distribution shows that in this case the probability of winning $25,000 is $.21 \times 10^{-8}$, that is, practically 0. A player would have to expect to play more than 476 million games at a total expense of $285 million to win the big prize!

The formula for the expected value of a r.v., which was given in the definition, is very suitable for numerical calculations. However, in some theoretical proofs another formula will prove more expedient. Let $\{A_1, A_2, \ldots, A_k\}$ be the partition induced by X. Then

$$E(X) = \sum_{i=1}^{k} x_i f(x_i) = \sum_{i=1}^{k} x_i P(A_i) = \sum_{i=1}^{k} x_i P\{\omega: X(\omega) = x_i\} = \sum_{\omega \in \Omega} X(\omega) P(\{\omega\}).$$

We have proved the following:

→ **THEOREM 1:** Let $X(\omega)$ be a r.v. on a probability measure space (Ω,P). Then

$$E(X) = \sum_{\omega \in \Omega} X(\omega)P(\{\omega\}).$$

EXAMPLE 4

Consider tossing a coin twice. Then a sample space Ω and the r.v. X denoting the number of tails are as follows:

ω	$X(\omega)$
HH	0
HT	1
TH	1
TT	2

The distribution of X is

$$f(0) = \frac{1}{4}, \quad f(1) = \frac{1}{2}, \quad f(2) = \frac{1}{4}$$

hence

$$E(X) = \sum_{i=1}^{4} x_i f(x_i) = 0 \cdot \frac{1}{4} + 1 \cdot \frac{1}{2} + 2 \cdot \frac{1}{4} = 1.$$

But on the other hand,

$$\begin{aligned} E(X) &= \sum_{\omega \in \Omega} X(\omega)P(\{\omega\}) \\ &= X(HH) \cdot P(\{HH\}) + X(HT) \cdot P(\{HT\}) + X(TH) \cdot P(\{TH\}) \\ &\quad + X(TT) \cdot P(\{TT\}) \\ &= 0 \cdot \frac{1}{4} + 1 \cdot \frac{1}{4} + 1 \cdot \frac{1}{4} + 2 \cdot \frac{1}{4} = 1. \end{aligned}$$

EXAMPLE 5

Let $A \subset \Omega$ be a subset of Ω. In Chapter X, we defined the indicator random variable I_A as follows:

$$\begin{aligned} I_A(\omega) &= 0 \qquad \text{if } \omega \notin A. \\ &= 1 \qquad \text{if } \omega \in A. \end{aligned}$$

Hence the r.v. $I_A(\omega)$ attains only the values 0 and 1 and its distribution is

$$f(0) = P(\{\omega\colon I_A(\omega) = 0\}) = P(A')$$

$$f(1) = P(\{\omega: I_A(\omega) = 1\}) = P(A).$$

Therefore $E(X) = 0 \cdot f(0) + 1 \cdot f(1) = 0 \cdot P(A') + 1 \cdot P(A) = P(A)$.

If X and Y are r.v.'s defined on a probability measure space (Ω,P) and c is a real number, then c can be understood as a r.v. (assigning c to every point $\omega \in \Omega$) and also cX, and $X + Y$ are r.v.'s. We can investigate their expected values.

THEOREM 2: Let X and Y be two random variables defined on a probability measure space (Ω,P) and let c be a real number. Then
(a) $E(c) = c$.
(b) $E(cX) = cE(X)$.
(c) $E(X + Y) = E(X) + E(Y)$.
(d) If for every $\omega \in \Omega$ we have $X(\omega) \geq 0$, then also $E(X) \geq 0$.

Proof: (a) We can understand c as a r.v. such that $Z(\omega) = c$ for every $\omega \in \Omega$. Using Theorem 1, we get

$$E(c) = \sum_{\omega\in\Omega} Z(\omega)\, P(\{\omega\}) = \sum_{\omega\in\Omega} c\, P(\{\omega\}) = c \sum_{\omega\in\Omega} P(\{\omega\}) = c \cdot 1 = c.$$

(b) We shall use Theorem 1:

$$E(cX) = \sum_{\omega\in\Omega} cX(\omega)P(\{\omega\})) = c \sum_{\omega\in\Omega} X(\omega)P(\{\omega\}) = cE(X).$$

(c) Using Theorem 1,

$$E(X + Y) = \sum_{\omega\in\Omega} (X(\omega) + Y(\omega))P(\{\omega\}) = \sum_{\omega\in\Omega} X(\omega)P(\{\omega\}) + \sum_{\omega\in\Omega} Y(\omega)P(\{\omega\}) = E(X) + E(Y).$$

(d) If for every $\omega \in \Omega$ we have $X(\omega) \geq 0$, then also $X(\omega)P(\{\omega\}) \geq 0$, hence $E(X) = \sum_{\omega\in\Omega} X(\omega)P(\{\omega\}) \geq 0$. q.e.d.

COMMENT. The statement of Theorem 2 (c) can be obviously generalized to any n r.v.'s $X_1,X_2, \ldots ,X_n$. Then $E(X_1 + X_2 + \ldots + X_n) = E_1(X) + E_2(X) + \ldots + E(X_n)$.

EXAMPLE 5

Let three coins be flipped; let X mean the number of heads and Y mean the number of runs. The joint probability distribution of X and Y was found in Chapter XI.

$X \backslash Y$	1	2	3	f
3	$\frac{1}{8}$	0	0	$\frac{1}{8}$
2	0	$\frac{2}{8}$	$\frac{1}{8}$	$\frac{3}{8}$
1	0	$\frac{2}{8}$	$\frac{1}{8}$	$\frac{3}{8}$
0	$\frac{1}{8}$	0	0	$\frac{1}{8}$
g	$\frac{2}{8}$	$\frac{4}{8}$	$\frac{2}{8}$	1

From the marginal distributions, we get

$$E(X) = 3 \cdot \frac{1}{8} + 2 \cdot \frac{3}{8} + 1 \cdot \frac{3}{8} + 0 \cdot \frac{1}{8} = \frac{12}{8} = 1.5.$$

$$E(Y) = 1 \cdot \frac{2}{8} + 2 \cdot \frac{4}{8} + 3 \cdot \frac{2}{8} = \frac{16}{8} = 2.$$

From the joint probability distribution, we can get the probability distribution of the r.v. $Z = X + Y$:

$z = x + y$	1	3	4	5
$f(z)$	$\frac{1}{8}$	$\frac{2}{8}$	$\frac{4}{8}$	$\frac{1}{8}$

$$E(Z) = 1 \cdot \frac{1}{8} + 3 \cdot \frac{2}{8} + 4 \cdot \frac{4}{8} + 5 \cdot \frac{1}{8} = \frac{28}{8} = 3.5.$$

Obviously $$E(Z) = E(X) + E(Y).$$

Further, $$E(X) \geq 0, \quad E(Y) \geq 0 \quad \text{and} \quad E(Z) \geq 0.$$

EXAMPLE 6

Let 10 fair coins be tossed and let the r.v. X denote the number of heads minus the number of tails. Find the expected value of X.

SOLUTION. Let Y_1 denote the number of heads, Y_2 denote the number of tails. Obviously, $E(Y_1) = E(Y_2)$. Further,

$$X = Y_1 - Y_2.$$

Using Theorem 2, we have

$$E(X) = E(Y_1 - Y_2) = E(Y_1 + (-1) \cdot Y_2) = E(Y_1) + E((-1) \cdot Y_2)$$
$$= E(Y_1) + (-1) \cdot E(Y_2) = E(Y_1) - E(Y_2) = 0.$$

EXAMPLE 7

A fair coin is tossed until a head or five consecutive tails occur. Let X denote the number of tosses performed. Find the expected value of X.

SOLUTION. An appropriate sample space with the probability measure and the distribution of X is as follows:

$$f(1) = P(H) = \frac{1}{2}$$

$$f(2) = P(TH) = \frac{1}{4}$$

$$f(3) = P(TTH) = \frac{1}{8}$$

$$f(4) = P(TTTH) = \frac{1}{16}$$

$$f(5) = P(TTTTH) + P(TTTTT) = \frac{1}{32} + \frac{1}{32} = \frac{1}{16}.$$

Hence, $$E(X) = 1 \cdot \frac{1}{2} + 2 \cdot \frac{1}{4} + 3 \cdot \frac{1}{8} + 4 \cdot \frac{1}{16} + 5 \cdot \frac{1}{16} = 1.9375.$$

In probability theory, we say that a game of chance is *fair* if the expected gain is 0.

EXAMPLE 8

A player tosses three fair coins. He wins \$8 if three heads occur, \$3 if two heads occur and \$1 if one head occurs. If the game is to be fair, how much should he lose if no heads occur?

SOLUTION. Denote by x the payment if no heads occur. Then the expected gain must be 0 if the game is to be fair.

$$8 \cdot \frac{1}{2^3} + 3\binom{3}{2}\frac{1}{2^3} + 1\binom{3}{1}\frac{1}{2^3} + (-x)\frac{1}{2^3} = 0. \quad x = 20.$$

He should lose \$20 if the game is to be fair.

EXAMPLE 9

The Petersburg "Paradox." An owner of a casino in czarist Petersburg tried to design a game as follows: Each player will toss a fair coin until it falls heads; if this occurs at the r-th throw the player receives 2^r rubles. The player receives nothing if no decision is reached in N tosses. How much should the player pay as an initial bet or "admission fee" in order for the game to be fair?

SOLUTION. The probability of getting the first head on the r-th toss is equal to the probability of getting $(r - 1)$ tails and then a head, and it is equal to $\frac{1}{2^r}$. The expectation is, hence, $2^1 \cdot \frac{1}{2^1} + 2^2 \cdot \frac{1}{2^2} + 2^3 \cdot \frac{1}{2^3} + \ldots + 2^N \cdot \frac{1}{2^N} = N.$

Therefore each player should pay N rubles as an admission fee. The owner of the casino, however, was not happy about the rule that the game must be stopped after N tosses. How much should the fee be if the game is to be continued until the coin falls heads without any further restriction? If the game were to be stopped after N tosses, then the fee should be N rubles. If N increases, then so should the fee. If N is unlimited, then regardless of the sum required for the admission fee, the bank of the casino would be completely ruined in such a game. A similar "paradox" was earlier considered by Daniel Bernoulli (1700–1782).

EXAMPLE 10

Sometimes we can calculate the expected value of a r.v. without the tedious calculations required by Definition 1 or by Theorem 1. Let us consider the experiment of drawing a bridge hand of 13 cards from a standard deck of 52 cards, and let us assume all hands to be equally likely. According to bridge theoretician Charles Goren, a simplified *honor count* of such a hand, say X, is calculated by assigning four points to each ace, three points to a king, two points to a queen and one point to a jack. To find the distribution of X is quite simple and straightforward but very tedious. We can find $E(X)$ using the following trick.

Imagine that all cards have been dealt to each of four players. Let X_1, X_2, X_3 and X_4 be the honor counts of the four hands. The total honor count of the complete deck is 40, hence

$$X_1 + X_2 + X_3 + X_4 = 40.$$

Using the comment at the end of the proof of Theorem 2, we get

$$E(X_1) + E(X_2) + E(X_3) + E(X_4) = 40.$$

However, we assumed that all hands are equally likely, hence X_1, X_2, X_3 and X_4 have the same distribution as X, and

$$E(X_1) = E(X_2) = E(X_3) = E(X_4) = E(X).$$

Therefore,
$$4E(X) = 40$$

and finally
$$E(X) = 10.$$

If the reader has some misgivings about the result, then we can find $E(X)$ using another trick. Let I_1, I_2, I_3 and I_4 be indicator random variables of the events that the hand contains the ace of spades, hearts, diamonds and clubs, respectively. Similarly, I_5, I_6, I_7 and I_8 are the indicators for the king of spades, hearts, diamonds and clubs, respectively, and so on up to I_{16}, which is the indicator that the hand contains the jack of clubs. Then

$$X = 4\sum_{j=1}^{4} I_j + 3\sum_{j=5}^{8} I_j + 2\sum_{j=9}^{12} I_j + \sum_{j=13}^{16} I_j.$$

Further,
$$E(X) = 4\sum_{j=1}^{4} E(I_j) + 3\sum_{j=5}^{8} E(I_j) + 2\sum_{j=9}^{12} E(I_j) + \sum_{j=13}^{16} E(I_j).$$

From the previous investigation, we know that

$$E(I_j) = P(\{I_j = 1\}) = \frac{1}{4}$$

because the event $\{I_j = 1\}$ means that the hand contains a specified card, and its probability is

$$\frac{\binom{51}{12}}{\binom{52}{13}} = \frac{1}{4}.$$

Hence
$$E(X) = 4 + 3 + 2 + 1 = 10.$$

Let X be a r.v. on a probability measure space and let F be a function of a real variable, that is $F: R^{\#} \longrightarrow R^{\#}$. Then, according to Chapter X, the composite function $F(X)$ is also a random variable and we shall often have to calculate its expected value $E(F(X))$. Let us consider first several simple situations.

EXAMPLE 11

Let X be a r.v. with the following probability distribution

x_i	-2	-1	0	1	2
$f(x_i)$	.1	.2	.3	.2	.2

(a) The distribution of the r.v. $Z = F(X) = 2X + 3$ is

z_i	-1	1	3	5	7
$g(z_i)$	.1	.2	.3	.2	.2

We can easily find $E(Z) = E(F(X)) = E(2X + 3)$ as follows:

$$E(Z) = E(F(X)) = (-1) \cdot (.1) + 1 \cdot (.2) + 3 \cdot (.3) + 5 \cdot (.2) + 7 \cdot (.2) = 3.4.$$

However, we could have calculated $E(F(X))$ by taking all the values of X with corresponding probabilities and evaluating Z always according to the formula $Z = 2X + 3$.

$$\begin{aligned} E(Z) &= E(F(X)) \\ &= [2 \cdot (-2) + 3] \cdot (.1) + [2 \cdot (-1) + 3] \cdot (.2) + [2 \cdot (0) + 3] \cdot (.3) \\ &\quad + [2 \cdot (1) + 3] \cdot (.2) + [2 \cdot (2) + 3] \cdot (.2) \\ &= 3.4. \end{aligned}$$

In this case of a very simple function F, it is absolutely obvious that both methods will give the same result. Let us consider a more complicated function of X.

(b) Let $Z = F(X) = X^2$. Find the distribution and the expectation of the r.v. $Z = F(X)$.

z_i	0	1	4
$f(z_i)$	.3	.4	.3

$$E(Z) = E(F(X)) = 0 \cdot (.3) + 1 \cdot (.4) + 4 \cdot (.3) = 1.6.$$

However, we can get the same result by taking all the values of X with corresponding probabilities and evaluating Z always according to the formula $Z = X^2$.

$$\begin{aligned} E(Z) &= E(F(X)) \\ &= (-2)^2 \cdot (.1) + (-1)^2 \cdot (.2) + 0^2 \cdot (.3) + 1^2 \cdot (.2) + 2^2 \cdot (.2) \\ &= 1.6. \end{aligned}$$

Let us consider now an even more complicated function F.

(c) Let $Z = F(X) = X^3 - 2X^2 - X + 2$. Find the probability distribution and the expectation of the r.v. $Z = F(X)$.

z_i	-12	0	2
$g(z_i)$	.1	.6	.3

In this case $g(0) = P(\{Z = 0\}) = P(\{X = -1\}) + P(\{X = 1\}) + P(\{X = 2\}) = .2 + .2 + .2 = .6$.

$$E(Z) = E(F(X)) = -12 \cdot (.1) + 0 \cdot (.6) + 2 \cdot (.3) = -.6.$$

However, we can get the same result as follows (see the distribution of X):

$$\begin{aligned} E(Z) &= E(F(X)) \\ &= [(-2)^3 - 2 \cdot (-2)^2 - (-2) + 2] \cdot (.1) + [(-1)^3 - 2 \cdot (-1)^2 - (-1) + 2] \cdot (.2) \\ &\quad + [(0)^3 - 2 \cdot (0)^2 - (0) + 2] \cdot (.3) + [1^3 - 2 \cdot (1)^2 - 1 + 2] \cdot (.2) \\ &\quad + [2^3 - 2 \cdot (2)^2 - 2 + 2] \cdot (.2) \\ &= -.6. \end{aligned}$$

These examples show intuitively that the theorem given below might be true. We went into such detailed calculations only to give the reader a real feeling for the thing, because the proof of the theorem is not easy to understand. For this reason, we give two proofs. Some people prefer the first one, while others find the second proof more comprehensible.

→ **THEOREM 3:** Let X be a r.v. with probability distribution $f(x_i)$, $i = 1,2, \ldots, k$ and let F be a real valued function of a real variable. Then the expected value of the r.v. $F(X)$ is equal to

$$E(F(X)) = \sum_{i=1}^{k} F(x_i)\, f(x_i).$$

Proof 1: Using Theorem 1, we get

$$E(F(X)) = \sum_{\omega \in \Omega} F(X(\omega))P(\{\omega\}) = \sum_{i=1}^{k} \sum_{\{\omega: X(\omega) = x_i\}} F(X(\omega))P(\{\omega\})$$

In the preceding step, we have put together into one set all those ω yielding the same value $X(\omega) = x_i$, hence $F(X(\omega)) = F(x_i)$.

$$\begin{aligned} E(F(X)) &= \sum_{i=1}^{k} \sum_{\{\omega: X(\omega) = x_i\}} F(x_i)P(\{\omega\}) = \sum_{i=1}^{k} F(x_i) \sum_{\{\omega: X(\omega) = x_i\}} P(\{\omega\}) \\ &= \sum_{i=1}^{k} F(x_i)P(\{\omega: X(\omega) = x_i\}) = \sum_{i=1}^{k} F(x_i) f(x_i) \end{aligned}$$

q.e.d.

Proof 2: If z_j are all possible distinct values which the r.v. $F(X)$ can attain then, using Definition 1, we get

$$\begin{aligned} E(F(X)) &= \sum_j z_j P(\{\omega: F(X(\omega)) = z_j\}) = \sum_j z_j \sum_{\{x_i: F(x_i) = z_j\}} P(\{\omega: X(\omega) = x_i\}) \\ &= \sum_j \sum_{\{x_i: F(x_i) = z_j\}} z_j P(\{\omega: X(\omega) = x_i\}) = \sum_j \sum_{\{x_i: F(x_i) = z_j\}} F(x_i) f(x_i) = \sum_{i=1}^{k} F(x_i) f(x_i). \end{aligned}$$

q.e.d.

Let us consider now two random variables X and Y, and let G be a function of two real variables, $G: R^{\#} \times R^{\#} \longrightarrow R^{\#}$ (see the last part of Chapter II, dealing with composite functions). Then also the composite function $G(X,Y)$ can be considered as a r.v. and we are interested in its expected value. The simplest case of $G(X,Y) = X + Y$ was considered in Example 5.

EXAMPLE 12

Let the joint distribution of the r.v.'s X and Y be given as follows:

$X \backslash Y$	1	2	3	f
0	0	$\frac{1}{6}$	$\frac{1}{6}$	$\frac{2}{6}$
1	$\frac{1}{6}$	$\frac{2}{6}$	$\frac{1}{6}$	$\frac{4}{6}$
g	$\frac{1}{6}$	$\frac{3}{6}$	$\frac{2}{6}$	1

(a) Let the r.v. Z be given as $Z = G(X,Y) = X + Y$. Find the distribution and expectation of Z.

z_i	2	3	4
$g(z_i)$	$\frac{2}{6}$	$\frac{3}{6}$	$\frac{1}{6}$

$$E(Z) = 2 \cdot \frac{2}{6} + 3 \cdot \frac{3}{6} + 4 \cdot \frac{1}{6} = \frac{17}{6}.$$

However, we can get $E(Z)$ also by taking all possible combinations of values of X and Y with corresponding probabilities, and evaluating Z from the formula $Z = X + Y$.

$$\begin{aligned} E(Z) &= E(G(X,Y)) \\ &= (0 + 1) \cdot 0 + (0 + 2) \cdot \frac{1}{6} + (0 + 3) \cdot \frac{1}{6} + (1 + 1) \cdot \frac{1}{6} + (1 + 2) \cdot \frac{2}{6} + (1 + 3) \cdot \frac{1}{6} \\ &= \frac{17}{6}. \end{aligned}$$

(b) Let the r.v. Z be given as $Z = G(X,Y) = X\,Y$. Find the probability distribution and expectation of Z.

z_i	0	1	2	3
$g(z_i)$	$\frac{2}{6}$	$\frac{1}{6}$	$\frac{2}{6}$	$\frac{1}{6}$

$$E(Z) = 0 \cdot \frac{2}{6} + 1 \cdot \frac{1}{6} + 2 \cdot \frac{2}{6} + 3 \cdot \frac{1}{6} = \frac{8}{6}.$$

However, we can get $E(Z)$ also as follows (see the joint distribution of X and Y):

$$E(Z) = E(G(X,Y))$$
$$= (0 \cdot 1) \cdot 0 + (0 \cdot 2) \cdot \frac{1}{6} + (0 \cdot 3) \cdot \frac{1}{6} + (1 \cdot 1) \cdot \frac{1}{6} + (1 \cdot 2) \cdot \frac{2}{6} + (1 \cdot 3) \cdot \frac{1}{6}$$
$$= \frac{8}{6}.$$

We can see that the expected value of the composite r.v. $G(X,Y)$ can be calculated in a way similar to what was done in Theorem 3 using the original distribution of X and Y. This fact motivates the following:

→ **THEOREM 4:** Let X and Y be two r.v.'s with the joint probability distribution $h(x_i,y_j)$, $i = 1,2, \ldots ,k$; $j = 1,2, \ldots ,l$. Let G be a real valued function of two real variables. Then the expected value of the r.v. $G(X,Y)$ is equal to

$$E(G(X,Y)) = \sum_{i=1}^{k} \sum_{j=1}^{l} G(x_i,y_j)h(x_i,y_j).$$

Proof: Using Theorem 1, we get

$$E(G(X,Y)) = \sum_{\omega \in \Omega} G(X(\omega), Y(\omega))P(\{\omega\}) = \sum_{i=1}^{k} \sum_{j=1}^{l} \sum_{\{\omega: X(\omega)=x_i, Y(\omega)=y_j\}} G(X(\omega), Y(\omega))P(\{\omega\}) = \sum_{i=1}^{k} \sum_{j=1}^{l} \sum_{\{\omega: X(\omega)=x_i, Y(\omega)=y_j\}} G(x_i,y_j)P(\{\omega\})$$

$$= \sum_{i=1}^{k} \sum_{j=1}^{l} G(x_i,y_j) \sum_{\{\omega: X(\omega)=x_i, Y(\omega)=y_j\}} P(\{\omega\}) = \sum_{i=1}^{k} \sum_{j=1}^{l} G(x_i,y_j)h(x_i,y_j).$$ q.e.d.

EXAMPLE 13

Let r.v.'s X and Y have the following joint probability distribution:

X \ Y	1	2	f
0	$\frac{1}{5}$	$\frac{2}{5}$	$\frac{3}{5}$
1	$\frac{2}{5}$	0	$\frac{2}{5}$
g	$\frac{3}{5}$	$\frac{2}{5}$	1

(a) Find $E(X)$ and $E(Y)$.

$$E(X) = 0 \cdot \frac{3}{5} + 1 \cdot \frac{2}{5} = \frac{2}{5}, \qquad E(Y) = 1 \cdot \frac{3}{5} + 2 \cdot \frac{2}{5} = \frac{7}{5}.$$

(b) If $G(X,Y) = XY$, find $E(G(X,Y))$.

$$E(XY) = (0 \cdot 1) \cdot \frac{1}{5} + (0 \cdot 2) \cdot \frac{2}{5} + (1 \cdot 1) \cdot \frac{2}{5} + (2 \cdot 1) \cdot 0 = \frac{2}{5}.$$

(c) If $G(X,Y) = (X + 1)^2 + X(Y - 2)$, find $E(G(X,Y))$.

$$E(G(X,Y)) = E[(X + 1)^2 + X(Y - 2)]$$
$$= [(0 + 1)^2 + 0 \cdot (1 - 2)] \cdot \frac{1}{5} + [(0 + 1)^2 + 0 \cdot (2 - 2)] \cdot \frac{2}{5}$$
$$+ [(1 + 1)^2 + 1 \cdot (1 - 2)] \cdot \frac{2}{5} + [(1 + 1)^2 + 1 \cdot (2 - 2)] \cdot 0$$
$$= \frac{9}{5}.$$

If we investigate Examples 12 and 13 closely, we can notice that

$$E(XY) \neq E(X)E(Y)$$

and also that X and Y are dependent. However, one might be interested in whether the relation $E(XY) = E(X)E(Y)$ can occur, and under what conditions.

THEOREM 5: Let X and Y be independent r.v.'s. Then

$$E(XY) = E(X)E(Y).$$

Proof: Let $h(x_i,y_j)$ be the joint distribution of X and Y, $f(x_i)$ be the distribution of X and $g(y_j)$ be the distribution of Y, $i = 1,2, \ldots ,k$; $j = 1,2, \ldots ,l$. From the assumption of independence of X and Y, we get

$$h(x_i,y_j) = f(x_i)g(y_j) \quad i = 1,2, \ldots ,k; \quad j = 1,2, \ldots ,l.$$

Using Theorem 4, we get

$$E(XY) = \sum_{i=1}^{k}\sum_{j=1}^{l} x_iy_jh(x_i,y_j) = \sum_{i=1}^{k}\sum_{j=1}^{l} x_iy_jf(x_i)g(y_j) = \sum_{i=1}^{k} x_if(x_i)\sum_{j=1}^{l} y_jg(y_j) = E(X)E(Y).$$ q.e.d.

EXAMPLE 14

Let the r.v.'s X and Y be as follows:

X \ Y	0	1	f
0	$\frac{1}{4}$	$\frac{1}{4}$	$\frac{1}{2}$
1	$\frac{1}{4}$	$\frac{1}{4}$	$\frac{1}{2}$
g	$\frac{1}{2}$	$\frac{1}{2}$	

We can consider X and Y as indicators of whether a head appeared on the first or second toss, respectively, when two coins were flipped. Obviously, X and Y are independent,

$$E(X) = \frac{1}{2}, \quad E(Y) = \frac{1}{2}, \quad E(XY) = \frac{1}{4}$$

$$E(XY) = E(X)E(Y).$$

> WARNING: If the r.v.'s X and Y are dependent, then the relation
>
> $$E(XY) = E(X)E(Y)$$
>
> may be, but does not necessarily have to be, satisfied. If X and Y are independent, then it is always satisfied.

EXAMPLE 15

Let us consider an example of dependent r.v.'s for which $E(XY) = E(X)E(Y)$. Toss a coin three times and let X be the number of heads and Y be the number of runs (see Example 5). Then X and Y are dependent: $E(X) = 1.5$, $E(Y) = 2$, $E(XY) = 3$, hence $E(XY) = E(X)E(Y)$.

EXAMPLE 16

Let us investigate one more example of random variables X and Y which are dependent but for which $E(XY) = E(X)E(Y)$. Let X be a r.v. with distribution *symmetrical* around 0, i.e., $P(\{X = x\}) = P(\{X = -x\})$ and let $Y = X^2$. It can be easily seen that $E(X) = 0$ and that X and Y are dependent (provided X attains more than one value). Furthermore, $E(XY) = E(X^3) = 0$ because the distribution of X^3 is again symmetrical. Hence $E(XY) = 0 = 0E(Y) = E(X)E(Y)$.

EXAMPLE 17

You order a wall-to-wall carpet for your 9×15 foot rectangular den. The rug should be cut in one piece from a very wide (more than 9 feet) and long roll of material. It is known that the workman cutting the carpet never gives less than the ordered amount but usually makes a "random error" in the measurement of each linear dimension, which is, on average, 2% above the required length. Assuming the errors in measuring the length and width of a carpet to be independent, what is the expected area of the delivered carpet?

SOLUTION. Let the r.v. X respective to Y denote the actual percentage error in cutting the length respective to the width of your carpet. Then X and Y are independent. The error of the length of the carpet is $15X$ feet, therefore the carpet will be $15(1 + X)$ feet long and similarly, we can see that it will be $9(1 + Y)$ feet wide. Let the r.v. Z denote the actual area of the carpet. Then

$$Z = 15(1 + X)9(1 + Y) = 135(1 + X + Y + XY).$$

The expected area of the carpet is

$$E(Z) = 135(1 + E(X) + E(Y) + E(XY)).$$

We know that $E(X) = E(Y) = .02$ and because of the independence of X and Y, using Theorem 5 we get $E(XY) = E(X)E(Y) = (.02)^2 = .0004$. Hence $E(Z) = 135(1 + .02 + .02 + .0004) = 140.454$ square feet.

Until now the examples of expected values of random variables were fairly simple. Let us finish this section with two examples that are more sophisticated.

EXAMPLE 18

According to the time schedule, exactly one bus stops at a bus station within every hour (i.e., one bus between 1 P.M. and 2 P.M., one bus between 2 P.M. and 3 P.M., etc.). With probability $\frac{1}{6}$ it arrives 10 minutes past the hour, with probability $\frac{1}{2}$ it arrives half an hour past the hour and with probability $\frac{1}{3}$ it arrives 10 minutes to the hour.

(a) You come to the bus station on the hour. How long do you expect to wait for the bus?

(b) What is the expected waiting time if you arrive 20 minutes past the hour?

(c) What is the best time to arrive at the station, i.e., when is the expected waiting time shortest?

SOLUTION. (a). Let the r.v. X denote the waiting time. Then the distribution of X is as follows:

x	10	30	50
$f(x)$	$\frac{1}{6}$	$\frac{3}{6}$	$\frac{2}{6}$

The expected waiting time is hence

$$E(X) = 10 \cdot \frac{1}{6} + 30 \cdot \frac{3}{6} + 50 \cdot \frac{2}{6} = \frac{200}{6} \text{ minutes.}$$

(b) Denoting the waiting time by X, we have the distribution of X as follows:

x	10	30	50	70	90
$f(x)$	$\frac{3}{6}$	$\frac{2}{6}$	$\frac{1}{6} \cdot \frac{1}{6}$	$\frac{1}{6} \cdot \frac{3}{6}$	$\frac{1}{6} \cdot \frac{2}{6}$

In the above table, for example, $P(\{X = 70\}) = P(\{\text{bus missed within the first hour}\})$ $P(\{\text{bus arrives 30 minutes past the second hour/bus missed within the first hour}\}) = \frac{1}{6} \cdot \frac{3}{6}$. Hence the expected waiting time is

$$E(X) = 10 \cdot \frac{3}{6} + 30 \cdot \frac{2}{6} + 50 \cdot \frac{1}{36} + 70 \cdot \frac{3}{36} + 90 \cdot \frac{2}{36} = \frac{980}{36} \text{ minutes.}$$

(c) Assume that you arrive t minutes past the hour and let X be your waiting time.

(i) If $0 \leq t \leq 10$, then the distribution of X is

x	$10 - t$	$30 - t$	$50 - t$
$f(x)$	$\frac{1}{6}$	$\frac{3}{6}$	$\frac{2}{6}$

hence $E(X) = \frac{1}{6} \cdot (10 - t) + \frac{3}{6} \cdot (30 - t) + \frac{2}{6} \cdot (50 - t)$. The minimum of $E(X)$ within the given time range is reached for $t = 10$ when

$$E(X) = \frac{3}{6} \cdot 20 + \frac{2}{6} \cdot 40 = \frac{140}{6} = \frac{840}{36} \text{ minutes.}$$

(ii) If $10 < t \leq 30$, then the distribution of X is

x	$30 - t$	$50 - t$	$70 - t$	$90 - t$	$110 - t$
$f(x)$	$\frac{3}{6}$	$\frac{2}{6}$	$\frac{1}{6} \cdot \frac{1}{6}$	$\frac{1}{6} \cdot \frac{3}{6}$	$\frac{1}{6} \cdot \frac{2}{6}$

The expected waiting time is

$$E(X) = \frac{3}{6} \cdot (30 - t) + \frac{2}{6} \cdot (50 - t) + \frac{1}{36} \cdot (70 - t) + \frac{3}{36} \cdot (90 - t) + \frac{2}{6} \cdot (110 - t).$$

The minimum of $E(X)$ within the given time range is reached for $t = 30$ when $E(X) = \frac{620}{36}$ minutes.

(iii) Similarly, for $30 < t \leq 50$ the distribution of X is

x	$50 - t$	$70 - t$	$90 - t$	$110 - t$
$f(x)$	$\frac{2}{6}$	$\frac{4}{6} \cdot \frac{1}{6}$	$\frac{4}{6} \cdot \frac{3}{6}$	$\frac{4}{6} \cdot \frac{1}{6}$

The minimum of $E(X)$ within the given time range is reached for $t = 50$ when $E(X) = \frac{800}{36}$ minutes.

(iv) If $50 < t \leq 60$, then you have to wait, on average, more than in the case of (i).

The best arrival time at the bus station for the shortest expected waiting time is half an hour past the hour.

EXAMPLE 19

Rumor has it that the president of the United Cloud Blowers has a special bargain for the members of his union. He proposed a new labor contract for each factory, which defines a day off as any day that is a birthday of at least one worker of the factory, while the rest of the year are working days. The ratified package, however, leaves a free hand to the boards of directors to maximize the number of workdays. Find the optimal number of workers for a factory.

SOLUTION. To clarify the expressions: a working day is any day when people work. The number of workdays is equal to the number of working days times the number of workers. Hence if we assume that a year has 365 days, then one worker will yield 364 workdays. If two workers have distinct birthdays, then they yield $2 \times (365 - 2) = 726$ workdays. The number of workdays, therefore, initially increases with the increasing number of workers. However, if a factory employs too many workers (say a million), then there will be hardly any working day left because most likely every day of the year will be somebody's birthday. We have to find the number of workers that will maximize the expected number of workdays each year. Let $N = 365$ be the number of days and n be the number of workers in a factory. The probability p that nobody has a birthday on a particular day is

$$p = \left(\frac{N - 1}{N}\right)^n$$

and this is true for any day of the year. Therefore, the expected number of working days each year is

$$Np = N\left(\frac{N-1}{N}\right)^n$$

giving a total of

$$nN\left(\frac{N-1}{N}\right)^n$$

workdays. We have to find $n = n_0$, which will give a maximum of the expected number of workdays, i.e.,

$$(n_0 + 1)N\left(1 - \frac{1}{N}\right)^{n_0+1} \leq n_0 N\left(1 - \frac{1}{N}\right)^{n_0} \geq (n_0 - 1)N\left(1 - \frac{1}{N}\right)^{n_0-1}.$$

After a simplification, we get

$$(n_0 + 1)\left(1 - \frac{1}{N}\right) \leq n_0 \quad \text{and} \quad (n_0 - 1) \leq n_0\left(1 - \frac{1}{N}\right).$$

Solving these inequalities, we easily obtain

$$n_0 \leq N \leq n_0 + 1.$$

We can see that for

$$n = n_0 = N \quad \text{we get} \quad N^2\left(1 - \frac{1}{N}\right)\left(1 - \frac{1}{N}\right)^{N-1} = N(N-1)\left(1 - \frac{1}{N}\right)^{N-1} \text{ workdays,}$$

$$n = n_0 = N - 1 \quad \text{we get} \quad N(N-1)\left(1 - \frac{1}{N}\right)^{N-1} \text{ workdays.}$$

Therefore, each factory should employ $n_0 = N - 1 = 364$ workers, giving on the average $N(N-1)\left(1 - \frac{1}{N}\right)^{N-1}$ workdays each year. If the workers would work every day of the year, there would be $N(N-1)$ workdays. Therefore, the employees will work on the average $\left(1 - \frac{1}{N}\right)^{N-1} \cdot 100\% = 37\%$ of the days of each year.

VARIANCE, COVARIANCE AND CORRELATION

In the previous section on expectations, we emphasized that the expected value $E(X)$ in some sense is a *measure of location* or *central tendency* of the probability distribution of the random variable X. However we would like to be able to find also a simple measure of *variability, dispersion* or *spread* of such a distribution.

EXAMPLE 20

Imagine a competition shooter who is using a perfect heavy bench rest stand and an electronic trigger mechanism. He does not even have to touch the rifle stock with his shoulder, and the gun is practically always perfectly vertically leveled on the target. The only variability of the point of impact in such an ideal situation is along a horizontal straight line. Consider the following hypothetical targets of two different shooters:

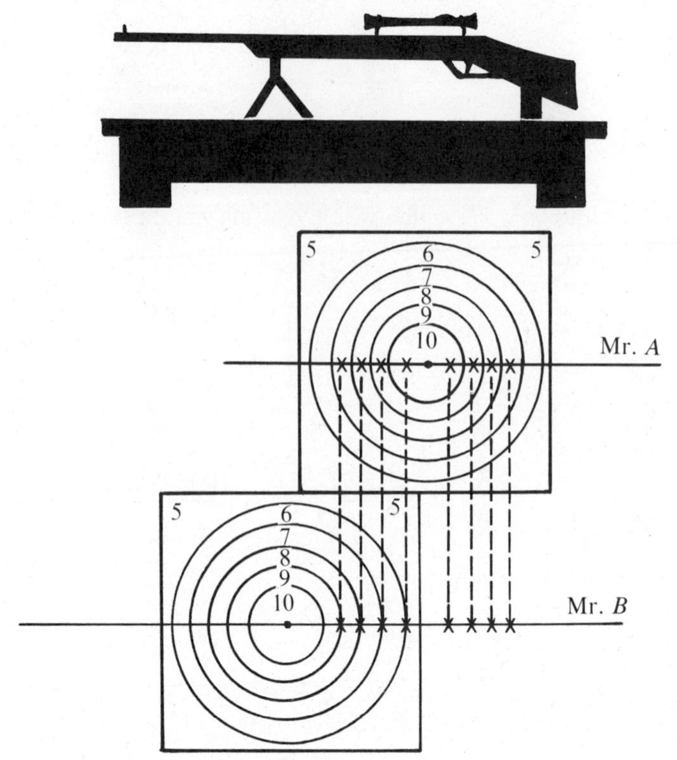

Let the r.v. X denote the score of Mr. A, and Y be the score of Mr. B on a particular shot. Obviously, both r.v.'s X and Y can attain values of 10, 9, 8, 7, 6, 5 and 0. The picture above shows several realizations of X and Y. Obviously X and Y are similarly distributed, however, they are *shifted* with respect to each other; they differ in their *location* or, in other words, $E(X) \neq E(Y)$. The expected point of impact $E(X)$ of Mr. A is the bull's eye. His gun is perfectly *centered* on the number 10, while the gun of Mr. B is misaligned. His expected point of impact $E(Y)$ is far to the right, due perhaps to a wrong windage (right-left) adjustment of his sights. However, the *spread* or *grouping* of both shooters is *equal.*

Consider now two other hypothetical targets:

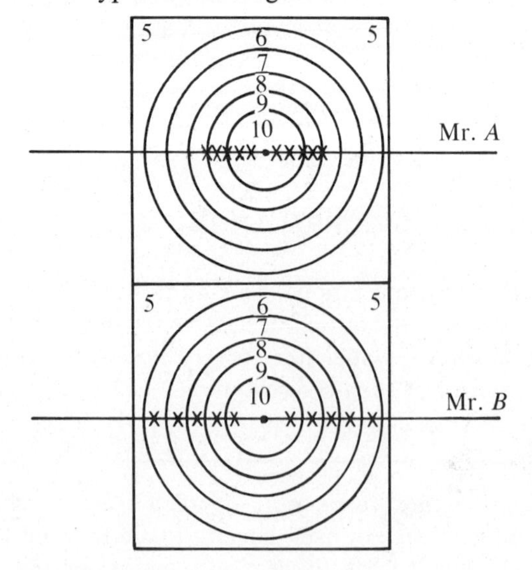

Obviously, both Mr. A and Mr. B have guns perfectly centered on the bull's eye; both expected points of impact are equal to 10, $E(X) = E(Y) = 10$. The main difference this time is in the *spread, variability* or *dispersion* of these two groups. Mr. A is obviously doing a much better job than Mr. B, even though both have their guns perfectly zeroed in on the bull's eye. In order to express the different degrees of dispersion, we should measure the *deviation* of the point of impact X from its expected value $E(X)$, which is equal to $(X - E(X))$.

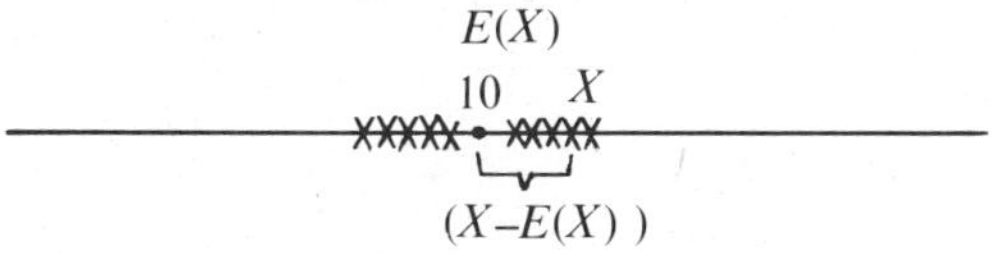

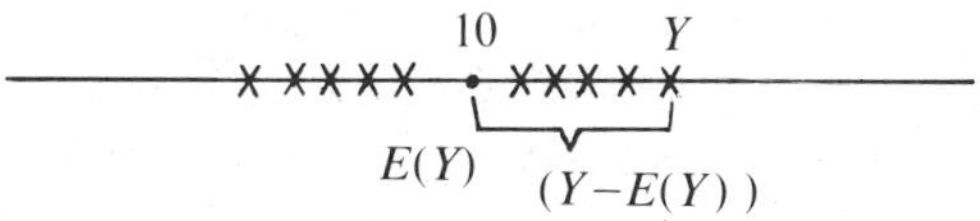

$E(X)$
10 X
XXXXX • XXXXX
$(X-E(X))$

10 Y
X XXXX • XXXX X
$E(Y)$ $(Y-E(Y))$

This deviation $(X - E(X))$ is again a r.v. and we should consider its *expectation* $E(X - E(X))$ as a reasonable measure of dispersion. Unfortunately this value is always *zero,* because (by Theorem 2)

$$E(X - E(X)) = E(X) - E(E(X)) = E(X) - E(X) = 0.$$

The problem is in the fact that values of X larger than $E(X)$ will give a positive deviation $(X - E(X))$, while the values of X smaller than $E(X)$ will give a negative deviation $(X - E(X))$, and these deviations will cancel each other. We could consider the *absolute values* of the deviations (regardless of a sign) $|X - E(X)|$ but it would be mathematically very awkward and ill-adapted for handling. In order to get rid of the signs of deviations, it is mathematically better to square them: the bigger the absolute value of the deviation $|X - E(X)|$ is, the bigger is its square $(X - E(X))^2$ and vice versa. Hence, it seems plausible and reasonable to consider the expected value of the r.v. $(X - E(X))^2$ as a measure of spread, dispersion or variability of X. In the above example, we may expect that Mr. B will have bigger deviations of his points of impact from the bull's eye than Mr. A, therefore

$$E[(X - E(X))^2] < E[(Y - E(Y))^2].$$

This inequality expresses the fact that the dispersion of the group of Mr. B is bigger than that of Mr. A. There are other possible measures of variability of a r.v. X than $E[(X - E(X))^2]$ but this measure, usually called the *variance,* is most frequently accepted and used.

DEFINITION 2: Let X be a r.v. Then the variance of X, denoted by $V(X)$, is defined as

$$V(X) = E[(X - E(X))^2].$$

Obviously $(X - E(X))^2 \geq 0$, hence $V(X) = E[(X - E(X))^2] \geq 0$. In calculating the variance, we square each deviation of X from $E(X)$. In order to compensate for squares we sometimes consider a measure of dispersion which is a square root of $V(X)$, and which is called the *standard deviation of X.*

DEFINITION 3: Let X be a r.v. with the variance $V(X)$. Then the standard deviation $\sigma(X)$ is given as

$$\sigma(X) = \sqrt{V(X)}.$$

→ **THEOREM 6:** Let X be a r.v. with the probability distribution $f(x_i)$, $i = 1,2, \ldots, k$. Then the variance $V(X)$ and the standard deviation $\sigma(X)$ of X are equal to

$$V(X) = \sum_{i=1}^{k} (x_i - E(X))^2 f(x_i).$$

$$\sigma(X) = \sqrt{\sum_{i=1}^{k} (x_i - E(X))^2 f(x_i)}.$$

Proof: In order to get the variance of X, we have to find the expected value of a function F of X, which is given as

$$F(X) = (X - E(X))^2.$$

According to Theorem 4, we have

$$V(X) = E(F(X)) = E[(X - E(X))^2] = \sum_{i=1}^{k} (x_i - E(X))^2 f(x_i)$$

and

$$\sigma(X) = \sqrt{V(X)} = \sqrt{\sum_{i=1}^{k} (x_i - E(X))^2 f(x_i)}.$$

q.e.d.

COMMENT. The formulae in Theorem 6 are useful for theoretical purposes, but they are often useless for actual numerical calculations. In such a case, it is better to use the formula in the following theorem.

→ **THEOREM 7:** For any r.v. X,

$$V(X) = E(X^2) - (E(X))^2.$$

Proof:
$$V(X) = \sum_{i=1}^{k} (x_i - E(X))^2 f(x_i) = \sum_{i=1}^{k} [x_i^2 - 2x_iE(X) + (E(X))^2] f(x_i)$$

$$= \sum_{i=1}^{k} x_i^2 f(x_i) - 2\sum_{i=1}^{k} E(X)x_i f(x_i) + (E(X))^2 \sum_{i=1}^{k} f(x_i) = E(X^2) - 2E(X)E(X) + (E(X))^2$$

$$= E(X^2) - (E(X))^2.$$

q.e.d.

EXAMPLE 21

Let a pair of fair dice be tossed, and let X denote the maximum of the numbers obtained. Then

$$f(1) = P\{(1,1)\} = \frac{1}{36}.$$

$$f(2) = P\{(1,2), (2,2), (2,1)\} = \frac{3}{36}.$$

$$f(3) = P\{(1,3), (2,3), (3,3), (3,2), (3,1)\} = \frac{5}{36}.$$

Similarly, $f(4) = \dfrac{7}{36}.$

$$f(5) = \frac{9}{36}.$$

$$f(6) = \frac{11}{36}.$$

$$E(X) = \sum_{i=1}^{k} x_i f(x_i) = 1 \cdot \frac{1}{36} + 2 \cdot \frac{3}{36} + 3 \cdot \frac{5}{36} + 4 \cdot \frac{7}{36} + 5 \cdot \frac{9}{36} + 6 \cdot \frac{11}{36} = \frac{161}{36}.$$

$$E(X^2) = 1^2 \frac{1}{36} + 2^2 \frac{3}{36} + 3^2 \frac{5}{36} + 4^2 \frac{7}{36} + 5^2 \frac{9}{36} + 6^2 \frac{11}{36} = \frac{791}{36}.$$

$$V(X) = \frac{791}{36} - \left(\frac{161}{36}\right)^2 = 1.9716. \qquad \sigma(X) = \sqrt{1.9716} = 1.4042.$$

THEOREM 8: For any r.v. X and any constant c:
(a) $V(X + c) = V(X)$.
(b) $V(cX) = c^2V(X)$.

Proof: (a) $V(X + c) = E[(X + c - E(X + c))^2] = E[(X + c - E(X) - c)^2] = E[(X - E(X))^2] = V(X)$.
(b) $V(cX) = E[(cX - E(cX))^2] = E[(cX - cE(X))^2] = E[c^2(X - E(X))^2] = c^2E[(X - E(X))^2] = c^2V(X)$.
q.e.d.

EXAMPLE 22

Let X be a r.v. and let a be a constant. Show that $E[(X - a)^2]$ is minimized by $a = E(X)$.

SOLUTION.

$$\begin{aligned} E[(X - a)^2] &= E[((X - E(X)) + (E(X) - a))^2] \\ &= E[(X - E(X))^2 + 2(X - E(X))(E(X) - a) + (E(X) - a)^2] \\ &= E[(X - E(X))^2] + E[2(X - E(X))(E(X) - a)] + E[(E(X) - a)^2 \\ &= V(X) + 2(E(X) - a)E[(X - E(X))] + (E(X) - a)^2 \\ &= V(X) + (E(X) - a)^2 \quad \text{because} \quad E[(X - E(X))] = 0. \end{aligned}$$

If $a \neq E(X)$, then $(E(X) - a)^2 > 0$, hence $E[(X - a)^2] \geq V(X)$. In other words, the expression $E[(X - a)^2]$ is minimized in the case of $a = E(X)$. That is, the expected value of the squared deviation of the r.v. X from any value a is at least as large as its variance.

EXAMPLE 23

Uniformly Distributed Random Variable. Let X be a number selected at random from among the numbers $\{1,2, \ldots ,n\}$. Find $E(X)$ and $V(X)$.

COMMENT. If $n = 6$, then this model represents tossing a fair die.

SOLUTION. The r.v. X attains the values $1,2, \ldots ,n$ with equal probabilities $\frac{1}{n}$, i.e.,

$$f(i) = \frac{1}{n} \qquad i = 1,2, \ldots ,n.$$

By definition,
$$E(X) = \sum_{i=1}^{n} i f(i) = \frac{1}{n} \sum_{i=1}^{n} i = \frac{n + 1}{2}.$$

$$E(X^2) = \sum_{i=1}^{n} i^2 f(i) = \frac{1}{n} \sum_{i=1}^{n} i^2 = \frac{1}{n} \frac{n(n + 1)(2n + 1)}{6} = \frac{(n + 1)(2n + 1)}{6}.$$

$$V(X) = E(X^2) - (E(X))^2 = \frac{(n+1)(2n+1)}{6} - \frac{(n+1)^2}{4} = \frac{n^2-1}{12}.$$

We used a known fact that $\sum_{i=1}^{n} i = \frac{n(n+1)}{2}$ and $\sum_{i=1}^{n} i^2 = \frac{n(n+1)(2n+1)}{6}$.

EXAMPLE 24

A manufacturer of paper clips would like to buy a new line of automatic filling machines that will put 100 clips into each box. There are two different types of machine on the market; both are equally expensive and equally suitable in their physical dimensions. It is well known that these machines fill 100 clips per box most of the time, but sometimes they make a small error by putting 98, 99, 101 or 102 clips into a box. Before making a decision the manufacturer requests a trial demonstration of each type, and estimates that the numbers of clips per box are distributed as follows:

Type A machine

No. of clips A	98	99	100	101	102
Probability	.02	.06	.85	.04	.03

Type B machine

No. of clips B	98	99	100	101	102
Probability	.01	.07	.84	.07	.01

Which type of machine should he buy?

SOLUTION. Let us calculate first the expected values of the number of clips per box.

$$E(A) = 98 \cdot (.02) + 99 \cdot (.06) + 100 \cdot (.85) + 101 \cdot (.04) + 102 \cdot (.03) = 100.$$
$$E(B) = 98 \cdot (.01) + 99 \cdot (.07) + 100 \cdot (.84) + 101 \cdot (.07) + 102 \cdot (.01) = 100.$$

Both types of machine thus in a long run fill on the average 100 clips per box and satisfy the requirement equally well. Let us try to evaluate the fluctuations in the number of clips per box. Using the definition of variance, we get for the variances of A and B the following values:

$$V(A) = 2^2 \cdot (.02) + 1^2 \cdot (.06) + 0^2 \cdot (.85) + 1^2 \cdot (.04) + 2^2 \cdot (.03) = .30.$$
$$V(B) = 2^2 \cdot (.01) + 1^2 \cdot (.07) + 0^2 \cdot (.84) + 1^2 \cdot (.07) + 2^2 \cdot (.01) = .22.$$

Therefore, type B machines have a smaller variability in the number of clips per box, which will result in smaller fluctuations in the contents of boxes. If the manufacturer is concerned about filling the boxes as accurately as possible, then he should buy type B machines.

The full meaning and usefulness of the notion of variance will appear only gradually, particularly in connection with limit theorems in Chapters XIII and XVIII. We should like to give now a measure of the joint behavior of two random variables.

DEFINITION 4: Let X and Y be r.v.'s with the expected values $E(X)$ and $E(Y)$. The covariance of X and Y, denoted by $COV(X,Y)$, is defined as

$$COV(X,Y) = E[(X - E(X))(Y - E(Y))].$$

THEOREM 9: Let X and Y be r.v.'s with the joint probability distribution $h(x_i, y_j)$, $i = 1,2, \ldots, k$; $j = 1,2, \ldots, l$. Then the covariance of X and Y is equal to

$$COV(X,Y) = \sum_{i=1}^{k} \sum_{j=1}^{l} (x_i - E(X))(y_j - E(Y))h(x_i,y_j).$$

Proof: In order to get the covariance of X and Y, we have to find the expected value of a function G of X and Y given as

$$G(X,Y) = (X - E(X))(Y - E(Y)).$$

According to Theorem 5, we have

$$COV(X,Y) = E(G(X,Y)) = E[(X - E(X))(Y - E(Y))] = \sum_{i=1}^{k} \sum_{j=1}^{l} (x_i - E(X))(y_j - E(Y))h(x_i,y_j). \quad \text{q.e.d.}$$

COMMENT 1. The covariance of X and Y is a generalization of the notion of variance. Let $X = Y$, then

$$COV(X,X) = E[(X - E(X))(X - E(X))] = E[(X - E(X))^2] = V(X).$$

COMMENT 2. In the case of the variance, the definition itself was inappropriate for numerical calculations, therefore we derived Theorem 7. Similarly, in the case of the covariance the actual numerical evaluation is usually done using the following:

THEOREM 10: For any two random variables X and Y,

$$COV(X,Y) = E(XY) - E(X)E(Y).$$

Proof:
$$\begin{aligned} COV(X,Y) &= E[(X - E(X))(Y - E(Y))] = E[XY - XE(Y) - YE(X) + E(X)E(Y)] \\ &= E(XY) - E[XE(Y)] - E[YE(X)] + E[E(X)E(Y)] \\ &= E(XY) - E(Y)E(X) - E(X)E(Y) + E(X)E(Y) = E(XY) - E(X)E(Y). \end{aligned} \quad \text{q.e.d.}$$

COMMENT. The formula for the covariance given above corresponds in the case of $X = Y$ to the formula of Theorem 7.

EXAMPLE 25

A fair coin is tossed three times. Let Y denote the number of heads, and X be 0 or 1 according to whether a head or a tail occurs in the first toss. Then the joint distribution of X and Y is as follows:

X \ Y	0	1	2	3	
0	0	$\frac{1}{8}$	$\frac{2}{8}$	$\frac{1}{8}$	$\frac{1}{2}$
1	$\frac{1}{8}$	$\frac{2}{8}$	$\frac{1}{8}$	0	$\frac{1}{2}$
	$\frac{1}{8}$	$\frac{3}{8}$	$\frac{3}{8}$	$\frac{1}{8}$	

$$E(Y) = 0 \cdot \frac{1}{8} + 1 \cdot \frac{3}{8} + 2 \cdot \frac{3}{8} + 3 \cdot \frac{1}{8} = \frac{3}{2}.$$

$$E(X) = 0 \cdot \frac{1}{2} + 1 \cdot \frac{1}{2} = \frac{1}{2}.$$

The distribution of the product is

$z = xy$	0	1	2	3
$f(z)$	$\frac{5}{8}$	$\frac{2}{8}$	$\frac{1}{8}$	0

Hence, $$E(XY) = 0 \cdot \frac{5}{8} + 1 \cdot \frac{2}{8} + 2 \cdot \frac{1}{8} + 3 \cdot 0 = \frac{1}{2}.$$

Finally, $$COV(X,Y) = E(XY) - E(X)E(Y) = \frac{1}{2} - \frac{1}{2} \cdot \frac{3}{2} = -\frac{1}{4}.$$

→ **THEOREM 11:** For any two random variables X and Y,

$$V(X + Y) = V(X) + V(Y) + 2COV(X,Y).$$

Proof: $$\begin{aligned} V(X + Y) &= E[(X + Y - E(X + Y))^2] = E[((X - E(X)) + (Y - E(Y)))^2] \\ &= E[(X - E(X))^2 + (Y - E(Y))^2 + 2(X - E(X))(Y - E(Y))] \\ &= E[(X - E(X))^2] + E[(Y - E(Y))^2] + 2E[(X - E(X))(Y - E(Y))] \\ &= V(X) + V(Y) + 2COV(X,Y). \end{aligned}$$ q.e.d.

→ **THEOREM 12:** If X and Y are independent random variables, then
(a) $COV(X,Y) = 0$.
(b) $V(X + Y) = V(X) + V(Y)$.

Proof: (a) If X and Y are independent, then $E(XY) = E(X)E(Y)$ by Theorem 5. Hence by Theorem 9,

$$COV(X,Y) = E(XY) - E(X)E(Y) = E(X)E(Y) - E(X)E(Y) = 0.$$

(b) By Theorem 10, we have

$$V(X + Y) = V(X) + V(Y) + 2COV(X,Y) = V(X) + V(Y).$$ q.e.d.

EXAMPLE 26

Consider the r.v.'s X and Y from Example 25.

Then $$E(X^2) = 0^2 \cdot \frac{1}{2} + 1^2 \cdot \frac{1}{2} = \frac{1}{2}.$$

$$V(X) = E(X^2) - (E(X))^2 = \frac{1}{2} - \left(\frac{1}{2}\right)^2 = \frac{1}{4}.$$

$$E(Y^2) = 0^2 \cdot \frac{1}{8} + 1^2 \cdot \frac{3}{8} + 2^2 \cdot \frac{3}{8} + 3^2 \cdot \frac{1}{8} = \frac{24}{8} = 3.$$

$$V(Y) = E(Y^2) - (E(Y))^2 = 3 - \left(\frac{3}{2}\right)^2 = \frac{3}{4}.$$

$$COV(X,Y) = -\frac{1}{4}.$$

Let us find the distribution of the r.v. $Z = X + Y$.

z_i	1	2	3
$g(z_i)$	$\frac{2}{8}$	$\frac{4}{8}$	$\frac{2}{8}$

$$E(Z) = E(X + Y) = 1 \cdot \frac{2}{8} + 2 \cdot \frac{4}{8} + 3 \cdot \frac{2}{8} = \frac{16}{8} = 2.$$

$$E(Z^2) = E[(X + Y)^2] = 1^2 \cdot \frac{2}{8} + 2^2 \cdot \frac{4}{8} + 3^2 \cdot \frac{2}{8} = \frac{36}{8} = \frac{9}{2}.$$

$$V(X + Y) = E[(X + Y)^2] - (E(X + Y))^2 = \frac{9}{2} - 2^2 = \frac{1}{2}.$$

However, $$V(X) + V(Y) = \frac{1}{4} + \frac{3}{4} = 1 \neq \frac{1}{2} = V(X + Y).$$

Further, $$V(X) + V(Y) + 2COV(X,Y) = \frac{1}{4} + \frac{3}{4} + 2\left(-\frac{1}{4}\right) = \frac{1}{2} = V(X + Y).$$

WARNING: If X and Y are dependent r.v.'s, then the relation

$$V(X + Y) = V(X) + V(Y)$$

may be but does not necessarily have to be true. However, if X and Y are independent, then it is always true.

COMMENT. Using a similar method as in Theorems 11 and 12, we can easily prove that for any n r.v.'s $X_1, X_1, \ldots, X_n$:

$$V\left(\sum_{i=1}^{n} X_i\right) = \sum_{i=1}^{n} V(X_i) + \sum_{i=1}^{n} \sum_{\substack{j=1 \\ i \neq j}}^{n} COV(X_i, X_j).$$

Hence, if the X_i's are independent, then

$$V\left(\sum_{i=1}^{n} X_i\right) = \sum_{i=1}^{n} V(X_i).$$

EXAMPLE 27

Sample Mean. Let X be a r.v. on a probability measure space (Ω, P) – imagine, for example, drawing k balls from an urn containing m white and n black balls, and X is the number of white balls obtained. Perform the corresponding experiment once, and let X_1 be the value of the r.v. which is observed, that is, let X_1 be the first realization of X. Then perform the experiment *independently* a second time and let X_2 be the value of the observed r.v., that is, let X_2 be the second independent realization of X. Now repeat this experiment independently n times till you get *n independent and identically distributed r.v.'s* $X_1, X_2, \ldots, X_n$. These r.v.'s are called an *independent random sample* of the r.v. X. Obviously, all the r.v.'s X_i have the same distribution as X, hence

$$E(X_i) = E(X), \qquad V(X_i) = V(X) \qquad i = 1,2, \ldots n.$$

The *sample mean* $\overline{X}$ is defined as the arithmetic mean of $X_1, X_2, \ldots, X_n$, that is,

$$\overline{X} = \frac{1}{n}(X_1 + X_2 + \ldots + X_n) = \frac{1}{n}\sum_{i=1}^{n} X_i.$$

The sample mean is a function of the r.v.'s X_i and it is therefore a r.v. We may ask, what is the expectation of the sample mean and what is its variance equal to? Using the comment after Theorem 2, we get

$$E(\overline{X}) = E\left(\frac{1}{n}\sum_{i=1}^{n} X_i\right) = \frac{1}{n}E\left(\sum_{i=1}^{n} X_i\right) = \frac{1}{n}\sum_{i=1}^{n} E(X_i) = \frac{1}{n}\sum_{i=1}^{n} E(X) = \frac{1}{n}\, n\, E(X) = E(X).$$

Therefore, we can conclude that the *expected value of the sample mean is equal to the mean of the original distribution.* Furthermore, using the properties of the variance and the comment after Theorem 12,

$$V(\overline{X}) = V\left(\frac{1}{n}\sum_{i=1}^{m} X_i\right) = \frac{1}{n^2}V\left(\sum_{i=1}^{n} X_i\right) = \frac{1}{n^2}\sum_{i=1}^{n} V(X_i) = \frac{1}{n^2}\sum_{i=1}^{n} V(X) = \frac{1}{n^2}\, n\, V(X) = \frac{V(X)}{n}.$$

We can conclude that the *variance of the sample mean is equal to the variance of the original distribution divided by the sample size n.*

This result, even though it might be surprising to some people, is actually very plausible. Imagine that X is the height in inches of a randomly selected student. This r.v. has a certain variability—one may get by chance a 5′1″ girl studying arts as well as a gigantic 6′8″ phys. ed. student. However, if one selects an independent random sample of 100 students and calculates its mean, then the sample mean certainly will vary from sample to sample, but this variability will be much smaller; it will never reach extremely high or low values. And the variability of the means from samples of size 1000 or 10,000 will be practically negligible.

DEFINITION 5: For any two random variables X and Y such that $V(X) \neq 0$, $V(Y) \neq 0$, we define the correlation coefficient $R(X,Y)$ as follows:

$$R(X,Y) = \frac{\text{COV}(X,Y)}{\sqrt{V(X)\,V(Y)}}$$

→ **THEOREM 13:** For any two random variables X and Y,

$$-1 \leq R(X,Y) \leq 1.$$

The equality $R(X,Y) = \pm 1$ holds if and only if there are constants a and $b \neq 0$ such that

$$Y(\omega) = a + bX(\omega)$$

for all points $\omega \in \Omega$ with the exception of at most a null set.

Proof: For every real number λ, we have

$$\begin{aligned} 0 \leq E[((Y - E(Y)) + \lambda COV(X,Y)(X - E(X)))^2] &= E[(Y - E(Y))^2] \\ &\quad + 2\lambda\, COV(X,Y)E[(Y - E(Y))(X - E(X))] \\ &\quad + \lambda^2[COV(X,Y)]^2 E[(X - E(X))^2] \\ &= V(Y) + 2\lambda[COV(X,Y)]^2 + \lambda^2[COV(X,Y)]^2\, V(X) \\ &= c\lambda^2 + d\lambda + e \quad \text{where} \quad c = [COV(X,Y)]^2 V(X), \\ d = 2[COV(X,Y)]^2,\quad e = V(Y). \end{aligned}$$

The quadratic polynomial in λ cannot have two distinct real roots because then for some λ_0, we would have

$$c\lambda_0^2 + d\lambda_0 + e < 0$$

and this is impossible. Hence the discriminant must be nonpositive,

$$d^2 - 4ce \leq 0.$$

This means that $4[COV(X,Y)]^4 - 4[COV(X,Y)]^2 V(X)V(Y) \leq 0$

or $[COV(X,Y)]^2 \leq V(X)V(Y)$

hence
$$[R(X,Y)]^2 = \frac{[COV(X,Y)]^2}{V(X)V(Y)} \leq 1.$$

Thus
$$-1 \leq R(X,Y) \leq +1.$$

The equality $R(X,Y) = \pm 1$ holds if and only if the discriminant is 0. This is equivalent to the condition that there is a double real root λ_0 of the quadratic polynomial

$$c\lambda_0^2 + d\lambda_0 + e = 0$$

or, equivalently,

$$E[((Y - E(Y)) + \lambda_0 COV(X,Y)(X - E(X)))^2] = 0.$$

But the r.v. inside the brackets [ ] is nonnegative, and Theorem 1 implies that

$$[Y(\omega) - E(Y)) + \lambda_0 COV(X,Y)(X(\omega) - E(X))] = 0$$

for every $\omega \in \Omega$ for which $P\{\omega\} > 0$. Therefore $R(X,Y) = \pm 1$ if and only if there is λ_0 such that

$$Y(\omega) = \lambda_0 COV(X,Y)(X(\omega) - E(X)) + E(Y)$$

for all $\omega \in \Omega$, with the exception of at most a null set. Taking

$$a = -\lambda_0 COV(X,Y)E(X) + E(Y) \quad \text{and} \quad b = \lambda_0 COV(X,Y),$$

we get the condition which was to be proved. q.e.d.

THEOREM 14: (*Schwarz-Buniakowski Inequality*): For any two random variables X,Y,

$$|COV(X,Y)| \leq \sqrt{V(X)V(Y)}.$$

Proof: From Theorem 13, we have $|R(X,Y)| = \dfrac{|COV(X,Y)|}{\sqrt{V(X)V(Y)}} \leq 1$ hence $|COV(X,Y)| \leq \sqrt{V(X)V(Y)}$.

q.e.d.

COMMENT 1. Inequalities of this kind are very important in pure mathematics.

COMMENT 2. The correlation coefficient $R(X,Y)$ is very closely related to the concept of independence of the r.v.'s X and Y. If X and Y are independent, then $COV(X,Y) = 0$, hence $R(X,Y) = 0$. On the other hand, if $R(X,Y) = \pm 1$, then one r.v. is a function of the other, which can be considered as an opposite of independence. In this sense, the *correlation coefficient measures the level of interdependence of two r.v.'s.* We could develop a whole theory of correlation and interdependence between random variables using $R(X,Y)$, $COV(X,Y)$, $V(X)$ and $V(Y)$, but this is somewhat out of the range of this text.

EXAMPLE 28

Let us put three identical balls into three numbered cells so that all $3^3 = 27$ distributions are equally likely. Let X be the number of empty cells and Y be the number of balls in the first cell. Find $R(X,Y)$.

SOLUTION. The joint probability distribution is:

$X \backslash Y$	0	1	2	3	f
0	0	$\frac{6}{27}$	0	0	$\frac{6}{27}$
1	$\frac{6}{27}$	$\frac{6}{27}$	$\frac{6}{27}$	0	$\frac{18}{27}$
2	$\frac{2}{27}$	0	0	$\frac{1}{27}$	$\frac{3}{27}$
g	$\frac{8}{27}$	$\frac{12}{27}$	$\frac{6}{27}$	$\frac{1}{27}$	1

Obviously X and Y are dependent. Further, $E(X) = \frac{8}{9}$; $E(Y) = 1$; $E(XY) = \frac{8}{9}$. Therefore,

$$COV(X,Y) = E(XY) - E(X)E(Y) = \frac{8}{9} - \frac{8}{9} \cdot 1 = 0; \quad R(X,Y) = \frac{COV(X,Y)}{\sqrt{V(X)V(Y)}} = 0.$$

In this example, X and Y are dependent but $R(X,Y) = 0$. We usually call such r.v.'s *uncorrelated* r.v.'s.

EXAMPLE 29

Let us toss a coin three times and let X be the number of heads and Y the number of tails. Find $R(X,Y)$.

SOLUTION. According to the second part of Theorem 13, the correlation coefficient $R(X,Y)$ must be either 1 or -1 because

$$Y = 3 - X.$$

However, as a check let us calculate it in full. The joint distribution of X and Y is:

$X \backslash Y$	0	1	2	3	f
0	0	0	0	$\frac{1}{8}$	$\frac{1}{8}$
1	0	0	$\frac{3}{8}$	0	$\frac{3}{8}$
2	0	$\frac{3}{8}$	0	0	$\frac{3}{8}$
3	$\frac{1}{8}$	0	0	0	$\frac{1}{8}$
g	$\frac{1}{8}$	$\frac{3}{8}$	$\frac{3}{8}$	$\frac{1}{8}$	1

$$E(X) = E(Y) = 1.5; \quad E(X^2) = 0^2 + \frac{1}{8} + 1^2\frac{3}{8} + 2^2\frac{3}{8} + 3^2\frac{1}{8} = \frac{24}{8} = 3.$$

$$V(X) = V(Y) = 3 - (1.5)^2 = .75.$$

$$E(XY) = 0 \cdot 3 \cdot \frac{1}{8} + 1 \cdot 2 \cdot \frac{3}{8} + 2 \cdot 1 \cdot \frac{3}{8} + 3 \cdot 0 \cdot \frac{1}{8} = \frac{12}{8} = 1.5.$$

$$COV(X,Y) = E(XY) - E(X)E(Y) = 1.5 - (1.5)^2 = -.75; \quad R(X,Y) = \frac{-.75}{\sqrt{(.75)^2}} = -1.$$

EXPECTED VALUE AND VARIANCE OF THE BINOMIAL DISTRIBUTION

The binomial random variable X attains values $0,1, \ldots ,n$ with probabilities

$$P(\{X = i\}) = \binom{n}{i} p^i q^{n-i} \quad i = 0,1, \ldots ,n,$$

where n is an integer and $p + q = 1, p \geq 0, q \geq 0$.

Therefore,
$$E(X) = \sum_{i=0}^{n} i \binom{n}{i} p^i q^{n-i} = \sum_{i=1}^{n} i \binom{n}{i} p^i q^{n-i}.$$

But
$$i\binom{n}{i} = \frac{in!}{i!(n-i)!} = \frac{n(n-1)!}{(i-1)![(n-1)-(i-1)]!} = n\binom{n-1}{i-1}.$$

Therefore,
$$E(X) = \sum_{i=1}^{n} n \binom{n-1}{i-1} p^i q^{n-i} = np \sum_{i=1}^{n} \binom{n-1}{i-1} p^{i-1} q^{(n-1)-(i-1)}.$$

Substituting $j = i - 1$, the limits of summation will be from $j = 0$ to $j = n - 1$; hence

$$E(X) = np \sum_{j=0}^{n-1} \binom{n-1}{j} p^j q^{(n-1)-j} = np(p + q)^{n-1} = np.$$

In order to get $V(X)$, let us first find $E(X^2)$.

$$E(X^2) = \sum_{i=0}^{n} i^2 \binom{n}{i} p^i q^{n-i} = \sum_{i=0}^{n} (i(i-1)+i) \binom{n}{i} p^i q^{n-i}$$

$$= \sum_{i=2}^{n} i(i-1) \binom{n}{i} p^i q^{n-i} + \sum_{i=1}^{n} i \binom{n}{i} p^i q^{n-i}.$$

As shown above, the second term is $E(X) = np$. As for the first one,

$$i(i-1)\binom{n}{i} = \frac{i(i-1)n!}{i!(n-i)!} = \frac{n(n-1)(n-2)!}{(i-2)![(n-2)-(i-2)]!} = n(n-1)\binom{n-2}{i-2}.$$

Hence
$$E(X^2) = \sum_{i=2}^{n} n(n-1)\binom{n-2}{i-2} p^i q^{n-i} + np$$

$$= n(n-1)p^2 \sum_{i=2}^{n} \binom{n-2}{i-2} p^{i-2} q^{(n-2)-(i-2)} + np.$$

Substituting $j = i - 2$, the limits of summation will be from $j = 0$ to $j = n - 2$, hence

$$E(X^2) = n(n-1)p^2 \sum_{j=0}^{n-2} \binom{n-2}{j} p^j q^{(n-2)-j} + np$$

$$= n(n-1)p^2(p+q)^{n-2} + np = n(n-1)p^2 + np$$

$$= n^2p^2 + npq.$$

Finally,
$$V(X) = E(X^2) - (E(X))^2 = n^2p^2 + npq - (np)^2 = npq.$$

This proves the following:

→ **THEOREM 15:** For the binomial r.v. X with parameters n and p, we have

$$E(X) = np \qquad V(X) = npq \qquad \sigma(X) = \sqrt{npq}$$

EXAMPLE 30

The probability of an item being defective is .02. A shipment of 10,000 items is sent to a warehouse. Find the expected number of defective items and the standard deviation.

SOLUTION. Let X denote the number of defectives. Assuming that the quality of one item does not influence the quality of another item, we can conclude that X is a binomial r.v.

$$E(X) = np = 10{,}000 \times (.02) = 200 \text{ items.}$$

$$\sigma(X) = \sqrt{npq} = \sqrt{10{,}000 \times (.02) \times (.98)} = \sqrt{196} = 14 \text{ items.}$$

We shall show in Example 31 that we may reasonably expect that the number of defective items will be within the interval $(200 - 3 \cdot 14)$ and $(200 + 3 \cdot 14)$, that is, within the interval 158 and 242.

The calculations of the expected value and variance of the binomial r.v. preceding Theorem 15 were rather tedious and cumbersome. We can get the same answer in a much more comfortable way. The binomial r.v. X means the number of successes in n independent repetitions of an experiment with the alternative outcomes a success (with the probability p) and a failure (with the probability $q = 1 - p$). Let the r.v. X_i, $i = 1, 2, \ldots, n$ be defined as follows:

$$X_i = 0 \text{ if the outcome of the } i\text{-th trial is a failure.}$$
$$= 1 \text{ if the outcome of the } i\text{-th trial is a success.}$$

The random variable X_i is called the *alternative* r.v., and we can see that $X = \sum_{i=1}^{n} X_i$. Therefore,

$$E(X) = E\left(\sum_{i=1}^{n} X_i\right) = \sum_{i=1}^{n} E(X_i).$$

Further, any two r.v.'s X_i and X_j, $i \neq j$, are independent, hence (using the comments after Theorem 12), we have

$$V(X) = V\left(\sum_{i=1}^{n} X_i\right) = \sum_{i=1}^{n} V(X_i).$$

But $\quad P(\{X_i = 0\}) = q \qquad P(\{X_i = 1\}) = p.$

Thus $\quad E(X_i) = 0 \cdot q + 1 \cdot p = p \qquad E(X_i^2) = 0^2 \cdot q + 1^2 \cdot p = p$

$$V(X_i) = p - p^2 = p(1 - p) = pq.$$

Hence $\quad E(X) = \sum_{i=1}^{n} E(X_i) = np, \qquad V(X) = \sum_{i=1}^{n} V(X_i) = npq.$

As a matter of fact, there is still another nice and fast way to calculate $E(X)$. However, it requires calculus, and readers who are not familiar with differentiation should skip the next several lines. Let p and q be two arbitrary real numbers. Using the Binomial Theorem, we have

$$(p + q)^n = \sum_{i=0}^{n} \binom{n}{i} p^i q^{n-i}.$$

Imagine that q is a constant and p is a variable. Then, differentiating both sides of the above equality with respect to p, we get

$$\frac{d}{dp}(p + q)^n = n(p + q)^{n-1} = \sum_{i=0}^{n} \binom{n}{i} i p^{i-1} q^{n-i}.$$

However, if we set now $q = 1 - p$, then

$$\sum_{i=0}^{n} i \binom{n}{i} p^{i-1} q^{n-i} = n.$$

Let us multiply both sides by p, then obviously,

$$E(X) = \sum_{i=0}^{n} i \binom{n}{i} p^i q^{n-i} = np.$$

We can calculate $V(X)$ similarly.

EXAMPLE 31

Let us toss a fair coin n times and let X denote the number of heads. Calculate the probability $P(\{E(X) - 3\sigma(X) \le X \le E(X) + 3\sigma(X)\})$ for different values of $n = 10, 20, 30$, etc.

SOLUTION. The probability of getting a head on a single toss is $\frac{1}{2}$ and therefore X is a binomial r.v. with

$$E(X) = np = \frac{n}{2} \qquad \sigma(X) = \sqrt{npq} = \frac{\sqrt{n}}{2}.$$

For $n = 10$, we have $E(X) = 5 \qquad \sigma(X) = \frac{\sqrt{10}}{2} = 1.58.$

$$E(X) - 3\sigma(X) = 5 - 4.74 = .026.$$

$$E(X) + 3\sigma(X) = 5 + 4.74 = 9.74.$$

$$P(\{0.26 \le X \le 9.74\}) = P\{1 \le X \le 9\} = \sum_{i=1}^{9} \binom{10}{i} \left(\frac{1}{2}\right)^i \left(\frac{1}{2}\right)^{10-i} = .9980.$$

For $n = 20$, $E(X) = 10 \qquad \sigma(X) = \frac{\sqrt{20}}{2} = 2.24.$

$$E(X) \pm 3\sigma(X) = 10 \pm 6.72 = \begin{cases} 16.72 \\ 3.28 \end{cases}.$$

$$P(\{3.28 \le X \le 16.72\}) = P(\{4 \le X \le 16\}) = \sum_{i=4}^{16} \binom{20}{i} \left(\frac{1}{2}\right)^i \left(\frac{1}{2}\right)^{20-i} = .9974.$$

Similarly for $n = 30$, $E(X) = 15 \quad \sigma(X) = 2.74$

$$P(\{7 \le X \le 23\}) = .9986.$$

Let us denote the smallest integer that is bigger than or equal to $[E(X) - 3\sigma(X)]$ by LB (for lower bound) and the largest integer that is smaller or equal to $[E(X) + 3\sigma(X)]$ by UB (for upper bound). Then

$$P(\{E(X) - 3\sigma(X) \le X \le E(X) + 3\sigma(X)\}) = P(\{LB \le X \le UB\}).$$

We can summarize our calculations in the following table. The first column contains the number of tosses. The second contains the lower bound and the third contains the upper bound, as defined above. The fourth column contains the probability that the binomial r.v. X will be within these bounds. If we toss the coin n times, we can get $(n + 1)$ possible outcomes, i.e., the number of heads obtained can be 0,1,2, . . . ,n. Out of these values, only a certain percentage falls within the limits $LB \leq X \leq UB$. This percentage is given in the last column. (For example, for $n = 20$ we have 21 possible values, while we have 13 in the interval 4 to 16, the percentage being $\frac{100}{21} \cdot 13 = 62\%$.)

n	Lower Bound	Upper Bound	$P(\{LB \leq X \leq UB\})$	Percentage of Values $OB \leq X \leq UB$
10	1	9	.9980	82%
20	4	16	.9974	62%
30	7	23	.9986	55%
40	11	29	.9978	46%
50	15	35	.9974	41%
70	23	47	.9974	35%
100	35	65	.9982	31%
150	57	93	.9976	24%
300	125	175	.9968	17%

We can see that with a probability of more than 99%, the binomial random variable X lies within the interval $(E(X) - 3\sigma(X))$ and $(E(X) + 3\sigma(X))$. There are many possible values of X outside of this interval (from $n = 40$, it is a majority of values), however their probability taken together is always less than 1%. This fact is true not only in the case of the binomial r.v. but in many other cases as well. The interval $(E(X) \pm 3\sigma(X))$ gives practical bounds on the values of the r.v. X that have a reasonable chance to appear. Any value outside of this interval is *practically impossible* to appear. This clearly shows the tremendous importance of the expected value $E(X)$, and of the standard deviation $\sigma(X)$. Incidentally, the calculation of the probabilities was very tedious in this example, and we shall show an easy way to do it in the last chapter.

EXPECTED VALUE AND VARIANCE OF THE HYPERGEOMETRIC DISTRIBUTION

In Chapter XI, we defined for any integers $n \leq N$ and $k \leq N$ the hypergeometric r.v. X by the condition

$$P(\{X = r\}) = \frac{\binom{k}{r}\binom{N-k}{n-r}}{\binom{N}{n}} \quad \text{for } 0 \leq r \leq k, r \leq n, r \geq n + k - N.$$

We can calculate the expected value $E(X)$ directly from the definition using the probability distribution of X, but this method is very tedious and cumbersome. Let us return to the model for the hypergeometric r.v. We have an urn containing k black balls and $(N - k)$ white balls, and we draw n balls at random. The r.v. X means the number of black balls obtained. We can imagine that we successively draw one ball after another without replacement. Let $X_i = 1$ if the i-th draw yields a black ball and $X_i = 0$ otherwise. The r.v.'s X_i, $i = 1,2, \ldots ,n$ are *not independent*. We know that

$$X = \sum_{i=1}^{n} X_i$$

hence $$E(X) = \sum_{i=1}^{n} E(X_i), \quad V(X) = \sum_{i=1}^{n} V(X_i) + \sum_{\substack{i=1 \\ i \neq j}}^{n} \sum_{j=1}^{n} COV(X_i, X_j)$$

as follows from the comments after Theorem 2 and Theorem 12. Let us investigate the probability $P(\{X_i = 1\})$. In Chapter XI, we found that the sample space has $\frac{N!}{(N-n)!}$ equally likely points. The r.v. $X_i(\omega) = 1$ if and only if we get a black ball in the i-th trial (there are k ways), and the outcomes of the remaining $(n - 1)$ trials are arbitrarily selected out of the remaining $(N - 1)$ balls. (Sample points are ordered n-tuples and hence there are $\frac{(N-1)!}{(N-1-(n-1))!}$ ways.) Hence,

$$P(\{\omega: X_i(\omega) = 1\}) = \frac{k\dfrac{(N-1)!}{(N-n)!}}{\dfrac{N}{(N-n)!}} = \frac{k}{N}$$

Obviously, $$P(\{\omega: X_i(\omega) = 0\}) = 1 - \frac{k}{N}.$$

Further, we need the probability $P(\{\omega X_i(\omega) = 1 \text{ and } X_j(\omega) = 1\})$. We can draw a black ball on the i-th draw in k ways, another black ball on the j-th draw in $(k - 1)$ ways and the remaining $(n - 2)$ balls in $\frac{(N-2)!}{(N-2-(n-2))!}$ ways.

Hence $$P(\{\omega: X_i(\omega) = 1 \text{ and } X_j(\omega) = 1\}) = \frac{k(k-1)\dfrac{(N-2)!}{(N-n)!}}{\dfrac{N!}{(N-n)!}} = \frac{k(k-1)}{N(N-1)}.$$

Let us calculate the expected values, variances and covariances.

$$E(X_i) = 0 \cdot P(\{\omega: X_i(\omega) = 0\}) + 1 \cdot P(\{\omega: X_i(\omega) = 1\}) = \frac{k}{N}.$$

Similarly, $E(X_i^2) = \frac{k}{N}.$

$$V(X_i) = E(X_i^2) - (E(X_i))^2 = \frac{k}{N} - \frac{k^2}{N^2} = \frac{k}{N}\left(1 - \frac{k}{N}\right).$$

$$E(X_iX_j) = 1 \cdot P(\{\omega: X_i(\omega) = 1 \quad \text{and} \quad X_j(\omega) = 1\}) = \frac{k(k-1)}{N(N-1)}.$$

Therefore, $$COV(X_iX_j) = E(X_iX_j) - E(X_i)E(X_j) = \frac{k(k-1)}{N(N-1)} - \frac{k^2}{N^2} = -\frac{k(N-k)}{N^2(N-1)}.$$

Therefore, $$E(X) = \sum_{i=1}^{n} E(X_i) = \frac{nk}{N}.$$

Further, $\sum_{i=1}^{n} V(X_i) = n\frac{k(N-k)}{N^2}$, $\sum_{\substack{i=1\\i\neq j}}^{n}\sum_{j=1}^{n} COV(X_i,X_j) = -n(n-1)\frac{k(N-k)}{N^2(N-1)}$.

Finally, $V(X) = \sum_{i=1}^{n} V(X_i) + \sum_{\substack{i=1\\i\neq j}}^{n}\sum_{j=1}^{n} COV(X_i,X_j) = \frac{nk(N-k)(N-n)}{N^2(N-1)}$.

THEOREM 16: For the hypergeometric r.v. X, we have

$$E(X) = \frac{nk}{N}, \qquad V(X) = \frac{nk(N-k)(N-n)}{N^2(N-1)}.$$

EXAMPLE 32

A shipment of 30 semiconductors contains 8 defectives. Six semiconductors are selected at random for testing. Let X be the number of discovered defectives. Find $E(X)$ and $\sigma(X)$.

SOLUTION. $E(X) = \frac{6\cdot 8}{30} = 1.6 \qquad \sigma(X) = \sqrt{\frac{6\cdot 8\cdot(30-8)(30-6)}{30^2(30-1)}} = .99$

EXPECTED VALUE AND VARIANCE OF THE NUMBER OF RUNS

In Chapter XI, we investigated the distribution of the number of runs. We could calculate its expected value and variance from the definition but it would be immensely tedious. Let us use the same method as in the case of the hypergeometric distribution. The sample space Ω consists of all sequences of m 0's and n 1's. Let the r.v. X denote the number of runs in each sequence. Further, for $k = 2,3, \ldots ,n$, let:

$$X_k = \begin{cases} 1 \text{ if the } k\text{-th element of the sequence differs from the } (k-1) \text{ element} \\ 0 \text{ otherwise} \end{cases}$$

Obviously,
$$X = 1 + \sum_{k=2}^{m+n} X_k.$$

Then X_k is an alternative r.v., and

$$P(\{X_k = 1\}) = \frac{2mn}{(m+n)(m+n-1)}.$$

Therefore,
$$E(X_k) = E(X_k^2) = E(X_2) = p = \frac{2mn}{(m+n)(m+n-1)}.$$

Hence, $E(X) = 1 + \sum_{k=2}^{m+n} E(X_k) = 1 + \frac{2mn}{m+n}$.

$$V(X) = V\left(\sum_{k=2}^{m+n} X_k\right) = \sum_{k=2}^{m+n} V(X_k) + \sum\sum_{2\leq j\neq k\leq m+n} COV(X_j,X_k)$$

$$= (m+n-1)E(X_2^2) + \sum\sum_{2\leq j\neq k\leq m+n} E(X_jX_k) - (m+n-1)^2(E(X_2))^2.$$

To evaluate $(m + n - 1)(m + n - 2)$ terms $E(X_j, X_k)$ for $j \neq k$, we shall split them into two groups:

(a) For $2(m + n - 2)$ terms where $j = k - 1$ or $j = k + 1$, we have

$$E(X_j X_k) = \frac{mn(m-1) + nm(n-1)}{(m+n)(m+n-1)(m+n-2)} = \frac{mn}{(m+n)(m+n-1)}.$$

(b) For the remaining $(m + n - 1)(m + n - 2) - 2(m + n - 2) = (m + n - 2)(m + n - 3)$ terms, we have

$$E(X_j X_k) = \frac{4mn(m-1)(n-1)}{(m+n)(m+n-1)(m+n-2)(m+n-3)}.$$

Adding all these expressions, we get

$$V(X) = \frac{2mn}{m+n} + \frac{2(m+n-2)mn}{(m+n)(m+n-1)} + \frac{4mn(m-1)(n-1)}{(m+n)(m+n-1)} - \frac{4m^2n^2}{(m+n)^2}$$
$$= \frac{2mn(2mn - m - n)}{(m+n)^2(m+n-1)}.$$

Blackjack in a London casino. (Courtesy of Fox Photos, Ltd.)

THEOREM 17: Let the r.v. X be the number of runs in a sequence of m 0's and n 1's. Then ←

$$E(X) = 1 + \frac{2mn}{m+n}, \qquad V(X) = \frac{2mn(2mn - m - n)}{(m+n)^2(m+n-1)}.$$

EXAMPLE 33

Chains are manufactured from two kinds of links, each kind being made of steel of a different chemical composition but of the same tensile strength. However, there is a suspicion that chemical reactions at those joints consisting of two different kinds of steel might cause accelerated aging and weakening of the chains. Find the expected number and the variance of the number of dangerous joints if the chains consist of 90 links, of which 35 are of the first kind, and if the links are connected at random.

SOLUTION. Let Y be the number of dangerous joints and X be the number of runs of the links of two types. We can see that $Y = X - 1$. We have $m = 35$ links of the first kind, and $n = 55$ links of the second kind, assembled at random.

Hence, $E(X) = 1 + \dfrac{2 \cdot 35 \cdot 55}{35 + 55} = 43.78$; $E(Y) = E(X) - 1 = 42.78$.

$$V(Y) = V(X) = \frac{2 \cdot 35 \cdot 55(2 \cdot 35 \cdot 55 - 35 - 55)}{(35+55)^2(35+55-1)} = 20.08.$$

EXERCISES

(Answers in Appendix to exercises marked)*

*1. A game is played in a casino as follows: A number is drawn at random from $\{1,2,3,4,5,6,7\}$. If the number is even, the player wins the number of dollars equal to the number drawn. Otherwise the player pays \$3, and rolls two dice. If the sum is not 6, he pays \$6; if the sum is equal to 6, he gets \$12. What is the probability that he will make money if he plays once? What is his expected payoff?

*2. A hunter finds a rock ptarmigan sitting in a tree. (It is well known that this bird will not fly away even if it is being shot at.) He has only 3 shells in his automatic shotgun (and no other ammunition). If the probability of hitting the bird with each shot is .4, find the distribution of the r.v. X, which means the number of shots fired by the hunter. Find also $E(X)$ and $V(X)$. (Assume that the hunter keeps shooting till he gets the bird or has no shells left.)

3. You play a game as follows: You pay \$2 and toss a coin. If it is heads, you are paid \$11, otherwise you pay \$4 more and roll a die. If the die shows 1, you are paid \$12. If it does not, you pay \$6 and draw a ball from an urn containing two black, two white and three red balls. If you draw a black or a red ball you are paid \$9, otherwise you pay \$8. Find the expected payoff.

4. Students at the College of Yraglac are allowed to attempt every course up to five times in order to get credit. Knowing this fact, students do not apply themselves exorbitantly to their books, thus having an equal chance of passing or failing on the first attempt. However, owing to the oddities of the learning processes of "habitual repetends," one knows that chances of a pass are 10% higher on each successive trial. Find the expected value of the number of trials taken by the students to pass a course.

5. Consider the following game. You toss a fair die, and if the number is 2, 4 or 6 you win the corresponding number of dollars. If the number is odd, you draw a card from a standard deck of 52 cards. If it is a heart you win \$10, otherwise you toss a fair coin again, winning \$3 for heads and \$4 for tails. How much should you pay as an admission to play the game if it is to be fair?

*6. A lumber yard can make 6 cents a board foot selling a certain variety of ungraded lumber. If it separates the lumber into grades 1 and 2, it can make 10 cents a board on grade 1 lumber and 4 cents on grade 2. It is estimated that about half of the lumber is grade 1 and separation costs $\frac{1}{2}$ cent per board foot. Does it pay the yard to separate and sell graded lumber?

7. Investigate the following form of a game called Russian roulette. A player puts one live round of ammunition into the empty cylinder of a six-gun. (This jargon means that he puts one loaded cartridge into a standard-type revolver with a capacity of six cartridges in the cylinder.) Then he spins the cylinder at random, cocks the gun, puts the muzzle of the barrel against his temple and pulls the trigger. If by chance he survives, then he is expected to repeat the spinning and shooting four more times. Find the expected number of shots he is able to fire.

*8. A greedy hunter shoots at a duck sitting on the pond. The probability of a hit by any shot is $\frac{1}{3}$. If the duck is missed, then it is as likely to fly away, scared by the noise of the gun, as to stay sitting on the pond. The hunter has three cartridges and he fires till the duck is hit or it flies away (he does not shoot at a flying duck). Find the expected number of shots and the variance.

9. Twelve balls have been drawn from the BINGO blower (see Problem 59, page 123): three of them for the letter B, two of them for each of the letters ING. How many numbers can you expect to be able to cover up on your BINGO card?

10. Susan likes Italian Chianti red table wine very much, therefore she buys a bottle every month. Her problem is that she likes Chianti Antinori just as much as Chianti Brolio, and she can never decide which one to get: to make it simple she always tosses a coin to select the wine of the month. Let the r.v. X denote the difference between the numbers of bottles of Antinori and Brolio which Susan bought during the last year (12 months). Find $E(X)$.

*11. A casino owner decided to charge $2 for playing the following game: Three cards are drawn randomly from an ordinary deck of 52 cards. If no hearts turn up, the gambler wins nothing. If one heart shows up, the gambler is paid $1 and if two hearts show up, he gets $5. How high can he afford to make the jackpot reward for three hearts, if he wants the house to average a 50¢ profit per game?

12. Some roulette wheels have only numbers 0,1,2, . . . ,36 on them and 00 is omitted. Assume that the payoffs are the same as in the ordinary roulette game (see page 176). Calculate the expected payoffs and house advantages for the standard bets.

13. You are dealt 2 cards for blackjack from a complete deck of 52 cards out of which a king has been "burned" (turned face up so that it cannot be used). The sum X of these cards is of top importance for the game. Calculate its probability distribution and its expectation $E(X)$. (For the game of blackjack, each number card counts its face value, all face cards count 10 and, for the sake of simplicity, assume that an ace counts 11.) What is the probability of getting a "blackjack" (the sum of 21) in these two cards?

*14. An organizer of a lottery is sure that he can sell all 1000 tickets. He wants to award one first prize worth $100, two second prizes each worth $30 and five third prizes each worth $10. How much should he charge per ticket so that the expectation of the prize per ticket is equal to two thirds of its costs?

*15. A space capsule contains three safety devices D_1, D_2 and D_3. Let the probabilities of their failure during a flight be p_1, p_2 and p_3, respectively. Find the expected number of devices that can fail during a single flight.

16. Let X be a r.v. so that $E(X) = 2$, $V(X) = 1$. Find the expected value and variance of the following r.v.'s:

*(a) $X + 2$.
(b) $2X + 3$.
(c) $\frac{3X - 7}{2}$.
(d) Find $E[2X^2 + 3X - 1]$.
*(e) Find $E[(X - 1)^2 + 7 - X]$.

17. Let X be a r.v. such that $E(X) = 5$ and $V(X) = 0$. Find the probability distribution of X.

18. Is it possible that for some random variable X
(a) $E(X) = 5$ and $E(X^2) = 15$?
(b) $E(X) = .5$ and $E(X^2) = .3$?
Give reasons for your answers.

19. Let X be a r.v., and define a r.v. Y as $Y = aX^2 + bX + c$. Show that $E(Y) = a[E(X)]^2 + bE(X) + c + aV(X)$.

20. Let X and Y be random variables. Suppose that $E(X) = E(Y) = 0$, $V(X) = V(Y) = 1$, $COV(X,Y) = 1$. Let $U = 3X - Y$, $T = Y + 3X$. Find
*(a) $V(T)$, $V(U)$.
(b) $COV(U,T)$.
*(c) $E[(X - 2Y + 1)^2]$.

21. Let two random variables X and Y be given as follows:

X \ Y	1	2	3
−1	.2	.1	0
0	.1	0	.3
1	.1	.1	.1

(a) Find the marginal distribution of X and Y. Are X and Y independent?
(b) Find $E(X)$, $E(Y)$, $V(X)$, $V(Y)$, $COV(X,Y)$, $R(X,Y)$.

22. Let X and Y be r.v.'s from the previous problem. For each of the following pairs of r.v.'s $T = G(X,Y)$, $U = H(X,Y)$, find $E(T)$, $E(U)$, $V(T)$, $V(U)$, $COV(T,U)$.
(a) $G(x,y) = \min(x,y) \quad H(x,y) = \frac{1}{2}(x + y)$.
(b) $G(x,y) = \frac{x}{y} \quad H(x,y) = \max(x,y)$.
(c) $G(x,y) = (x - y)^2 \quad H(x,y) = 2x + 3y$.

23. Let the r.v.'s X and Y be given by the following joint probability distribution:

X \ Y	1	$\frac{5}{3}$	2
−1	$\frac{1}{16}$	$\frac{1}{16}$	$\frac{2}{16}$
0	$\frac{1}{16}$	$\frac{2}{16}$	$\frac{2}{16}$
1	$\frac{2}{16}$	$\frac{1}{16}$	$\frac{4}{16}$

(a) Are X and Y independent?
(b) Find $E(X)$, $E(Y)$, $V(X)$, $V(Y)$.
(c) Find $COV(X,Y)$ and comment on your answer from (a).
(d) Find the distribution of $(X + Y)$, $E(X + Y)$ and $V(X + Y)$. Compare with your answers in (a) and (b).

24. Is it possible that for some random variables X,Y: $E(X) = 1$, $E(X^2) = 10$, $E(Y) = 2$, $E(Y^2) = 20$, and $E(XY) = 15$? Give a reason for your answer.

*25. An urn contains three red and five black balls. Four balls are drawn at random, and let X denote the number of red balls drawn. Find $E(X)$ and $V(X)$.

26. A die is tossed, and if 1 or 2 appears then three balls are drawn from a bag containing three red and three blue balls. Otherwise the same number of balls is drawn from another bag containing four red and five blue balls. Let X denote the number of red balls drawn. Calculate $E(X)$ and $V(X)$.

*27. Ten fair dice are tossed. If X is the total number showing, find $E(X)$ and $V(X)$.

28. Thirteen cards are dealt to you from an ordinary deck of 52 cards. You receive \$2 for each spade in your hand, but you have to pay \$6 for playing the game. If X is your net profit after playing the game eight times, calculate $E(X)$ and $\sigma(X)$.

29. A coin is tossed until we get two tails in a row. After seven times we shall stop the game, even if no such event has occurred. Find the distribution of the number of tosses, its expected value and variance.

*30. A service station offers a routine car checkup before winter, which consists of testing four important parts of the vehicle. Assume that part i will fail the test with the probability

$$p_i = .1 + .05i$$

and that the failures of the parts are independent. Find the expected number and variance of the number of parts which fail the test.

31. A ski rental shop had five pairs of new and six pairs of used EDGER men's ski boots No. 10 before the beginning of the season. On the first day of the season, two people rented these boots (they were chosen at random from the 11 pairs available). On the second day, three pairs of these boots were rented out. Let X be the number of pairs of new boots rented on the first day and Y be the number of pairs of new boots rented on the second day. Find the joint distribution of X and Y, $E(X)$, $E(Y)$, $V(X)$, $V(Y)$ and $COV(X,Y)$. (Consider new boots rented for a day as used, and assume that all selections are at random.)

32. Let a fair coin be tossed four times, and let the r.v.'s X_1,X_2,X_3,X_4 be given as follows:

$X_i = 1$ if i-th toss results in a head

$= 0$ if i-th toss results in a tail,

$i = 1,2,3,4.$

Find $E(X_i)$ and $V(X_i)$. Now let the r.v. $\overline{X}$ denote the "average number of heads" in four tosses, that is,

$$\overline{X} = \frac{X_1 + X_2 + X_3 + X_4}{4}.$$

Find the probability distributions of $\overline{X}$, $E(\overline{X})$ and $V(\overline{X})$. Compare them with the expected value and the variance of X_i's and comment on your findings. (See Example 23, page 199).

33. Let $X_1,X_2, \ldots ,X_n$ be an independent random sample and $\overline{X}$ be the sample mean (see Example 27, page 203). The r.v.'s X_i are identically distributed and let us denote

$E(X_i) = \mu, \quad V(X_i) = \sigma^2, \quad i = 1,2, \ldots ,n.$

The random variable S^2, defined by

$$S^2 = \frac{1}{n-1}\sum_{i=1}^{n}(X_i - \overline{X})^2$$

is called the *sample variance.*
(a) Show that S^2 can be written as

$$S^2 = \frac{1}{n-1}\sum_{i=1}^{n}(X_i - \mu)^2 - \frac{n}{n-1}(\overline{X} - \mu)^2.$$

(b) Using this fact, show that

$$E(S^2) = \sigma^2.$$

34. Consider an urn containing R red and W white marbles. The marbles are drawn one after another sequentially till nothing is left in the urn. Let the r.v.'s X and Y be given as follows:

X is the number of red marbles drawn before the first white is obtained.

Y is the number of runs consisting of red marbles.

Show that if h is the joint distribution of X and Y, and f,g are the distributions of X and Y, respectively, then

$$h(0,m) = \frac{\binom{R-1}{m-1}\binom{W}{m}}{\binom{R+W}{R}} \qquad m \geq 1$$

$$h(R,1) = \frac{1}{\binom{R+W}{R}}$$

$$h(n,m) = \frac{\binom{R-n-1}{m-2}\binom{W}{m-1}}{\binom{R+W}{R}} \text{ otherwise}$$

$$f(n) = \frac{\binom{R+W-n-1}{R-n}}{\binom{R+W}{R}}$$

$$g(m) = \frac{\binom{R-1}{m-1}\binom{W+1}{m}}{\binom{R+W}{R}}$$

$$E(X) = \frac{R}{W+1} \quad E(Y) = \frac{R(W+1)}{R+W}$$

$$E(XY) = \frac{R^2}{R+W}$$

hence X and Y are dependent, and $COV(X,Y) = 0$.

*35. A statistics instructor comes to his class on time four out of five times (his arrivals being independent). How many times would you expect him to be late in a half course of 12 weeks (3 lectures weekly)?

36. A warehouse contains 10,000 LP records on the Colophonia label and 7,000 LP's on the Cape Tall label. It is known that Colophonia produces about 3% of warped records, while Cape Tall has a 4% defective output due to warping. Find the expectation and the variance of the number of defective records due to warping in the warehouse.

37. A salesman figures that each contract results in a sale with probability $\frac{1}{2}$. During a day, he contacts five prospective clients. Find the probability distribution of the number of clients who sign a sales contract, its expected value and standard deviation.

*38. In a binomial experiment, the mean number of successes is 10 and the variance is 6. What is the probability of a single success and the number of repetitions? Assume that you repeat this experiment again, this time four times. What is the probability of getting more failures than successes?

39. The probability that an aircraft trapped in an "air pocket" (or "air hole," a nearly vertical air current) will lose suddenly more than 100 feet of altitude is strictly positive. Let the r.v. X be 0 if the plane does not lose more than 100 feet of altitude upon entering an air hole, and let X be 1 otherwise. Show that there is an upper bound for the variance of X, namely $V(X) \leq \frac{1}{4}$. Imagine that the plane has to fly through a stormy and turbulent area, and hits 16 such air holes (which act independently). Let Y be the number of sudden drops in altitude of over 100 feet. Show that there is an upper bound for the standard deviation of Y, namely $\sigma(Y) \leq 2$.

*40. An oil refinery has a contract with a large independent gas station chain, specifying that all regular gasoline must have an octane number of at least 90, and at least 80% of the shipments should have an octane number of at least 92. The chain manager decided to test randomly 15 out of the next 120 shipments. Assuming that the manufacturer keeps the contract exactly (but is not generous, that is, exactly 80% of the shipments are of high quality, above 91 octanes), find the expected number and the variance of the number of tested shipments with an octane number under 92. (Assume successive shipments to be independent.)

41. A wholesaler buys films from a manufacturer in lots of 1000. Before sending them to retailers the wholesaler always opens 25 films from each lot for a technical inspection. It was found over a long period of time that on the average 2 out of these 25 films are in plastic containers while the rest have metal containers. Upon request, the manufacturer admitted that each lot contains a certain (always the same) number of films in plastic containers. Using the formula for the expected value of the hypergeometric distribution (why?), estimate the number of films in plastic containers in each lot. Notice that this method is very plausible and would be used by a person without any knowledge of probability.

LAW OF LARGE NUMBERS

A law of Nature is something that we being ignorant of may attain to the knowledge of by the use and due application of our natural Faculties.

John Locke, 1690

In Chapter IV, we attempted to define probability as a "limit" of relative frequencies in a sequence of trials. Perhaps the simplest example is the case of tossing a fair coin many times. From our experience, we know that almost always in long sequences of tosses the relative frequency of the occurrences of heads will closely approach one half. We already know that one cannot accept this observation as a definition of probability. At the same time, this fact is very interesting and worthy of further investigation. However, it is fascinating that we can *prove* within our axiomatic system statements of this type; e.g., in any sequence of independent trials of an experiment with outcomes both a success and a failure, the relative frequency of the occurrences of a success in a long run almost always must be very close to the probability of a success. Theorems of a similar kind are called *laws of large numbers.* Before we start with the investigation of a certain law of large numbers, we have to prove one auxiliary inequality.

THEOREM 1: *Chebyshev's Inequality:* Let X be a r.v. on a probability measure space (Ω,P) with the expected value $E(X)$ and the variance $V(X)$. Then for any real number $\varepsilon > 0$,

$$P(\{\omega: |X(\omega) - E(X)| \geq \varepsilon\}) \leq \frac{V(X)}{\varepsilon^2}.$$

Proof: Let $A = \{\omega: |X(\omega) - E(X)| \geq \varepsilon\}$. Then the left-hand side of the inequality is $P(A)$. From Theorem 1 of Chapter XII, we get

$$V(X) = E[(X - E(X))^2] = \sum_{\omega\in\Omega} (X(\omega) - E(X))^2 P\{\omega\}.$$

This sum is taken for all $\omega \in \Omega$ but we can split it into two parts, namely, into sums where $\omega \in A$ and where $\omega \in A'$.

Hence
$$V(X) = \sum_{\omega\in A} (X(\omega) - E(X))^2 P\{\omega\} + \sum_{\omega\in A'} (X(\omega) - E(X))^2 P\{\omega\}.$$

The second sum is always nonnegative and if we omit it, we decrease the right-hand side:

$$V(X) \geq \sum_{\omega\in A} (X(\omega) - E(X))^2 P\{\omega\}.$$

But for every $\omega \in A$, we have $(X(\omega) - E(X))^2 \geq \varepsilon^2$, hence

$$V(X) \geq \sum_{\omega\in A} \varepsilon^2 P\{\omega\} = \varepsilon^2 P(A).$$

Therefore
$$P(A) \leq \frac{V(X)}{\varepsilon^2}.$$
q.e.d.

COMMENT. Chebyshev's inequality can be written in another equivalent way if we take the probability of the complement of the investigated event, namely,

$$P(\{\omega: |X(\omega) - E(X)| < \varepsilon\}) \geq 1 - \frac{V(X)}{\varepsilon^2}.$$

EXAMPLE 1

Let the r.v. X have the following distribution:

x_i	-1	1
$f(x_i)$	.5	.5

Then $$E(X) = 0, \qquad V(X) = 1.$$

Take $\varepsilon = 1$ in Chebyshev's inequality, then

$$P(\{\omega: |X - 0| \geq 1\}) = P(\{-1,1\}) = 1 \leq \frac{V(X)}{\varepsilon^2} = \frac{1}{1^2} = 1.$$

In this case we get an equality, and in this sense it is impossible to improve the statement of Chebyshev's inequality.

EXAMPLE 2

Toss a fair coin six times and estimate the probability of the number of heads being between two and four inclusive, using Chebyshev's inequality.

SOLUTION. If X is the number of heads, then

$$E(X) = np = 3. \qquad V(X) = npq = 1.5.$$

Choosing $\varepsilon = 2$ for the second version of Chebyshev's inequality, we get

$$P\{\omega: |X - 3| < 2\} \geq 1 - \frac{1.5}{2^2} = .6250.$$

Since $$P(\{\omega: |X - 3| < 2\}) = P(\{2,3,4\}) = \sum_{i=2}^{4} \binom{6}{i}\left(\frac{1}{2}\right)^i\left(\frac{1}{2}\right)^{6-i} = .7812,$$

we can see that we got a fairly good estimation of the required probability. A merit (or rather a demerit) of Examples 1 and 2 is that *we know the distribution* of the r.v. X and therefore *can calculate exactly* the required probability. However, in many situations *we do not know the distribution,* but only $E(X)$ and $V(X)$ and the only way to find anything about the required probability is to *estimate it approximately,* using Chebyshev's inequality.

EXAMPLE 3

A r.v. X attaining natural numbers as its values has an expected value $E(X) = 12$ and variance $V(X) = 3$. Estimate the probability that X will attain any value between 10 and 14 inclusive.

SOLUTION. We know nothing about the distribution of X, and have to use Chebyshev's inequality. For $\varepsilon = 3$, we get

$$P(\{10,11,12,13,14\}) = P(\{\omega: |X - 12| < 3\}) \geq 1 - \frac{3}{3^2} = \frac{2}{3}.$$

Therefore this probability is at least $\frac{2}{3}$.

EXAMPLE 4

A r.v. X, attaining natural numbers as its values, has an expected value $E(X) = 25$ and variance $V(X) = 1$. Find an upper and a lower bound so that with a probability of at least 99%, the r.v. X will be inside the interval with these bounds.

SOLUTION. The second version of Chebyshev's inequality can be written as

$$P(\{\omega: E(X) - \varepsilon < X < E(X) + \varepsilon\}) \geq 1 - \frac{V(X)}{\varepsilon^2}.$$

What should ε be equal to so that

$$1 - \frac{V(X)}{\varepsilon^2} = .99?$$

We get $\frac{1}{\varepsilon^2} = .01$ or $\varepsilon = \sqrt{100} = 10$.

Therefore $$P(\{\omega: 25 - 10 < X < 25 + 10\}) \geq .99$$

or $$P(\{\omega: 15 < X < 35\}) \geq .99.$$

Hence with a probability of at least 99%, the r.v. X is within the interval [16, 34]. (The brackets [ ] mean that 16 and 34 are included in the interval).

EXAMPLE 5

Estimate the probability that a r.v. X will differ from its expected value by less than three times its standard deviation (compare Example 31, page 210).

SOLUTION. Let us investigate the probability that X will differ from $E(X)$ by less than k times $\sigma(X)$, i.e.,

$$P(\{\omega: E(X) - k\sigma(X) < X < E(X) + k\sigma(X)\}).$$

Using $\varepsilon = k\sigma(X)$ in Chebyshev's inequality, we get

$$P(\{\omega: E(X) - k\sigma(X) < X < E(X) + k\sigma(X)\}) \geq 1 - \frac{V(X)}{k^2\sigma^2(X)} = 1 - \frac{1}{k^2}.$$

For a special selection $k = 3$, we get

$$P(\{\omega: E(X) - 3\sigma(X) < X < E(X) + 3\sigma(X)\}) \geq \frac{8}{9} = .88\overline{8}.$$

We can conclude that *any r.v. X will be with a probability of at least 88.8$\bar{8}$% within the interval $E(X) - 3\sigma(X)$ and $E(X) + 3\sigma(X)$.* This fact clearly shows the importance and the role of the expected value and of the standard deviation of a random variable. However, we shall see in the last chapter that this estimation is in many cases rather conservative; on the other hand, this estimation may be the only one available in some situations.

EXAMPLE 6

An oil well yields daily, on average, 1200 barrels with a standard deviation of the yield of 65 barrels. Estimate the range of the daily yield within which you will get at least 90% of the production.

SOLUTION. Using Chebyshev's inequality, we get

$$P(\{\omega: E(X) - \varepsilon < X < E(X) + \varepsilon\}) \geq 1 - \frac{V(X)}{\varepsilon^2} = .90.$$

The right-hand side of the inequality gives

$$\frac{(65)^2}{\varepsilon^2} = .01 \quad \text{or} \quad \varepsilon = 205.5.$$

Therefore, in at least 90% of cases, the daily yield of the crude oil will be within the range of 994.5 to 1405.5 barrels. It may be expected that the actual range will not be so wide because Chebyshev's inequality often gives rather conservative answers.

In Chapters XI and XII, we investigated the binomial r.v., its expected value and variance. We started with an experiment which had outcomes success and failure, hence we took $\Omega = \{S,F\}$. We assumed that the probability of success is p and the probability of failure is $q = 1 - p$. Then we investigated n independent repetitions of this experiment, and found that an appropriate sample space for this "new combined" experiment is

$$\Omega_n = \underbrace{\Omega \times \Omega \times \ . \ . \ . \ \times \Omega}_{n\,-\,times}.$$

We defined the binomial random variable S_n on Ω^n as the number of successes. Then we proved that

$$P^{(n)}(\{S_n = k\}) = \binom{n}{k} p^k q^{n-k} \qquad k = 0,1,2, \ . \ . \ . \ ,n,$$

$$E(S_n) = np, \qquad V(S_n) = npq.$$

→ **THEOREM 2:** *Bernoulli's Law of Large Numbers:* Let S_n be the binomial random variable, ε be a positive number. Then $P^{(n)}\left(\left\{\omega: \left|\frac{S_n}{n} - p\right| \geq \varepsilon\right\}\right)$ approaches arbitrarily close to 0 if n grows above all limits.

COMMENT. Let us have a sequence of nonnegative numbers, say $a_1,a_2,a_3 \ . \ . \ . \ .$ We say that this sequence approaches arbitrarily close to 0 if n grows above all limits, if the following is satisfied: Imagine that you pick an arbitrarily small number $\delta > 0$. Then starting from a certain n, say n_0, all the numbers a_n must be less than δ, that is,

$$0 \leq a_n < \delta \qquad \text{for all} \qquad n \geq n_0.$$

We can picture it as follows:

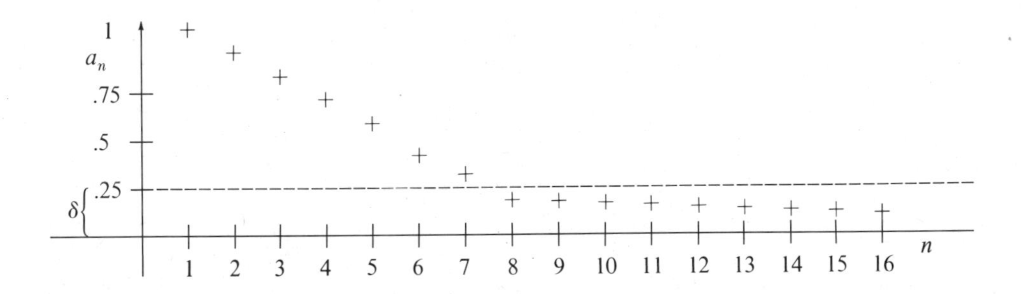

For given $\delta = .25$, we can see that $a_n < \delta$ for $n \geq n_0 = 8$. If $a_n = \frac{1}{n}$, then obviously the sequence a_n approaches arbitrarily close to 0 because it is of the form $\frac{1}{1}, \frac{1}{2}, \frac{1}{3}, \frac{1}{4}, \frac{1}{5}, \frac{1}{6}, \frac{1}{7}$, etc. But also the sequence $a_n = \frac{2}{n}$ approaches 0, and in general for any constant k, the sequence $a_n = \frac{k}{n}$ approaches arbitrarily close to 0 as n grows above all limits.

Proof: We know that

$$E(S_n) = np. \qquad V(S_n) = npq.$$

Using Theorem 2 of Chapter XII, we get

$$E\left(\frac{S_n}{n}\right) = \frac{E(S_n)}{n} = p.$$

$$V\left(\frac{S_n}{n}\right) = \frac{V(S_n)}{n^2} = \frac{pq}{n}.$$

Using Chebyshev's inequality, we get

$$P^{(n)}\left(\left\{\omega: \left|\frac{S_n}{n} - p\right| \geq \varepsilon\right\}\right) \leq \frac{V\left(\frac{S_n}{n}\right)}{\varepsilon^2} = \frac{pq}{\varepsilon^2} \cdot \frac{1}{n}.$$

But the right-hand side approaches the value 0 arbitrarily close with n increasing, and so therefore does the left-hand side because it is always less than or equal to the right-hand side and nonnegative. q.e.d.

The value $\frac{S_n}{n}$ is the relative frequency of successes in n independent repetitions. We can see from Bernoulli's Law of Large Numbers that regardless of how small ε is, *the probability that the relative frequency* $\frac{S_n}{n}$ *of successes will differ from the probability of a success p by more than* ε *can be made arbitrarily small if we repeat the experiment sufficiently many times.* In this sense we say that with an increasing number of repetitions, the relative frequency of successes approaches or converges to the probability p of a success.

EXAMPLE 7

How many times should we toss a fair coin so that with a probability of at least 90%, the relative frequency of heads (the number of heads divided by the number of tosses) will differ from $\frac{1}{2}$ by less than one hundredth?

SOLUTION. The last formula in the proof of the Law of Large numbers (if rewritten using complements) shows that

$$P^{(n)}\left(\left\{\omega: p - \varepsilon < \frac{S_n}{n} < p + \varepsilon\right\}\right) \geq 1 - \frac{pq}{n\varepsilon^2}$$

where $\frac{S_n}{n}$ is the frequency of heads, $p = \frac{1}{2}$ is the probability of heads and $q = 1 - p$. We need the right-hand side to be at least 90% or

$$1 - \frac{1}{4n\varepsilon^2} \geq .9.$$

The required accuracy is $\varepsilon = .01$, hence

$$1 - \frac{1}{4n(.01)^2} \geq .9.$$

Solving this inequality, we get $n \geq 25{,}000$. Therefore we should flip the coin at least 25,000 times.

EXAMPLE 8

A machine produces a certain unknown percentage of defective items. The only way to find this percentage is to test a large number of items and find how many of them are defective. What is the least number of items necessary to test so that the quality controller can be at least 80% confident that the unknown proportion will differ from the relative frequency of defectives among tested items by less than .05?

SOLUTION. Let the probability of getting a defective item be p, and assume that the production of successive items is independent. We have to find p. Proceeding as in Example 7, we get $\varepsilon = .05$ and

$$P^{(n)}\left(\left\{\omega: p - \varepsilon < \frac{S_n}{n} < p + \varepsilon\right\}\right) \geq 1 - \frac{pq}{n\varepsilon^2}.$$

Unfortunately, without any knowledge of p we cannot calculate the right-hand side of the inequality. However, we know that for $0 \leq p \leq 1$ and $p + q = 1$ the following inequalities hold:

$$\frac{1}{4} - pq = \frac{1}{4} - p(1 - p) = p^2 - p + \frac{1}{4} = \left(p - \frac{1}{2}\right)^2 \geq 0$$

hence

$$pq \leq \frac{1}{4}$$

Therefore $$1 - \frac{pq}{n\varepsilon^2} \geq 1 - \frac{1}{4n\varepsilon^2} = 1 - \frac{1}{4n(.05)^2} = .8$$

Solving for n, we get $$n = 500.$$

Therefore, the quality controller should test at least 500 items. In this case also the estimation might be rather conservative, and a smaller number of tests might suffice. We shall discuss more efficient methods in the last chapter.

In Chapter XII, we defined so-called alternating r.v.'s X_i on Ω^n. The r.v. X_i was 0 if we got a failure in the i-th experiment, and it was 1 if we got a success. We saw also that X_i's are independent, and that the binomial r.v. S_n can be expressed as

$$S_n = X_1 + X_2 + \ldots + X_n.$$

We also know that $E(X_1) = E(X_2) = \ldots = E(X_n) = p$ and if we denote $p = E(X)$, then Bernoulli's Law of Large Numbers investigates the behavior of the expression

$$P^{(n)}\left(\left\{\omega: \left|\frac{S_n}{n} - E(X)\right| \geq \varepsilon\right\}\right)$$

for a special kind of independent, identically distributed random variables $X_1, X_2, \ldots, X_n$.

We can try to generalize this situation. Let us have a probability measure space (Ω,P), and let X be a r.v. defined on (Ω,P). We saw already that an appropriate probability measure space for the investigation of n independent repetitions of an experiment is the probability measure space $(\Omega^n,P^{(n)})$ defined as follows:

$$\Omega^n = \underbrace{\Omega \times \Omega \times \ldots \times \Omega}_{n - times}.$$

Points of Ω^n are sequences of the length n of the elements from Ω, i.e., $\omega \in \Omega^n$ if and only if $\omega = (\omega_1,\omega_2, \ldots ,\omega_n)$ such that $\omega_i \in \Omega$ for $i = 1,2, \ldots ,n$. Let us define a probability measure P^n for $\{\omega\}$ as follows:

$$P^{(n)}\{\omega\} = P\{\omega_1\}. \quad P\{\omega_2\} \ldots P\{\omega_n\}.$$

It can be easily shown that $P^{(n)}\{\omega\}$ defines a probability measure for all subsets of Ω^n. Let us define further n r.v.'s on Ω^n as follows:

$$X_i(\omega) = X(\omega_i). \quad i = 1,2, \ldots ,n, \quad \omega \in \Omega^n.$$

Obviously, for an appropriate real number x,

$$P^{(n)}\{\omega: \omega \in \Omega^n, X_i(\omega) = x\} = P^{(n)}\{\omega: \omega \in \Omega^n, X(\omega_i) = x\} = P\{\omega: \omega \in \Omega, X(\omega) = x\}.$$

Therefore, the r.v.'s X_i on Ω^n, $i = 1,2, \ldots ,n$ have the same probability distribution, or as we say they are *identically distributed.* It can be shown that they are independent. We can now generalize Bernoulli's Law of Large Numbers.

→ **THEOREM 3:** *Weak Law of Large Numbers:* For any n, let $X_1, X_2, \ldots, X_n$ be independent, identically distributed r.v.'s on the probability measure space $(\Omega^n, P^{(n)})$. Let $\varepsilon > 0$, $S_n = \sum_{i=1}^{n} X_i$ and let E be the common expected value of X_i's.

Then $P^{(n)}\left(\left\{\omega: \left|\frac{S_n}{n} - E\right| \geq \varepsilon\right\}\right)$ approaches arbitrarily close to 0 as n grows above all limits.

Proof: Let $E(X_i) = E$ and $V(X_i) = V$, $i = 1,2, \ldots n$.

Then
$$E(S_n) = \sum_{i=1}^{n} E(X_i) = nE,$$
$$V(S_n) = \sum_{i=1}^{n} V(X_i) = nV,$$
$$E\left(\frac{S_n}{n}\right) = \frac{1}{n} E(S_n) = E,$$
$$V\left(\frac{S_n}{n}\right) = \frac{1}{n^2} V(S_n) = \frac{1}{n} V,$$

because r.v.'s X_i are pairwise independent. From Chebyshev's inequality, we get

$$P^{(n)}\left\{\omega: \left|\frac{S_n}{n} - E\right| \geq \varepsilon\right\} = P^{(n)}\left\{\omega: \left|\frac{S_n}{n} - E\left(\frac{S_n}{n}\right)\right| \geq \varepsilon\right\} \leq \frac{V\left(\frac{S_n}{n}\right)}{\varepsilon^2} = \frac{V}{\varepsilon^2} \cdot \frac{1}{n}.$$

The right-hand side approaches arbitrarily close to 0 as n grows above all limits, and so does the left-hand side because it is at most equal to the right-hand side and because the left-hand side is nonnegative. q.e.d.

EXAMPLE 9

Application of the Weak Law of Large Numbers in Statistics. Let us investigate the traffic passing through a certain busy intersection in a downtown area. In order to develop appropriate road allowances, it is necessary to establish first the expected number of cars passing through this intersection on a weekday during the peak traffic hour, 4 to 5 P.M. We assume that all weekdays have just about the same traffic. If we denote the number of cars passing through this intersection between 4 P.M. and 5 P.M. on a weekday by X, then X is a random variable, the expectation of which we have to find. We do not know the probability distribution of X, hence we have to use another method to find $E(X)$. However, we can actually count the number of cars passing through the intersection between 4 and 5 P.M. for several weekdays—say n times—thus getting a set of n observations. It is possible to assume that these observations represent n independent random variables $X_1, X_2, \ldots, X_n$, and that all of them have the same probability distribution as X. The Weak Law of Large Numbers then assures that regardless of how small we take ε (say $\varepsilon = 0.1$ or $\varepsilon = .001$ or less), the probability that the arithmetic means $\frac{1}{n}(X_1 + X_2 + \ldots + X_n)$ of successive observations will differ from the unknown value $E(X)$ by more than ε can be made arbitrarily small by taking a sufficiently large number of observations. In practical life, we loosely say that the expected value $E(X)$ is somewhere very close to the arithmetic mean $\frac{1}{n}(X_1 + X_2 + \ldots + X_n)$ or, in other words, we estimate $E(X)$ by $\frac{1}{n}(X_1 + X_2 + \ldots + X_n)$. The Weak Law of Large Numbers says we can be quite confident that we do not commit too large an error provided we take a sufficiently large number of observations.

EXAMPLE 10

Consistency of the Sample Mean. Let us investigate an independent random sample X_1, $X_2, \ldots, X_n$ of the size n (see Example 27, page 203; compare also Example 9 above). In our terminology, we can say that $X_1, X_2, \ldots, X_n$ are independent, identically distributed random variables such that

$$E(X_i) = E(X). \qquad i = 1,2, \ldots, n.$$

Assume that the expected value $E(X)$ is *unknown,* as frequently happens in applications of statistics. Then we can investigate the *sample mean* $\overline{X}$

$$\overline{X} = \frac{S_n}{n} = \frac{X_1 + X_2 + \ldots + X_n}{n}$$

and we have already shown that

$$E(\overline{X}) = E(X) \quad \text{and} \quad V(\overline{X}) = \frac{V(X)}{n}.$$

We would like to find the unknown value of $E(X)$. We can collect (observe) an actual sample $X_1, X_2, \ldots, X_n$, calculate its (known) sample mean $\overline{X}$ and use it as an *estimator* of unknown $E(X)$. The sample mean $\overline{X}$ will vary from one sample to another; it is a random variable, but its expected value is equal to the unknown estimated value $E(X)$. In this sense, we say that the sample mean $\overline{X}$ is *an unbiased estimator of E(X).*

However, using the Weak Law of Large Numbers, we can claim yet another good property of $\overline{X}$ as an estimator of the unknown value $E(X)$. We know that for any $\varepsilon > 0$, $P(\{\omega: |\overline{X} - E(X)| \geq \varepsilon\})$ must approach arbitrarily close to 0 with the sample size n increasing above all limits. It means that for any arbitrarily small ε positive, we can be practically sure that the (unknown) expected value $E(X)$ will differ from the (known) sample mean $\overline{X}$ by less ε, provided we take a sufficiently large sample. In this sense, the sample mean approaches to or converges on $E(X)$, and we say therefore that $\overline{X}$ is a *consistent estimator of* $E(X)$. (Compare this with the conclusion in Example 9.)

EXAMPLE 11

The Registrar's Office of a large university is required to find the average height of students, for designing the dimensions of new classroom furniture. It is impossible to measure all the registered students so a random sample to estimate the average height is suggested. The Registrar's Office would like to be at least 90% sure that the actual average will differ from the obtained sample average by less than one inch. How many students should be measured?

This problem is complicated by the fact that we have practically no information about the student population; its mean and standard deviation are unknown. If X denotes the height of a randomly selected student, then we have to estimate $E(X)$. However, first we have to get some idea about $\sigma(X)$, which is also unknown. From Example 5, it follows that at least $88.\overline{88}\%$ of students will have heights within $\pm 3\sigma(X)$ of $E(X)$. We may reasonably assume that a large majority of students will be between 5 feet and 7 feet tall, and that this interval of 24 inches corresponds approximately to the range of $6\sigma(X)$ from Example 5. This is a rather crude guess, but it is the best we can do when more detailed information is missing.

Hence, $$6\sigma(X) = 24, \qquad \sigma(X) = 4, \qquad V(X) = 16.$$

Assume that the Registrar's Office will measure a random sample of n (independent) students. If we denote the sample mean by $\overline{X}$, then from the discussion of Example 10 and from the proof of the Weak Law of Large Numbers, we can see that

$$P(\{\omega: |\overline{X} - E(X)| < \varepsilon\}) \geq 1 - \frac{V(X)}{\varepsilon^2} \cdot \frac{1}{n}.$$

Taking $\varepsilon = 1$, $V(X) = 16$, we require the right-hand side to be 90%, i.e.,

$$P(\{\omega: |\overline{X} - E(X)| < 1\}) \geq 1 - \frac{16}{n} = .9.$$

Equivalently, $$\frac{16}{n} = .1 \quad \text{or} \quad n = 160.$$

Therefore, the Registrar's Office should take an independent random sample of at least 160 students. If the student body consists of 30,000 students, the sampling technique represents a considerable saving and still yields a reasonably accurate estimation.

EXERCISES

(Answers in Appendix to exercises marked)*

Assume that all r.v.'s attain natural numbers as their values.

1. Let $E(X) = 11$, $V(X) = 2$. Use Chebyshev's inequality to estimate from below $P(\{\omega: X(\omega) = 9,10,11,12,13)$.

2. Let $E(X) = 20$, $\sigma(X) = 1.5$. Estimate from above $P(\{\omega: X(\omega) \geq 23 \text{ or } X(\omega) \leq 17\})$.

*3. Let $E(X) = 0$, $V(X) = 1$. Estimate from below $P\{\omega: -5 \leq X(\omega) \leq +5\}$.

*4. An average daily deposit at the local branch of the Bank of Quickbuck is $150, and the standard deviation of deposits is $20. Construct a lower and an upper bound of a range within which at least 85% of daily deposits is located.

5. Hunters harvest, on average, 11,500 mule deer each hunting season in a certain wildlife management area. The standard deviation of the number of deer taken is 20. Find a lower and an upper bound of a range within which you would expect the annual yield at least nine times in a decade.

*6. How many times should you toss a fair coin so that the probability of getting the relative frequency of heads within the interval .4 to .6 is at least 80%? How many times should you toss it if you want the interval to be .45 to .55?

7. Assume that you have a fair die. How many times should you roll it (what is the least number of tosses) so that with a probability of at least 85% the frequency of the 6's will differ from $\frac{1}{6}$ by less than .03?

8. You again have a die (see problem 7) but you do not know whether it is loaded or not. How many times should you roll it so that with a probability of at least 85% the frequency of the 6's will differ from the actual (but unknown) probability of getting a 6 by less than .03?

9. Pacific Poorfield Oil Company would like to estimate the average daily yield of a new oil well. It is known that all the wells in this area have daily yields with approximately the same standard deviation of 40 barrels. For how many days should the chief engineer request the reports of the yield so that he can be at least 80% sure that the actual expectation of daily yields will differ from the reported average by less than 10 barrels?

MARKOV CHAINS

In this chapter, we are going to investigate repeated trials in the case where successive repetitions are *dependent.* However, we shall assume that this dependence has a very simple structure that can be conveniently formalized. Let us begin with the following example.

EXAMPLE 1

A traveling salesman sells his products in three cities: Abadan, Belial and Cithaeron. If he is in A on a certain day, then he is equally likely to be in any one of the three cities the next day. He never stays in B or C for two days in a row. If he is in B, then he is equally likely to be in one of the two other cities the next day. However, if he is in C then the next day he is twice as likely to be in A as in B. He makes the city of Abadan his permanent residence, and his wife noticed that he will usually start his weekly sales route in this city. As a matter of fact, he is twice as likely to start on Monday in A than in B or C (the latter two being equally likely places of his Monday sales). Investigate the salesman's traveling habits.

SOLUTION. On any day the salesman can be in any one of the three cities A, B and C, and he stays there for a day, possibly moving to another city for the next day. We can capture his behavior using a probability tree. On Monday he can be in the cities A, B, C with the following probabilities:

$$p_1 = P(A) = \frac{1}{2}, \quad p_2 = P(B) = \frac{1}{4}, \quad p_3 = P(C) = \frac{1}{4}.$$

For the next two days he can travel as follows:

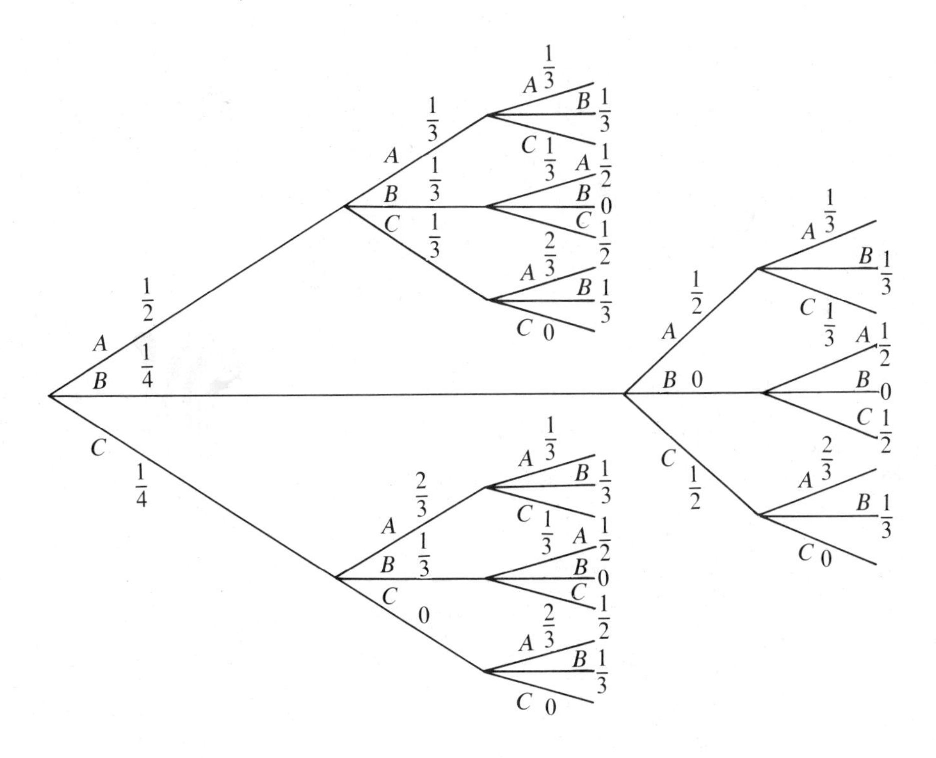

If we denote the cities Abadan by 1, Belial by 2 and Cithaeron by 3, then we can say that:

If he is in A on a certain day, then on the next day he will be in:

$$A \text{ with probability } p_{11} = \frac{1}{3}\cdot$$

$$B \text{ with probability } p_{12} = \frac{1}{3}\cdot$$

$$C \text{ with probability } p_{13} = \frac{1}{3}\cdot$$

If he is in B on a certain day, then on the next day he will be in:

$$A \text{ with probability } p_{21} = \frac{1}{2}\cdot$$

$$B \text{ with probability } p_{22} = 0.$$

$$C \text{ with probability } p_{23} = \frac{1}{2}\cdot$$

If he is in C on a certain day, on the next day he will be in:

$$A \text{ with probability } p_{31} = \frac{2}{3}\cdot$$

$$B \text{ with probability } p_{32} = \frac{1}{3}\cdot$$

$$C \text{ with probability } p_{33} = 0.$$

By the symbol p_{ij}, we have denoted the conditional probability that he will be in city j or he will "transit" into city j for the next day given the fact that he is in city i today. The probability tree is rather unwieldy even if we consider the salesman's travel only for two or three days. If we would go on for a longer period of time, then the tree would be practically unmanageable. We shall try to develop another method to help us cope with similar problems. First of all, we shall try to arrange all the given probabilities in the form of arrays. The *initial probabilities* for the beginning of his sales route on Monday are

$$(p_1, p_2, p_3) = \left(\frac{1}{2}, \frac{1}{4}, \frac{1}{4}\right).$$

The conditional probabilities of transition from city i to city j, or *transition probabilities,* can be arranged as follows

$$\begin{pmatrix} p_{11} & p_{12} & p_{13} \\ p_{21} & p_{22} & p_{23} \\ p_{31} & p_{32} & p_{33} \end{pmatrix} = \begin{array}{c} \\ A \\ B \\ C \end{array} \begin{array}{c} A \quad\; B \quad\; C \\ \begin{pmatrix} \frac{1}{3} & \frac{1}{3} & \frac{1}{3} \\ \frac{1}{2} & 0 & \frac{1}{2} \\ \frac{2}{3} & \frac{1}{3} & 0 \end{pmatrix} \end{array}$$

Note that we have placed the probability of a transition from city i into city j during one day, that is p_{ij}, into *row i and column j* of the array. Let us investigate the probability that the salesman will be in A on Monday, in C on Tuesday and B on Wednesday. This probability, denoted $P(A,C,B)$, is equal to

$$P(A,C,B) = \frac{1}{2} \cdot \frac{1}{3} \cdot \frac{1}{3} = p_1 p_{13} p_{32}$$

as we can easily see from the tree. Or the probability that he will make the following trip: Monday in B, Tuesday in A, Wednesday in C, Thursday in A can be expressed as

$$P(B,A,C,A) = \frac{1}{4} \cdot \frac{1}{2} \cdot \frac{1}{3} \cdot \frac{2}{3} = p_2 p_{21} p_{13} p_{31}.$$

This time the tree is not sufficient but it is quite easy to see how to get the corresponding probability. However, we might be interested in other important questions, such as: if he is in A on Monday, what is the probability that he will be in B on Wednesday; if he is in C on Monday, what is the probability that he will be in A on Friday? Such questions are quite interesting and we shall try to develop a formal method for their solution.

However, we have to observe two important facts related to the salesman's habits. First of all, the probability that he will be in a certain city on a certain day *given* the itinerary of his journey during the past several days depends *solely* on where he was the *very previous day* and not at all where he was two, three or more days before. For example, the probability of going to city 1 tomorrow is either $p_{11} = \frac{1}{3}$, $p_{21} = \frac{1}{2}$ or $p_{31} = \frac{2}{3}$, depending on whether the salesman is in city 1, city 2 or city 3 today, and it does not depend on his location yesterday or the day before yesterday. And second, the transition probability from a city, say A, on a certain day to another city, say B, on the next day is *always the same*, in our case $p_{12} = \frac{1}{3}$, whether the two mentioned days are Monday and Tuesday, or Tuesday and Wednesday or Thursday and Friday. In this sense, we can say that the transition probabilities do not change during time, or in other words they are *stationary*.

Let us consider an experiment with the sample space $\Omega = \{\omega_1, \omega_2, \ldots, \omega_n\}$. The outcomes of $\omega \in \Omega$ will sometimes be called *states of the system* in this chapter. We shall try to develop a probabilistic model for sequences of repetitions of this experiment, which are not necessarily independent but are of a rather simple nature.

If we investigate k repetitions of the experiment, then a corresponding sample space Ω^k to describe these repetitions is the set of all possible sequences of the length k consisting of elements of Ω, i.e.,

$$\Omega^k = \underbrace{\Omega \times \Omega \times \ldots \times \Omega}_{k\text{-times}}.$$

Let us have, for example, $k = 4$ repetitions, and let us investigate the probability that the outcome will be ω_5 in the first trial, ω_3 in the second, ω_7 in the third and ω_2 in the fourth trial.

Then, using the multiplication rule (page 89),

$$P(\omega_5,\omega_3,\omega_7,\omega_2) = P(\omega_5 \text{ in 1st trial}) \times P(\omega_3 \text{ in 2nd trial}/\omega_5 \text{ in 1st trial})$$
$$\times P(\omega_7 \text{ in 3rd trial}/\omega_5 \text{ in 1st and } \omega_3 \text{ in 2nd trial})$$
$$\times P(\omega_2 \text{ in 4th trial}/\omega_5 \text{ in 1st, } \omega_3 \text{ in 2nd and } \omega_7 \text{ in 3rd})$$

Let us assume now that (a) *the conditional probability of an outcome on any trial depends only on the outcome of the immediately preceding trial* but not on the outcomes of other previous trials, and (b) *the conditional probability of an outcome on any trial given the outcome of the immediately preceding trial does not depend on the successive ordinal number of the trial;* that is, for any k and l, the probability of occurrence of ω_j on the k-th trial given that ω_i has occurred on the $(k-1)^{\text{st}}$ trial is the same as the probability of occurrence of ω_j on the l-th trial given that ω_i has occurred on the $(l-1)^{\text{th}}$ trial. We shall denote this probability by p_{ij}. These conditions were considered first by *A. A. Markov,* hence chains of random events satisfying them are called *Markov chains.*

In our example, we get

$$P(\omega_7 \text{ in 3rd trial}/\omega_5 \text{ in 1st and } \omega_3 \text{ in 2nd trial}) = P(\omega_7 \text{ in 3rd trial}/\omega_3 \text{ in 2nd}) = p_{37}$$
$$P(\omega_2 \text{ in 4th trial}/\omega_5 \text{ in 1st, } \omega_3 \text{ in 2nd and } \omega_7 \text{ in 3rd trial}) = P(\omega_2 \text{ in 4th trial}/\omega_7 \text{ in 3rd trial}) = p_{72}$$

If we denote $P(\omega_5 \text{ in 1st trial}) = p_5$,

$P(\omega_3 \text{ in 2nd trial}/\omega_5 \text{ in 1st trial}) = p_{53}$,

then $P\{(\omega_5, \omega_3, \omega_7, \omega_2)\} = p_5 p_{53} p_{37} p_{72}$

One can observe how the probabilities are "linked" into a "chain."

Let us consider now a general case of k repetitions ($k = 2,3, \ldots$). We should like to find the probability that the system will be in the state ω_{i_1} on the first trial, in ω_{i_2} on the second trial, etc., in the state $\omega_{i_{k-3}}$ on the $(k-3)^{\text{rd}}$ trial, $\omega_{i_{k-2}}$ on the $(k-2)^{\text{nd}}$ trial, in $\omega_{i_{k-1}}$ on the $(k-1)^{\text{st}}$ trial and in the state ω_{i_k} on the k-th trial. In this notation $i_1,i_2, \ldots ,i_{k-1}i_k$ are some numbers chosen (with possible repetitions) from the numbers $1,2,3, \ldots ,n-1,n$, hence $\omega_{i_1},\omega_{i_2}, \ldots ,\omega_{i_{k-1}},\omega_{i_k}$ are some points chosen from Ω; e.g., if $i_1 = 3$, $i_2 = 5$, $i_3 = 1$, then $\omega_{i_1} = \omega_3$, $\omega_{i_2} = \omega_5$, $\omega_{i_3} = \omega_1$, and so on. Let us express first the conditional probability that the system will be in the state ω_{i_k} on the k-th trial given its whole "past history" from trial 1 to trial $(k-1)$.

$P(\omega_{i_k}$ on trial $k/\omega_{i_{k-1}}$ on trial $(k-1)$, $\omega_{i_{k-2}}$ on trial $(k-2)$, $\omega_{i_{k-3}}$ on trial $(k-3)$ . . . , ω_{i_2} on trial 2, ω_{i_1} on trial 1$) = P(\omega_{i_k}$ on trial $k/\omega_{i_{k-1}}$ on trial $(k-1)) = p_{i_{k-1}i_k}$. It follows from Markov's conditions (a) and (b) that this probability does not depend on the states at the trial $(k-2), (k-3) \ldots 3,2,1$ and furthermore it does not depend on k. It depends only on i_k and i_{k-1} and we denoted it therefore by $p_{i_{k-1}i_k}$. Obviously, $p_{i_{k-1}i_k}$ is the *transition probability* from state $\omega_{i_{k-1}}$ to state ω_{i_k} in *one step.* Let us denote the *initial* probability that the system will be in state ω_i on the first trial by p_i, that is,

$$P(\omega_i \text{ on first trial}) = p_i.$$

Obviously, $\sum_{i=1}^{n} p_i = 1$ because the system must be in one of its states in the beginning. Using the multiplication theorem for conditional probabilities, the required probability will be:

$$P(\omega_{i_1}, \omega_{i_2}, \omega_{i_3}, \ldots, \omega_{i_{k-2}}, \omega_{i_{k-1}}, \omega_{i_k}) = p_{i_1} p_{i_1 i_2} p_{i_2 i_3} \cdots p_{i_{k-2} i_{k-1}} p_{i_{k-1} i_k}$$

If we have all the probabilities p_i and p_{ij} for all the possible values $1 \leq i, j \leq n$, then we can completely describe the behavior of the system over an arbitrary period of time.

The probabilities $p_i (i = 1,2, \ldots, n)$ will be called the *initial probabilities* and the probabilities $p_{ij} (i,j = 1,2, \ldots, n)$ will be called the *transition probabilities.* We can arrange the transition probabilities in the following array, called a *matrix:*

$$\mathbf{P} = \begin{pmatrix} p_{11} & p_{12} & p_{13} & \cdots & p_{1n} \\ p_{21} & p_{22} & p_{23} & \cdots & p_{2n} \\ \cdot & \cdot & \cdot & \cdots & \cdot \\ p_{n1} & p_{n2} & p_{n3} & \cdots & p_{nn} \end{pmatrix}$$

In this matrix, the entry p_{ij}, denoting the transition probability from state ω_i to state ω_j *in one step,* is located in row i and in column j. The matrix P is called a *transition probability matrix.* Similarly, we can arrange the initial probabilities in the following array, called a *vector:*

$$\mathbf{p} = (p_1, p_2, \ldots, p_n).$$

EXAMPLE 2

Alice, Bob and Clint are playing ball. Alice likes Bob and always throws the ball to him, but Bob does not reciprocate and always throws the ball equally likely to Alice or Clint. Clint likes Alice, and throws the ball twice as likely to her than to Bob. Use a Markov chain to investigate their game.

SOLUTION. We can denote the probabilities p_{ij} as follows:

p_{11} is the probability that A keeps the ball; hence $p_{11} = 0$.

p_{12} is the probability that A throws the ball to B; hence $p_{12} = 1$.

p_{13} is the probability that A throws the ball to C; hence $p_{13} = 0$.

p_{21} is the probability that B throws the ball to A; hence $p_{21} = \frac{1}{2}$.

p_{22} is the probability that B keeps the ball; hence $p_{22} = 0$.

p_{23} is the probability that B throws the ball to C; hence $p_{23} = \frac{1}{2}$.

p_{31} is the probability that C throws the ball to A; hence $p_{31} = \frac{2}{3}$.

p_{32} is the probability that C throws the ball to B; hence $p_{32} = \frac{1}{3}$.

p_{33} is the probability that C keeps the ball; hence $p_{33} = 0$.

Therefore, the transition probability matrix P is

$$\mathbf{P} = \begin{matrix} & \begin{matrix} A & B & C \end{matrix} \\ \begin{matrix} A \\ B \\ C \end{matrix} & \begin{pmatrix} 0 & 1 & 0 \\ \frac{1}{2} & 0 & \frac{1}{2} \\ \frac{2}{3} & \frac{1}{3} & 0 \end{pmatrix} \end{matrix}$$

This is a Markov chain because, for example, whether A gets the ball or not after a certain throw depends only on who had the ball before that throw, and not at all on who had the ball two or more throws before that. Also, the preferences of A, B and C do not change during the game, yielding stationary transition probabilities.

EXAMPLE 3

A young man is well aware of the threat of pollution caused by car exhausts, therefore he frequently uses a bicycle. Not being exactly an athlete, he feels exhausted from the long pedaling to and from his office, therefore he never takes the bike two days in a row. However, if he takes the car on a certain day, then he is equally likely to take either the car or the bike the next day. Find the transition probability matrix.

SOLUTION. The states of this system are "bike" (denoted by b) and "car" (denoted by c). The transition probability matrix is

$$\mathbf{P} = \begin{array}{c} \\ b \\ c \end{array} \begin{array}{c} \begin{array}{cc} b & c \end{array} \\ \begin{pmatrix} 0 & 1 \\ \frac{1}{2} & \frac{1}{2} \end{pmatrix} \end{array}$$

Important questions in this case might be: If he takes the bike on Monday, what is the probability that he will take the car on Wednesday? If he drives on Monday, what is the probability that his wife will have the car for Friday's grocery shopping? We shall answer these questions later.

EXAMPLE 4

A cougar's territory consists of three regions A, B and C. It never hunts in the same region on successive days. If it hunts in region A, then next day it moves to region B. However, if it is in either B or C, then next day it is twice as likely to appear to region A as in the other region. The states of this system are A,B,C, and the transition probability matrix is

$$\mathbf{P} = \begin{array}{c} \\ A \\ B \\ C \end{array} \begin{array}{c} \begin{array}{ccc} A & B & C \end{array} \\ \begin{pmatrix} 0 & 1 & 0 \\ \frac{2}{3} & 0 & \frac{1}{3} \\ \frac{2}{3} & \frac{1}{3} & 0 \end{pmatrix} \end{array}$$

In each of these examples, rowwise sums in the transition matrix are always 1. This is a fundamental property of transition probability matrices. We know that if in a trial an event ω_i occurs, or as we sometimes say, the system is in the state ω_i, then with probability p_{i1} it will go to the state ω_1, with probability p_{i2} to the state ω_2, . . . , with probability p_{ij} to the state ω_j, . . . , and with probability p_{in} to the state ω_n. It is certain that in the next trial (or step, as we sometimes say) one of the outcomes $\omega_1,\omega_2, . . . ,\omega_j, . . . ,\omega_n$ must occur. We have proved:

→ **THEOREM 1:** $\sum_{j=1}^{n} p_{ij} = 1$ for every $i = 1,2, . . ,n$.

EXAMPLE 5

Random Walk with Reflecting Barriers. A happily intoxicated man walks home to his wife from a nearby beer parlor. In order to keep his balance, now and then he takes either a step forward with probability p or a step backward with probability $q = 1 - p$. When he finally reaches home, his angry wife throws him out immediately. If, by chance, he reaches the beer parlor again, he is also bounced out immediately. Investigate the behavior of this poor fellow. Assume that his house is only five steps from the beer parlor and that he always makes steps of equal length.

SOLUTION. This is an example of a random walk with reflecting barriers. The man is at an integral point on the axis between the origin (beer parlor) and point 5 (his wife). He takes a unit step with probability p to the right and with probability $q = 1 - p$ to the left, unless he is at the point 0 where he must go a unit step to the right, or at point 5 where he must go a unit step to the left. We can view the states of the system as $\{0,1,2,3,4,5\}$.

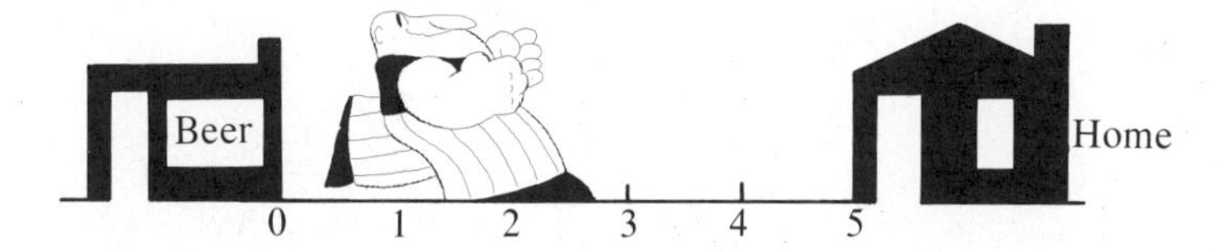

The transition matrix is

$$\mathbf{P} = \begin{array}{c} \\ 0 \\ 1 \\ 2 \\ 3 \\ 4 \\ 5 \end{array} \begin{array}{c} \begin{array}{cccccc} 0 & 1 & 2 & 3 & 4 & 5 \end{array} \\ \begin{pmatrix} 0 & 1 & 0 & 0 & 0 & 0 \\ q & 0 & p & 0 & 0 & 0 \\ 0 & q & 0 & p & 0 & 0 \\ 0 & 0 & q & 0 & p & 0 \\ 0 & 0 & 0 & q & 0 & p \\ 0 & 0 & 0 & 0 & 1 & 0 \end{pmatrix} \end{array}$$

EXAMPLE 6

Random Walk with Absorbing Barriers. Imagine that the man in the previous example has a more congenial wife and is befriended by the bouncer. He still stumbles between his house and the beer parlor: however, if he enters one of these places he is grabbed either by his wife (if at home) or by the bouncer (if in the tavern) and put to bed.

SOLUTION. The Markov chain describing the drunkard's behavior is similar to that in Example 5, except that his home and the beer parlor act as absorbing states: if he reaches one of them, he is trapped in it and his wandering stops.

$$\mathbf{P} = \begin{array}{c} \\ 0 \\ 1 \\ 2 \\ 3 \\ 4 \\ 5 \end{array} \begin{array}{c} \begin{array}{cccccc} 0 & 1 & 2 & 3 & 4 & 5 \end{array} \\ \begin{pmatrix} 1 & 0 & 0 & 0 & 0 & 0 \\ q & 0 & p & 0 & 0 & 0 \\ 0 & q & 0 & p & 0 & 0 \\ 0 & 0 & q & 0 & p & 0 \\ 0 & 0 & 0 & q & 0 & p \\ 0 & 0 & 0 & 0 & 0 & 1 \end{pmatrix} \end{array}$$

EXAMPLE 7

Diffusion of Gases. There are two containers C_1 and C_2. In the beginning, C_1 contains n black balls and C_2 contains n white balls. There is an opening between the containers that is occupied by a small creature, called the Maxwell demon (a hypothetical agent or device of arbitrarily small mass that is considered selectively to admit or block the passage of individual molecules from one compartment to another; named after *J. C. Maxwell*). Assume that this creature selects a ball at random from each container and interchanges these balls. Then he selects another two balls and interchanges them, and so on.

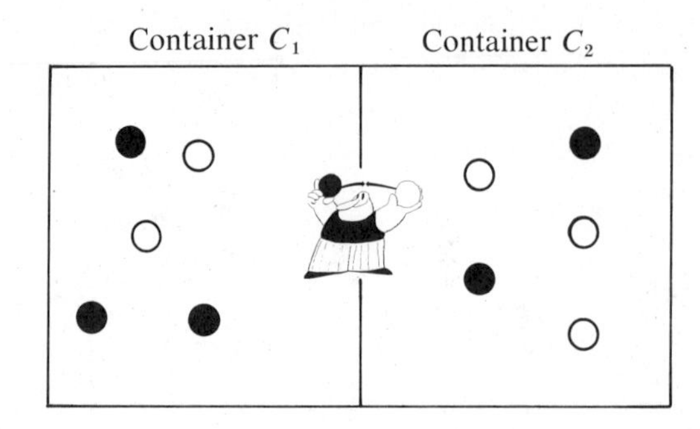

The state of this system is uniquely described if we know the number j of black balls in C_1. Then the number of white balls in C_1 must be $(n - j)$, the number of black balls in C_2 must be $(n - j)$ and the number of white balls in C_2 must be j. The states of the system are $\{0,1,2,3, \ldots, n\}$. Let us find $p_{j,j+1}$. This is the probability that if there are j black balls in C_1, then after one exchange there will be $j + 1$ of them. This can happen only if a white ball is selected from C_1 $\left(\text{the probability of which is } \frac{n-j}{n}\right)$, and a black ball is selected from C_2 $\left(\text{the probability of which is } \frac{n-j}{n}\right)$. Hence

$$p_{j,j+1} = \left(\frac{n-j}{n}\right)^2 \qquad j < n.$$

Similarly,

$$p_{j,j} = \frac{2j(n-j)}{n^2} \qquad 0 < j < n.$$

$$p_{j,j-1} = \left(\frac{j}{n}\right)^2 \qquad j > 0$$

$$p_{j,k} = 0. \qquad \text{otherwise}$$

If we have $n = 3$ balls in each container, then the transition probability matrix is

$$\mathbf{P} = \begin{array}{c} 0 \\ 1 \\ 2 \\ 3 \end{array} \overset{\begin{array}{cccc} 0 & 1 & 2 & 3 \end{array}}{\begin{pmatrix} 0 & 1 & 0 & 0 \\ \frac{1}{9} & \frac{4}{9} & \frac{4}{9} & 0 \\ 0 & \frac{4}{9} & \frac{4}{9} & \frac{1}{9} \\ 0 & 0 & 1 & 0 \end{pmatrix}}$$

This model describes diffusion of gases: one can imagine a molecule of oxygen as a black ball and a molecule of helium as a white ball. The first container is filled with oxygen, the second with helium and a hole is pierced in the common wall between them. The transition probability matrix then describes an idealized way of diffusion.

EXAMPLE 8

Queueing Model. Customers arrive at random at a service counter. Only one customer can be served at a time, but up to two more can be accommodated in a queue. We can imagine that there is a *waiting room* with a capacity of two people. If more than three customers at a time require service, then the fourth, the fifth, and so on, are turned away. Assume that within an interval of time a customer will arrive with probability q, and the customer who is being served will leave the counter with probability p. Assume also that the time interval is sufficiently small so that it is practically impossible that more than one customer will arrive or will leave the counter during this interval. Investigate the service using a Markov chain.

SOLUTION. Let the state of the system be the number of people accepted for the service (including the served customer). Then there are four possible states, 0,1,2,3.

p_{00} is the probability that if nobody is accepted for service, then after the interval of time there will still be nobody; it means nobody arrives, hence $p_{00} = 1 - q$.

p_{01} means that if there is nobody for service, then after the interval there will be one person, who arrives with probability q, hence $p_{01} = q$.

p_{10} means that if one person is served during the interval, he will leave and nobody will arrive, hence $p_{10} = p(1 - q)$.

p_{11} means the probability that either one person being served will leave and another customer will arrive, or the person being served will not leave and nobody will arrive, hence $p_{11} = pq + (1 - p)(1 - q)$.

p_{12} means the probability that one person being served will not leave and a second person will arrive, hence $p_{12} = q(1 - p)$.

p_{13} means the probability that if one person is served, then during the interval at least two more will arrive, hence $p_{13} = 0$.

Similarly, $p_{20} = 0,\quad p_{21} = p(1 - q),\quad p_{22} = pq + (1 - p)(1 - q),$

$p_{23} = q(1 - p),\quad p_{30} = p_{31} = 0,\quad p_{32} = p(1 - q).$

p_{32} means the probability that either one will leave the line and one will arrive, or nobody will leave, hence $p_{33} = pq + (1 - p)$.

Thus the transition probability matrix is

$$\mathbf{P} = \begin{pmatrix} (1-q) & q & 0 & 0 \\ p(1-q) & pq + (1-p)(1-q) & q(1-p) & 0 \\ 0 & p(1-q) & pq + (1-p)(1-q) & q(1-p) \\ 0 & 0 & p(1-q) & pq + (1-p) \end{pmatrix}$$

This simple example shows an approach to a solution of some problems in so-called *queueing theory*. Our model is very simple, but incorporates basic factors leading to some interesting practical results.

EXAMPLE 9

Cell Genetics. Some biometrical theories describe cell regeneration in the following way. Each cell has N certain particles, some of them type A and the others type B. The cell is said to be in the state i, if it contains exactly i particles of type A. The regeneration process runs as follows: first, all the particles of the mother cell replicate themselves; then N particles are chosen at random among these $2i$ particles of type A and $2(N - i)$ particles of type B in order to create the daughter cell; finally, the mother cell dies off and the daughter cell replaces her. Obviously, if the mother cell was in state i, then the probability of the daughter cell being in the state j after the regeneration is given by the hypergeometric distribution. The transition probabilities are therefore

$$p_{ij} = \frac{\binom{2i}{j}\binom{2(N-i)}{N-j}}{\binom{2N}{N}} \qquad i,j = 0,1,2, \ldots, N.$$

Clearly the states $i = 0$ and $i = N$ are absorbing states. If the cell reaches any one of them during its regeneration, then it can never leave them.

EXAMPLE 10

Incestuous Progeny. Some of the simplest models in genetics consider so-called brother-sister mating: two direct descendants from the same parents are chosen at random and mated again, their offspring are mated again at random, and so on. Consider the inheritance of genotypes determined by two alleles A and B. A genotype is any (unordered) pair of alleles, hence a member of the progeny can have one of the three genotypes AA, AB, BB. For any two parents, the genotype of the offspring is determined as follows: one allele is chosen at random from each parent, and the obtained pair is the child's genotype. If both parents have the AA genotype, then all their offspring also have the AA genotype and the next generation parents will be of the same type. The same is true if both parents have the BB genotype. If one parent has the AA genotype and the other has the BB, then both parents in the next generation will have the AB genotype. The situation is more complicated if one parent has the AA and the other the AB genotype. Then offspring will equally likely have the AA or AB genotype, and if two are selected at random for mating, according to the binomial distribution they will both have the AA genotype with probability $\frac{1}{4}$, one AA and one AB with probability $\frac{1}{2}$ and both AB with probability $\frac{1}{4}$.

This shows that we could consider the *type of mating* as a state of a Markov chain. There are six types of mating: $\omega_1 = AA \times AA$, $\omega_2 = AA \times AB$, $\omega_3 = AB \times AB$, $\omega_4 = AB \times BB$, $\omega_5 = BB \times BB$, $\omega_6 = AA \times BB$. We have just shown that

$$p_{11} = 1 \qquad p_{1j} = 0 \qquad j \neq 1$$
$$p_{55} = 1 \qquad p_{5j} = 0 \qquad j \neq 5$$
$$p_{63} = 1 \qquad p_{6j} = 0 \qquad j \neq 3$$

or, in other words, ω_1 and ω_5 are absorbing states. Further,

$$p_{21} = \frac{1}{4} \quad p_{22} = \frac{1}{2} \quad p_{23} = \frac{1}{4} \quad p_{2j} = 0. \quad j > 3$$

Similarly, we can calculate the remaining transition probabilities.

$$\mathbf{P} = \begin{pmatrix} 1 & 0 & 0 & 0 & 0 & 0 \\ \frac{1}{4} & \frac{1}{2} & \frac{1}{4} & 0 & 0 & 0 \\ \frac{1}{16} & \frac{1}{4} & \frac{1}{4} & \frac{1}{4} & \frac{1}{16} & \frac{1}{8} \\ 0 & 0 & \frac{1}{4} & \frac{1}{2} & \frac{1}{4} & 0 \\ 0 & 0 & 0 & 0 & 1 & 0 \\ 0 & 0 & 1 & 0 & 0 & 0 \end{pmatrix}$$

EXAMPLE 11

Coin Game. Two players A and B have five silver dollars distributed between them and they play the following game: each of them takes one coin and both flip the coins simultaneously. A keeps the coins displaying a head, B keeps the rest. The game stops when one of them is ruined (has no coins left).

SOLUTION. The coins need not be fair, but let us assume that they are identical and that the probability of getting H is p and the probability of getting T is $q = 1 - p$. Let the state of the system be the number of coins in A's possession. Then the transition probability matrix is

$$\mathbf{P} = \begin{pmatrix} 1 & 0 & 0 & 0 & 0 & 0 \\ q^2 & 2pq & p^2 & 0 & 0 & 0 \\ 0 & q^2 & 2pq & p^2 & 0 & 0 \\ 0 & 0 & q^2 & 2pq & p^2 & 0 \\ 0 & 0 & 0 & q^2 & 2pq & p^2 \\ 0 & 0 & 0 & 0 & 0 & 1 \end{pmatrix}$$

The first row corresponding to no coins and the last row corresponding to 5 coins in A's possession are absorbing states. It is intuitively clear that the game will sooner or later certainly end in one of them. We can modify the rules as follows: if one of the players has no coins left, then the other tosses a single coin alone and A keeps the coin if it displays a head, otherwise B gets it. In this case, the transition matrix is

$$\mathbf{P} = \begin{pmatrix} q & p & 0 & 0 & 0 & 0 \\ q^2 & 2pq & p^2 & 0 & 0 & 0 \\ 0 & q^2 & 2pq & p^2 & 0 & 0 \\ 0 & 0 & q^2 & 2pq & p^2 & 0 \\ 0 & 0 & 0 & q^2 & 2pq & p^2 \\ 0 & 0 & 0 & 0 & q & p \end{pmatrix}$$

All the previous examples gave a transition probability matrix in one step. Let us investigate now the actual behavior of a Markov chain over a longer period of time.

EXAMPLE 12

Consider again the traveling salesman in Example 1. On Monday, he starts his journey in one of the cities Abadan, Belial or Cithaeron with probabilities $\frac{1}{2}, \frac{1}{4}$ or $\frac{1}{4}$, respectively. What are his probabilities of being in the cities A, B or C on Tuesday?

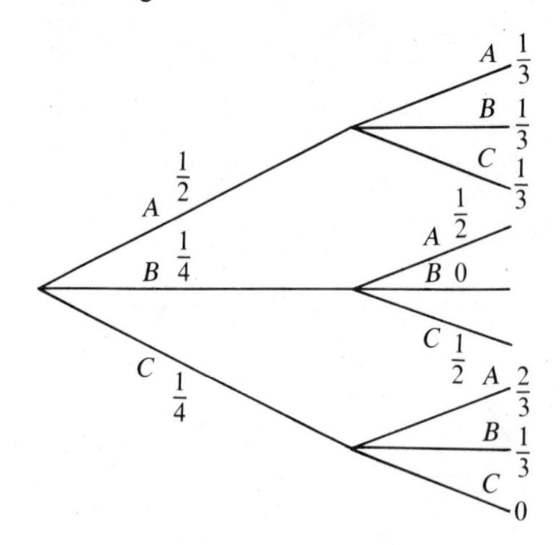

From this tree, we can see that he will be in A on Tuesday with probability

$$\frac{1}{2}\cdot\frac{1}{3}+\frac{1}{4}\cdot\frac{1}{2}+\frac{1}{4}\cdot\frac{2}{3}=\frac{11}{24}.$$

We can imagine his behavior schematically as follows:

Monday in	With probability	Transit to A for Tuesday with probability	Probabilities
A	$p_1=\frac{1}{2}$	$p_{11}=\frac{1}{3}$	$p_1p_{11}=\frac{1}{2}\cdot\frac{1}{3}$
B	$p_2=\frac{1}{4}$	$p_{21}=\frac{1}{2}$	$p_2p_{21}=\frac{1}{4}\cdot\frac{1}{2}$
C	$p_3=\frac{1}{4}$	$p_{31}=\frac{2}{3}$	$p_3p_{31}=\frac{1}{4}\cdot\frac{2}{3}$
		TOTAL: Probability of being in A on Tuesday:	$p_1p_{11}+p_2p_{21}+p_3p_{31}=\frac{1}{2}\cdot\frac{1}{3}+\frac{1}{4}\cdot\frac{1}{2}+\frac{1}{4}\cdot\frac{2}{3}=\frac{11}{24}$

Let us denote the probability of the salesman being in city j after one day (on Tuesday) by $p_j^{(1)}$. Then obviously,

$$p_1^{(1)}=p_1p_{11}+p_2p_{21}+p_3p_{31}=\frac{11}{24}.$$

Similarly, we can calculate the probability of him being in B on Tuesday:

$$p_2^{(1)} = p_1 p_{12} + p_2 p_{22} + p_3 p_{32} = \frac{1}{2} \cdot \frac{1}{3} + \frac{1}{4} \cdot 0 + \frac{1}{4} \cdot \frac{1}{3} = \frac{6}{24}.$$

Finally, the probability of being in C on Tuesday is

$$p_3^{(1)} = p_1 p_{13} + p_2 p_{23} + p_3 p_{33} = \frac{1}{2} \cdot \frac{1}{3} + \frac{1}{4} \cdot \frac{1}{2} + \frac{1}{4} \cdot 0 = \frac{7}{24}.$$

We shall note the *formal* way of getting these probabilities; for B, we get

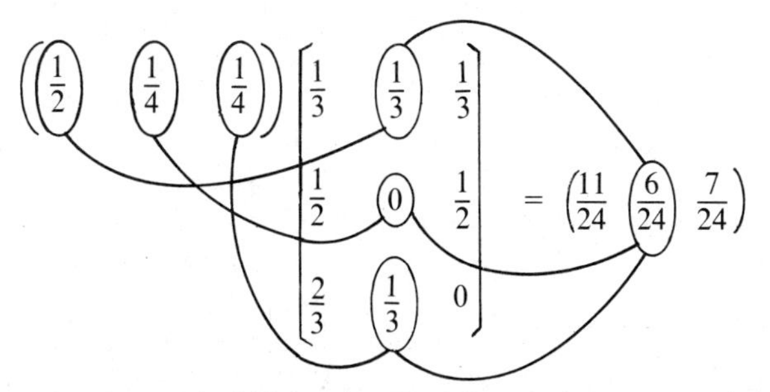

In order to get, for example, probabilities for Tuesday for the *second* city (B), we take the components of the vector of initial probabilities, multiply them by the entries in the *second* column, and add the products.

In general, we have n states, and the system can be initially in any one of them. The probability $p_j^{(1)}$ that the system will be in the state ω_j *after one step* can be calculated as

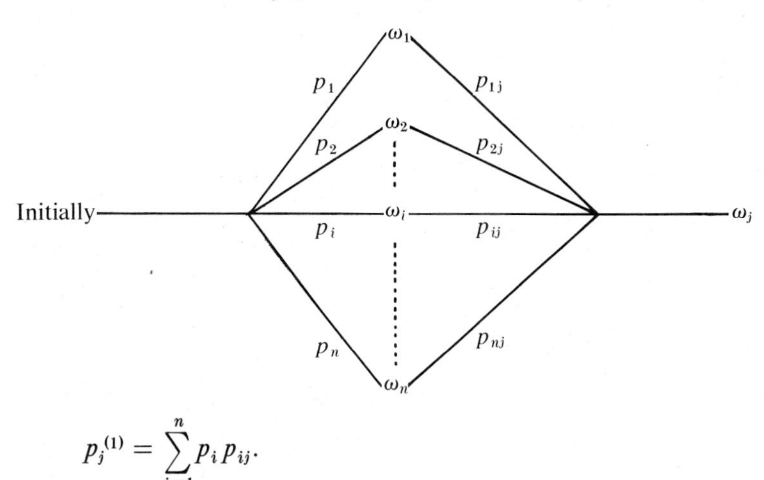

$$p_j^{(1)} = \sum_{i=1}^{n} p_i p_{ij}.$$

We have informally proved the following:

THEOREM 2: Let p be the vector of initial probabilities, and P be the transition probability matrix of a Markov chain. Then the probability $p_j^{(1)}$ that the system will be in state ω_j after one step is

$$p_j^{(1)} = \sum_{i=1}^{n} p_i p_{ij}.$$

COMMENT. The probabilities $p_j^{(1)}, j = 1,2, \ldots, n$ can be written again as a vector

$$p^{(1)} = (p_1^{(1)}, p_2^{(1)} \ldots, p_n^{(1)})$$

and we can easily see that

$$\sum_{j=1}^{n} p_j^{(1)} = \sum_{j=1}^{n} \sum_{i=1}^{n} p_i p_{ij} = \sum_{i=1}^{n} p_i \sum_{j=1}^{n} p_{ij} = \sum_{i=1}^{n} p_i = 1.$$

EXAMPLE 13

The three areas of the cougar's territory in Example 4 have been photographed from a helicopter, and evaluated by a wildlife biologist. It is impossible to tell exactly where the animal was located on the day of the survey but the aerial photos show that the animal was twice as likely to be in the heavily wooded area A as in the sparsely covered area B. It definitely could not be in the open rocky area C, otherwise the pictures would have shown it. Find the probabilities of the animal being in A, B and C, respectively, one day later.

SOLUTION. The initial vector of probabilities is $p = \left(\frac{2}{3}, \frac{1}{3}, 0\right)$, and we get the probabilities for the next day as

$$p_1^{(1)} = \frac{2}{3} \cdot 0 + \frac{1}{3} \cdot \frac{2}{3} + 0 \cdot \frac{2}{3} = \frac{2}{9}.$$

$$p_2^{(1)} = \frac{2}{3} \cdot 1 + \frac{1}{3} \cdot 0 + 0 \cdot \frac{1}{3} = \frac{6}{9}.$$

$$p_3^{(1)} = \frac{2}{3} \cdot 0 + \frac{1}{3} \cdot \frac{1}{3} + 0 \cdot \frac{1}{3} = \frac{1}{9}.$$

We can generalize the notion of a *vector* and of a *matrix* as follows. Let n be a natural number; then any ordered n-tuple of real numbers

$$v = (v_1, v_2, \ldots, v_n)$$

is called an n dimensional vector. Any array of real numbers of the form

$$\mathbf{M} = \begin{pmatrix} m_{11} & m_{12} & \ldots & m_{1n} \\ m_{21} & m_{22} & \ldots & m_{2n} \\ \vdots & \vdots & & \vdots \\ m_{n1} & m_{n2} & \ldots & m_{nn} \end{pmatrix}$$

is called an $n \times n$ square matrix. Motivated by Theorem 2, we can define the product of a vector v and a matrix $\mathbf{M}$.

DEFINITION: Let v be an n-dimensional vector and **M** be an $n \times n$ square matrix. Then the product of v and **M** is a vector u

$$v\mathbf{M} = \mathbf{u}$$

with components given as follows

$$u_j = \sum_{i=1}^{n} v_i M_{ij}.$$

COMMENT. Obviously the vector of probabilities $\mathbf{p}^{(1)}$ is equal to

$$\mathbf{p}^{(1)} = \mathbf{pP}.$$

TRANSITION PROBABILITIES OF HIGHER ORDERS

EXAMPLE 14

Let us continue the investigation of the salesman's journeys. (See Example 12.) We know that he will be in the three cities A,B,C after *one day,* i.e.,

$$\mathbf{p}^{(1)} = \left(\frac{11}{24}, \frac{6}{24}, \frac{7}{24}\right).$$

What are the probabilities of him being in A,B,C after *two days,* that is, on Wednesday? We can use a probability tree again but it is more or less clear that if we denote the vector of probabilities after two days by $p^{(2)}$, then we get

$$p^{(2)} = p^{(1)}\mathbf{P}.$$

$$p_1^{(2)} = p_1^{(1)}p_{11} + p_2^{(1)}p_{21} + p_3^{(1)}p_{31} = \frac{11}{24}\cdot\frac{1}{3} + \frac{6}{24}\cdot\frac{1}{2} + \frac{7}{24}\cdot\frac{2}{3} = \frac{68}{144}.$$

$$p_2^{(2)} = p_1^{(1)}p_{12} + p_2^{(1)}p_{22} + p_3^{(1)}p_{32} = \frac{11}{24}\cdot\frac{1}{3} + \frac{6}{24}\cdot 0 + \frac{7}{24}\cdot\frac{1}{3} = \frac{36}{144}.$$

$$p_3^{(2)} = p_1^{(1)}p_{13} + p_2^{(1)}p_{23} + p_2^{(1)}p_{33} = \frac{11}{24}\cdot\frac{1}{3} + \frac{6}{24}\cdot\frac{1}{2} + \frac{7}{24}\cdot 0 = \frac{40}{144}.$$

The reader is well advised to construct a probability tree, and see how the vector and matrix multiplication corresponds to the branches of the tree. The probability vector after two days is

$$p^{(2)} = \left(\frac{68}{144}, \frac{36}{144}, \frac{40}{144}\right).$$

However, we should like to ask now a very important question. We got the probabilities *after two steps* by multiplying the initial probabilities *two times by the transition probabilities in one step.* Could we not actually find *transition probabilities* from a state ω_i to a state ω_j *in two steps?* Imagine that the salesman is in A on Monday. What is the probability that he will be in C after two days, that is, on Wednesday?

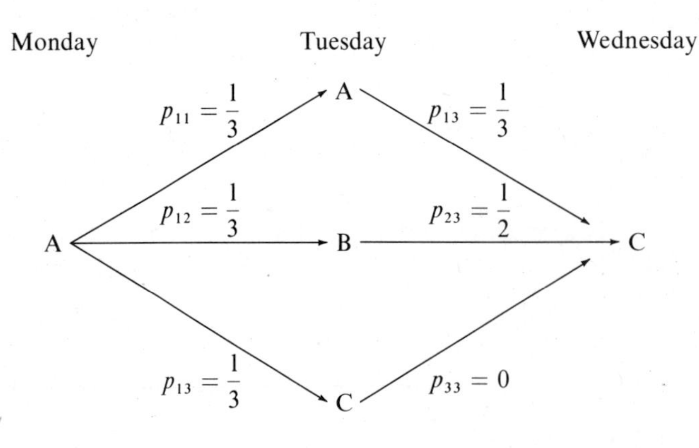

Probabilities

$$p_{11}p_{13} = \frac{1}{3} \cdot \frac{1}{3}$$
$$p_{12}p_{23} = \frac{1}{3} \cdot \frac{1}{2}$$
$$p_{13}p_{33} = \frac{1}{3} \cdot 0$$

$$p_{11}p_{13} + p_{12}p_{23} + p_{13}p_{33} = = \frac{1}{3} \cdot \frac{1}{3} + \frac{1}{3} \cdot \frac{1}{2} + \frac{1}{3} \cdot 0 = \frac{5}{18}$$

TOTAL: Probability of transition from A to C in two steps:

If we denote this probability $p_{13}^{(2)}$ (transition probability from A to C in two days), then we have

$$p_{13}^{(2)} = \frac{5}{18}.$$

Similarly, we can calculate all other probabilities, for example:

$$p_{11}^{(2)} = p_{11}p_{11} + p_{12}p_{21} + p_{13}p_{31} = \frac{1}{3} \cdot \frac{1}{3} + \frac{1}{3} \cdot \frac{1}{2} + \frac{1}{3} \cdot \frac{2}{3} = \frac{9}{18}.$$

$$p_{23}^{(2)} = p_{21}p_{13} + p_{22}p_{23} + p_{23}p_{33} = \frac{1}{2} \cdot \frac{1}{3} + 0 \cdot \frac{1}{2} + \frac{1}{2} \cdot 0 = \frac{3}{18}.$$

In general, if we have n states then the transition from ω_i to ω_j in *two steps* can happen as follows:

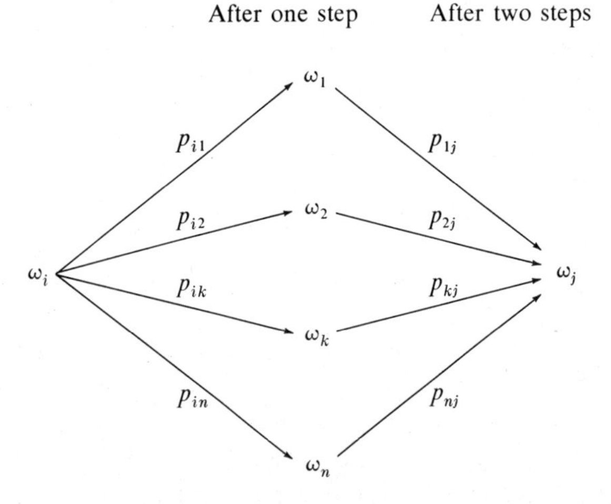

Probabilities

$$p_{i1}p_{1j}$$
$$p_{i2}p_{2j}$$
$$\vdots$$
$$p_{ik}p_{kj}$$
$$\vdots$$
$$p_{in}p_{nj}$$

$$\sum_{k=1}^{n} p_{ik}p_{kj}$$

TOTAL: Transition probability from i to j in two steps:

We have thus informally proved the following:

→ **THEOREM 3:** Let **P** be the transition probability matrix in one step. Then the transition probability matrix in two steps $\mathbf{P}^{(2)}$ consists of elements $p_{ij}^{(2)}$ given as

$$p_{ij}^{(2)} = \sum_{k=1}^{n} p_{ik}\, p_{kj}.$$

Motivated by Theorem 3, we can introduce the notion of multiplication of two $n \times n$ square matrices. This notion can be further generalized (so that, for example, matrix multiplication and vector-matrix multiplication become closely related) but we shall not attempt to do it here; students interested in these topics should refer to any book on linear algebra.

DEFINITION: Let **M** and **N** be two $n \times n$ square matrices consisting of elements m_{ij} and n_{ij}, $1 \leq i, j \leq n$, respectively. Then the product of **M** and **N** is an $n \times n$ square matrix **Q**

$$\mathbf{Q} = \mathbf{MN}$$

consisting of elements q_{ij} given as

$$q_{ij} = \sum_{k=1}^{n} m_{ik} m_{kj}.$$

COMMENT. The transition probability matrix $\mathbf{P}^{(2)}$ after two steps is, by Theorem 3, a product of the transition probability matrix **P** after one step with itself,

$$\mathbf{P}^{(2)} = \mathbf{PP}.$$

Let us now finish the calculation of the transition probability matrix in two days for the traveling salesman.

$$\mathbf{P}^{(2)} = \mathbf{PP} = \begin{pmatrix} \frac{1}{3} & \frac{1}{3} & \frac{1}{3} \\ \frac{1}{2} & 0 & \frac{1}{2} \\ \frac{2}{3} & \frac{1}{3} & 0 \end{pmatrix} \begin{pmatrix} \frac{1}{3} & \frac{1}{3} & \frac{1}{3} \\ \frac{1}{2} & 0 & \frac{1}{2} \\ \frac{2}{3} & \frac{1}{3} & 0 \end{pmatrix} = \begin{pmatrix} \frac{9}{18} & \frac{4}{18} & \frac{5}{18} \\ \frac{9}{18} & \frac{6}{18} & \frac{3}{18} \\ \frac{7}{18} & \frac{4}{18} & \frac{7}{18} \end{pmatrix}$$

Notice that, for example, the entry in the *second row and first column* of the matrix $\mathbf{P}^{(2)}$ is obtained by multiplication and addition of appropriate entries of the *second row and first column* of the matrix **P**, as shown by the joining lines:

$$p_{21}^{(2)} = p_{21}p_{11} + p_{22}p_{21} + p_{23}p_{31} = \frac{1}{2}\cdot\frac{1}{3} + 0\cdot\frac{1}{2} + \frac{1}{2}\cdot\frac{2}{3} = \frac{9}{18}.$$

In Example 14, we got the vector of probabilities that the salesman will be in A, B and C after *two steps,* that is, on Wednesday,

$$\mathbf{p}^{(2)} = \left(\frac{68}{144}, \frac{36}{144}, \frac{40}{144}\right)$$

by multiplication of the corresponding probabilities $p^{(1)}$ for Tuesday by the transition probability matrix in one step, P, that is,

$$\mathbf{p}^{(2)} = \mathbf{p}^{(1)}\mathbf{P}.$$

However, let us see what happens if we multiply the initial probabilities $\mathbf{p}$ for Monday by the transition probability matrix in two steps $\mathbf{P}^{(2)}$ (that is Monday–Tuesday–Wednesday):

$$\mathbf{pP}^{(2)} = \left(\frac{1}{2}, \frac{1}{4}, \frac{1}{4}\right) \begin{pmatrix} \frac{9}{18} & \frac{4}{18} & \frac{5}{18} \\ \frac{9}{18} & \frac{6}{18} & \frac{3}{18} \\ \frac{7}{18} & \frac{4}{18} & \frac{7}{18} \end{pmatrix} = \begin{pmatrix} \frac{68}{144} & \frac{36}{144} & \frac{40}{144} \end{pmatrix} = \mathbf{p}^{(2)}$$

We should notice that the vector of probabilities after two steps, i.e., on Wednesday, can be obtained *either* by multiplication of the vector of probabilities after one step, i.e., on Tuesday, by the transition probability matrix in one step, *or* by multiplication of the vector of initial probabilities, i.e., on Monday, by the transition probability matrix in two steps.

$$\mathbf{p}^{(2)} = \mathbf{p}^{(1)}\mathbf{P} = \mathbf{pP}^{(2)}.$$

EXAMPLE 15

Let us investigate the cougar's behavior once more. In Example 4, we found that if he is initially (say on Monday) in one of the three areas with probabilities $\mathbf{p} = \left(\frac{2}{3}, \frac{1}{3}, 0\right)$, then after one day the corresponding probabilities are $\mathbf{p}^{(1)} = \left(\frac{2}{9}, \frac{6}{9}, \frac{1}{9}\right)$. Let us calculate the transition probability matrix in two days.

$$\mathbf{P}^{(2)} = \mathbf{PP} = \begin{pmatrix} 0 & 1 & 0 \\ \frac{2}{3} & 0 & \frac{1}{3} \\ \frac{2}{3} & \frac{1}{3} & 0 \end{pmatrix} \begin{pmatrix} 0 & 1 & 0 \\ \frac{2}{3} & 0 & \frac{1}{3} \\ \frac{2}{3} & \frac{1}{3} & 0 \end{pmatrix} = \begin{pmatrix} \frac{2}{3} & 0 & \frac{1}{3} \\ \frac{2}{9} & \frac{7}{9} & 0 \\ \frac{2}{9} & \frac{6}{9} & \frac{1}{9} \end{pmatrix}$$

The probability vector of the cougar being respectively in A,B,C two days after the survey, i.e., on Wednesday, is equal to

$$\mathbf{p}^{(2)} = \mathbf{pP}^{(2)} = \left(\frac{2}{3}, \frac{1}{3}, 0\right) \begin{pmatrix} \frac{2}{3} & 0 & \frac{1}{3} \\ \frac{2}{9} & \frac{7}{9} & 0 \\ \frac{2}{9} & \frac{6}{9} & \frac{1}{9} \end{pmatrix} = \left(\frac{14}{27}, \frac{7}{27}, \frac{6}{27}\right) = \mathbf{p}^{(1)}\mathbf{P}$$

as can be easily verified. If we now multiply the transition probability matrix in one step $\mathbf{P}$ by the transition probability matrix in 2 steps $\mathbf{P}^{(2)}$, then we obviously get the transition probability matrix in three steps $\mathbf{P}^{(3)}$.

$$\mathbf{P}^{(3)} = \mathbf{P}\mathbf{P}^{(2)} = \begin{pmatrix} 0 & 1 & 0 \\ \frac{2}{3} & 0 & \frac{1}{3} \\ \frac{2}{3} & \frac{1}{3} & 0 \end{pmatrix} \begin{pmatrix} \frac{2}{3} & 0 & \frac{1}{3} \\ \frac{2}{9} & \frac{7}{9} & 0 \\ \frac{2}{9} & \frac{6}{9} & \frac{1}{9} \end{pmatrix} = \begin{pmatrix} \frac{2}{9} & \frac{7}{9} & 0 \\ \frac{14}{27} & \frac{6}{27} & \frac{7}{27} \\ \frac{14}{27} & \frac{7}{27} & \frac{6}{27} \end{pmatrix}$$

It can be easily verified that the following property is satisfied:

$$\mathbf{P}^{(3)} = \mathbf{P}\mathbf{P}^{(2)} = \mathbf{P}^{(2)}\mathbf{P}.$$

We can get the transition probability matrix in four steps $\mathbf{P}^{(4)}$ as follows:

$$\mathbf{P}^{(4)} = \mathbf{P}\mathbf{P}^{(3)} = \mathbf{P}^{(2)}\mathbf{P}^{(2)} = \mathbf{P}^{(3)}\mathbf{P} = \begin{pmatrix} \frac{14}{27} & \frac{6}{27} & \frac{7}{27} \\ \frac{26}{81} & \frac{49}{81} & \frac{6}{81} \\ \frac{26}{81} & \frac{48}{81} & \frac{7}{81} \end{pmatrix}$$

We can continue, and find the transition probability matrix after 5, 6, 7, 10 steps, etc. In general, we can use the following theorem.

THEOREM 4: *Chapman-Kolmogorov Equations.* Let $\mathbf{P}^{(k)}$ and $\mathbf{P}^{(l)}$ be the transition probability matrices in k and l steps, respectively. Then the transition probability matrix $\mathbf{P}^{(k+l)}$ in $(k + l)$ steps is equal to

$$\mathbf{P}^{(k+l)} = \mathbf{P}^{(k)}\mathbf{P}^{(l)}$$

or equivalently, $$p_{ij}^{(k+l)} = \sum_{m=1}^{n} p_{im}^{(k)} p_{mj}^{(l)}. \qquad i,j = 1,2, \ldots n.$$

COMMENT. The matrix $\mathbf{P}^{(k)}$ is sometimes called the k-th power of the matrix $\mathbf{P}$. In our notation, $\mathbf{P}^{(1)} = \mathbf{P}$ is the one step transition probability matrix. For every i and k we have $\sum_{j=1}^{n} p_{ij}^{(k)} = 1$ because the system which is in the state i in the beginning must be in one of the states $\omega_1, \omega_2, \omega_3, \ldots, \omega_n$ after k steps.

LIMIT BEHAVIOR

We have investigated the behavior of a Markov chain after 2, 3, 4, 5, 10 and more steps. Can we conclude anything about its behavior after a *very long period of time?* Sometimes the answer is quite simple. It is intuitively clear that if the chain has one or more absorbing states, as in Examples 6, 9 or 11, then with probability 1 the system will get trapped in one of them after a sufficiently long period of time (we shall not try to prove this intuitive statement mathematically). Let us investigate now the other types of chains without absorbing states.

EXAMPLE 16

Let us investigate the behavior of the Maxwell demon (see Example 7). Imagine that before he starts his regular activities, he fills the containers as follows: He picks three black and three white balls and rolls a die. If a 1, 2 or 3 appears, then he puts into the first container the number of black balls equal to the outcome on the die, otherwise he does not put any black balls there. Then he puts the remaining black balls into the second container, fills the containers with white balls so that each contains exactly three balls and starts his random exchanges. The initial probability vector is hence $\boldsymbol{p} = \left(\frac{1}{2}, \frac{1}{6}, \frac{1}{6}, \frac{1}{6}\right)$ and the one step transition matrix is given in Example 7. Let us calculate the probability vectors $\boldsymbol{p}^{(m)}$ after m steps for some values m.

$$
\begin{aligned}
\boldsymbol{p} &= (.5 \quad .1667 \quad .1667 \quad .1667) \\
\boldsymbol{p}^{(1)} &= (.0185 \quad .6482 \quad .3149 \quad .0185) \\
\boldsymbol{p}^{(2)} &= (.0720 \quad .4465 \quad .4465 \quad .0350) \\
\boldsymbol{p}^{(3)} &= (.0496 \quad .4689 \quad .4319 \quad .0496) \\
\boldsymbol{p}^{(5)} &= (.0500 \quad .4521 \quad .4480 \quad .0500) \\
\boldsymbol{p}^{(7)} &= (.0500 \quad .4503 \quad .4497 \quad .0500) \\
\boldsymbol{p}^{(10)} &= (.0500 \quad .4500 \quad .4500 \quad .0500) \\
\boldsymbol{p}^{(11)} &= (.0500 \quad .4500 \quad .4500 \quad .0500)
\end{aligned}
$$

$$\boldsymbol{p}^{(10)} = \boldsymbol{p}^{(11)} = \boldsymbol{p}^{(12)} = \boldsymbol{p}^{(m)}, m \geq 10$$

We can see that after 10 steps the distribution of black balls in the first urn "settled down," and remains unchanged with the increasing number of exchanges. About 5% of the time there will be no black ball, 45% of the time there will be one black ball, 45% of the time there will be two black balls and 5% of the time there will be three black balls in the first container. We have to ask, however: *does this interesting fact depend in any way on the initial distribution of the balls?* Imagine that the Maxwell demon will assign equally likely 0, 1, 2 or 3 black balls into the first urn in the beginning. What will be the behavior of the system now?

$$
\begin{aligned}
\boldsymbol{p} &= (.2500 \quad .2500 \quad .2500 \quad .2500) \\
\boldsymbol{p}^{(1)} &= (.0278 \quad .4722 \quad .4722 \quad .0278) \\
\boldsymbol{p}^{(2)} &= (.0525 \quad .4475 \quad .4475 \quad .0525) \\
\boldsymbol{p}^{(3)} &= (.0497 \quad .4503 \quad .4503 \quad .0497) \\
\boldsymbol{p}^{(5)} &= (.0500 \quad .4500 \quad .4500 \quad .0500)
\end{aligned}
$$

$$\boldsymbol{p}^{(5)} = \boldsymbol{p}^{(6)} = \boldsymbol{p}^{(7)} = \boldsymbol{p}^{(m)}, m \geq 5$$

In this case, the distribution of black balls settled after five steps, and is identical to the previous initial distribution. It looks as though this "*limit behavior*" *is independent of the initial probability distribution* and is an intrinsic property of the transition probability matrix itself. Let us investigate the transition probability matrix $\mathbf{P}^{(m)}$ in m steps for some values m.

$$\mathbf{P} = \begin{pmatrix} 0 & 1 & 0 & 0 \\ \frac{1}{9} & \frac{4}{9} & \frac{4}{9} & 0 \\ 0 & \frac{4}{9} & \frac{4}{9} & \frac{1}{9} \\ 0 & 0 & 1 & 0 \end{pmatrix} \qquad \mathbf{P}^{(2)} = \begin{pmatrix} .1111 & .4444 & .4444 & .0000 \\ .0494 & .5062 & .3951 & .0494 \\ .0494 & .3951 & .5062 & .0494 \\ .0000 & .4444 & .4444 & .1111 \end{pmatrix}$$

$$\mathbf{P}^{(3)} = \begin{pmatrix} .0494 & .5062 & .3951 & .0494 \\ .0562 & .4499 & .4499 & .0439 \\ .0439 & .4499 & .4499 & .0562 \\ .0494 & .3951 & .5062 & .0494 \end{pmatrix} \qquad \mathbf{P}^{(5)} = \begin{pmatrix} .0500 & .4562 & .4438 & .0500 \\ .0507 & .4500 & .4500 & .0493 \\ .0493 & .4500 & .4500 & .0507 \\ .0500 & .4438 & .4562 & .0500 \end{pmatrix}$$

$$\mathbf{P}^{(7)} = \begin{pmatrix} .0500 & .4507 & .4493 & .0500 \\ .0501 & .4500 & .4500 & .0499 \\ .0499 & .4500 & .4500 & .0501 \\ .0500 & .4493 & .4507 & .0500 \end{pmatrix} \qquad \mathbf{P}^{(10)} = \begin{pmatrix} .0500 & .4500 & .4500 & .0500 \\ .0500 & .4500 & .4500 & .0500 \\ .0500 & .4500 & .4500 & .0500 \\ .0500 & .4500 & .4500 & .0500 \end{pmatrix}$$

and

$$\mathbf{P}^{(m)} = \mathbf{P}^{(10)}, m \geq 10$$

It is striking to see that after only 10 steps, the transition probability matrix has settled and does not change anymore. Its rows are identical, and equal to the vector of probabilities of the system after a "very long period of time." We shall call this vector a *limit vector* and the corresponding matrix a *limit matrix* of transition probabilities. It is obvious that the limit vector (if it exists) is very important for the investigation of the behavior of the Markov chain because it tells us what will be the distribution of the states in a "very long run"; it determines the so-called *limit distribution* of the chain. We can always find the limit vector by multiplying the one step transition matrix again and again, and observing the matrix $\mathbf{P}^{(m)}$. However, this method is very tedious and inefficient. Assume that we can find a vector $\mathbf{v} = (v_1, v_2, \ldots, v_n)$ so that

$$\mathbf{vP} = \mathbf{v}; \quad v_i \geq 0 \quad i = 1,2, \ldots, n; \quad \sum_{i=1}^{n} v_i = 1$$

where $\mathbf{P}$ is the one step transition probability matrix. If we start the Markov chain with the vector $\mathbf{v}$ as its initial probability vector, then this vector does not change after one step. If we apply again the matrix $\mathbf{P}$, then we see that it does not change after two, three and more steps. Therefore $\mathbf{v}$ must be a limit vector of the chain, and this vector is sometimes called a *fixed vector*. Let us try to find such a vector $\mathbf{v}$ for the Markov chain in Example 16. Assume that

$$\mathbf{v} = (x, y, z, t).$$

Then necessarily,

$$x + y + z + t = 1$$

and further

$$(x,y,z,t)\begin{pmatrix} 0 & 1 & 0 & 0 \\ \frac{1}{9} & \frac{4}{9} & \frac{4}{9} & 0 \\ 0 & \frac{4}{9} & \frac{4}{9} & \frac{1}{9} \\ 0 & 0 & 1 & 0 \end{pmatrix} = (x,y,z,t)$$

We get the following system of linear equations:

$$0 \cdot x + \frac{1}{9} \cdot y + 0 \cdot z + 0 \cdot t = x.$$

$$1 \cdot x + \frac{4}{9} \cdot y + \frac{4}{9} \cdot z + 0 \cdot t = y.$$

$$0 \cdot x + \frac{4}{9} \cdot y + \frac{4}{9} \cdot z + 1 \cdot t = z.$$

$$0 \cdot x + 0 \cdot y + \frac{1}{9} \cdot z + 0 \cdot t = t.$$

From the first and last equations, we get

$$y = 9x \quad \text{and} \quad z = 9t.$$

Substituting into the second equation, we get

$$x + 4x + 4t = 9x \quad \text{or} \quad x = t.$$

Hence

$$1 = x + y + z + t = x + 9x + 9x + x = 20x$$

or

$$x = t = \frac{1}{20} = .0500.$$

$$y = z = 9 \cdot \frac{1}{20} = .4500.$$

The solution is, therefore,

$$v = (.05, \quad .45, \quad .45, \quad .05)$$

or exactly the same answer as before. Are the above observations only properties of the Markov chain describing the Maxwell demon, or do they have a wider validity?

EXAMPLE 17

Investigate the limit distribution of the Markov chain in Example 3.

SOLUTION.

$$\mathbf{P} = \begin{pmatrix} 0 & 1 \\ .5 & .5 \end{pmatrix} \qquad \mathbf{P}^{(2)} = \begin{pmatrix} .5 & .5 \\ .25 & .75 \end{pmatrix} \qquad \mathbf{P}^{(3)} = \begin{pmatrix} .2500 & .7500 \\ .3750 & .6250 \end{pmatrix}$$

$$\mathbf{P}^{(5)} = \begin{pmatrix} .3125 & .6875 \\ .3438 & .6563 \end{pmatrix} \qquad \mathbf{P}^{(7)} = \begin{pmatrix} .3281 & .6719 \\ .3359 & .6641 \end{pmatrix} \qquad \mathbf{P}^{(9)} = \begin{pmatrix} .3320 & .6680 \\ .3340 & .6660 \end{pmatrix}$$

$$\mathbf{P}^{(11)} = \begin{pmatrix} .3330 & .6670 \\ .3335 & .6665 \end{pmatrix} \qquad \mathbf{P}^{(15)} = \begin{pmatrix} .3333 & .6667 \\ .3333 & .6667 \end{pmatrix}$$

Hence $$\mathbf{P}^{(m)} = \begin{pmatrix} \frac{1}{3} & \frac{2}{3} \\ \frac{1}{3} & \frac{2}{3} \end{pmatrix} \qquad m \geq 15$$

Using linear equations for finding the limit vector, we get

$$\boldsymbol{\nu}\mathbf{P} = (x, y)\begin{pmatrix} 0 & 1 \\ .5 & .5 \end{pmatrix} = (x, y)$$

or $$.5y = x, x + .5y = y \quad \text{and further} \quad x + y = 1.$$

Hence $$.5y + y = 1 \quad \text{or} \quad y = \frac{2}{3}, \quad x = \frac{1}{3}.$$

The limit vector obtained by either method is

$$\boldsymbol{\nu} = \left(\frac{1}{3}, \frac{2}{3}\right).$$

It looks, therefore, as though the limit (fixed) vector for the limit distribution exists quite often, and we shall try to prove its existence under some general assumptions. However, we have to make a comment about our terminology: We investigated the probability vectors and matrices "after a very long period of time," noticed that they "settle" and "approach" some values and we called these values "a limit vector" and "a limit matrix." We demonstrated these facts on numerical examples, and we shall prove them using some intuitive ideas (as, for example, if $0 < x < 1$, then for increasing n, x^n "approaches" 0). However, our method is not really mathematically perfect, and we advise a reader interested in mathematically strict and precise treatment of the notion of limit to consult appropriate texts on calculus and analysis. On the other hand, our approach is mathematically consistent and complete, but it might appear a little nebulous, perhaps, to a reader not familiar with the notion of limit.

DEFINITION: A transition probability matrix is said to be regular if, for some m, its m-th power $\mathbf{P}^{(m)}$ has all its elements strictly positive.

COMMENT 1. It is intuitively clear what the regularity condition means: It says that there must exist a natural number m such that any state of the system is accessible from an arbitrary state in exactly m steps. In other words, regardless of the state the system begins in, it must be possible to reach the arbitrary state in exactly m steps (the number of steps being equal for all the states).

COMMENT 2. We do not have to calculate exactly the powers of $\mathbf{P}$ to find whether it is regular. It is enough to denote all the positive entries in $\mathbf{P}^{(m)}$ by a cross x. Then we formally multiply the matrices and realize that $x \cdot x = x, x \cdot 0 = 0, 0 \cdot 0 = 0, x + 0 = x, x + x = x$ and $0 + 0 = 0$. For example, in the case of the ball game (Example 2):

$$\mathbf{P} = \begin{pmatrix} 0 & 1 & 0 \\ \frac{1}{2} & 0 & \frac{1}{2} \\ \frac{2}{3} & \frac{1}{3} & 0 \end{pmatrix} = \begin{pmatrix} 0 & x & 0 \\ x & 0 & x \\ x & x & 0 \end{pmatrix}$$

$$\mathbf{P}^{(2)} = \begin{pmatrix} x & 0 & x \\ x & x & 0 \\ x & x & x \end{pmatrix} \qquad \mathbf{P}^{(3)} = \begin{pmatrix} x & x & 0 \\ x & x & x \\ x & x & x \end{pmatrix} \qquad \mathbf{P}^{(4)} = \begin{pmatrix} x & x & x \\ x & x & x \\ x & x & x \end{pmatrix}$$

We can see that the matrix **P** is regular, because after four steps all the transition probabilities are strictly positive. The best way is to find $\mathbf{P}^{(2)}$, then $\mathbf{P}^{(4)} = \mathbf{P}^{(2)}\mathbf{P}^{(2)}$, $\mathbf{P}^{(8)} = \mathbf{P}^{(4)}\mathbf{P}^{(4)}$, $\mathbf{P}^{(16)} = \mathbf{P}^{(8)}\mathbf{P}^{(8)}$, and so on, because using this method we can reach high powers of P fast. Sometimes we can see immediately that a matrix cannot be regular—for example, if it has a 1 on its diagonal.

→ **THEOREM 5:** Let **P** be a regular transition probability matrix. Then with m increasing above all limits, its powers $\mathbf{P}^{(m)}$ approach arbitrarily closely a matrix **Q** which has all its rows identical and equal to a vector $\mathbf{v} = (v_1, v_2, \ldots, v_n)$ such that

$$\mathbf{vP} = \mathbf{v}, \sum_{i=1}^{n} v_i = 1, \quad v_i > 0, \quad i = 1,2, \ldots, n.$$

COMMENT. The following proof does not require any background in calculus; however, it does require a high degree of sophistication and mathematical maturity. Students without these prerequisites are advised to do something pleasurable and omit reading the proof.

Proof: It is enough to prove the theorem for matrices which have all elements strictly positive. Let us denote

$$\min_{1 \le i, \; j \le n} p_{ij} = d.$$

Then obviously,

$$0 < d \le \frac{1}{2}, \quad 0 \le 1 - 2d < 1.$$

Let further $M_j^{(m)} = \max_{1 \le i \le n} p_{ij}^{(m)}$ be the maximum of the j-th column of $P^{(m)}$

$m_j^{(m)} = \min_{1 \le i \le n} p_{ij}^{(m)}$ be the minimum of the j-th column of $P^{(m)}$

Let us assume further that the maximum of the j-th column occurs in the row R, that is,

$$M_j^{(m)} = p_{Rj}^{(m)}$$

and the minimum of the j-th column occurs in the row r, that is

$$m_j^{(m)} = p_{rj}^{(m)}.$$

Clearly, both r and R depend on the value of m.

Then

$$p_{ij}^{(m)} = \sum_{l=1}^{n} p_{il} p_{lj}^{(m-1)} \ge m_j^{(m-1)} \sum_{l=1}^{n} p_{il} = m_j^{(m-1)}.$$

Therefore

$$m_j^{(m)} = \min_{1 \le i \le n} p_{ij}^{(m)} \ge m_j^{(m-1)}.$$

Similarly,

$$p_{ij}^{(m)} = \sum_{l=1}^{n} p_{il} p_{lj}^{(m-1)} \le M_j^{(m-1)} \sum_{l=1}^{n} p_{il} = M_j^{(m-1)}$$

and

$$M_j^{(m)} = \max_{1 \le i \le n} p_{ij}^{(m)} \le M_j^{(m-1)}.$$

So, therefore, the sequence of column minima is nondecreasing

$$0 \le m_j^{(1)} \le m_j^{(2)} \le m_j^{(3)} \le \ldots \le m_j^{(m-1)} \le m_j^{(m)} \le \ldots \le 1$$

and the sequence of column maxima is nonincreasing

$$1 \geq M_j^{(1)} \geq M_j^{(2)} \geq M_j^{(3)} \geq \ldots \geq M_j^{(m-1)} \geq M_j^{(m)} \geq \ldots \geq 0.$$

The sequence $M_j^{(m)}$ is nonincreasing and bounded from below by 0,

$$0 \leq M_j^{(m)} \quad \text{for} \quad m = 1,2,3, \ldots$$

Therefore, there must be a unique number $v_j > 0$ to which $M_j^{(m)}$ approaches arbitrarily close if m grows above all limits, or in other words,

$$(M_j^{(m)} - v_j)$$

must approach arbitrarily close to 0. We would like to prove that all the elements of the j-th column must approach this number v_j. We can see that this is the case if we show that the "span" between elements of the j-th column, that is, $(M_j^{(m)} - m_j^{(m)})$ approaches 0.

Let us assume that $p_{Rj}^{(m-1)} = M_j^{(m-1)}$ and that we can write

$$p_{iR} = d + t. \quad t \geq 0.$$

Then $$p_{ij}^{(m)} = \sum_{l=1}^{n} p_{il} p_{lj}^{(m-1)} = p_{iR} p_{Rj}^{(m-1)} + \sum_{l \neq R} p_{il} p_{lj}^{(m-1)} \geq (d + t) M_j^{(m-1)} + m_j^{(m-1)} \sum_{l \neq R} p_{il}$$

$$= (d + t) M_j^{(m-1)} + m_j^{(m-1)} (1 - (d + t))$$

$$= dM_j^{(m-1)} + (1 - d) m_j^{(m-1)} + t(M_j^{(m-1)} - m_j^{(m-1)}) \geq dM_j^{(m-1)} + (1 - d) m_j^{(m-1)}$$

because $$t \geq 0 \quad \text{and} \quad (M_j^{(m-1)} - m_j^{(m-1)}) \geq 0.$$

Therefore $$m_j^{(m)} = \min_{1 \leq i \leq n} p_{ij}^{(m)} \geq d\, M_j^{(m-1)} + (1 - d) m_j^{(m-1)}.$$

Similarly, we can estimate $M_j^{(m)}$ as follows: If $m_j^{(m-1)} = p_{rj}^{(m-1)}$, then assume that

$$p_{ir} = d + u \quad \text{where} \quad u \geq 0.$$

Thus $$p_{ij}^{(m)} = \sum_{l=1}^{n} p_{il} p_{lj}^{(m-1)} = p_{ir} p_{rj}^{(m-1)} + \sum_{l \neq r} p_{il} p_{lj}^{(m-1)} \leq (d + u) m_j^{(m-1)} + M_j^{(m-1)} \sum_{l \neq r} p_{il}$$

$$= (d + u) m_j^{(m-1)} + M_j^{(m-1)} (1 - (d + u))$$

$$= d\, m_j^{(m-1)} + (1 - d) M_j^{(m-1)} - u(M_j^{(m-1)} - m_j^{(m-1)}) \leq d\, m_j^{(m-1)} + (1 - d) M_j^{(m-1)}$$

Hence $$M_j^{(m)} = \max_{1 \leq i \leq n} p_{ij}^{(m)} \leq d\, m_j^{(m-1)} + (1 - d) M_j^{(m-1)}.$$

We can estimate the span as follows:

$$M_j^{(m)} - m_j^{(m)} \leq d\, m_j^{(m-1)} + (1 - d) M_j^{(m-1)} - d\, M_j^{(m-1)} - (1 - d) m_j^{(m-1)} = (1 - 2d)(M_j^{(m-1)} - m_j^{(m-1)}).$$

Using this fact recursively, we get $0 \leq (M_j^{(m)} - m_j^{(m)}) \leq (1 - 2d)(M_j^{(m-1)} - m_j^{(m-1)}) \leq (1 - 2d)^2 (M_j^{(m-2)} - m_j^{(m-2)} \leq \ldots \leq (1 - 2d)^{m-1} (M_j^{(1)} - m_j^{(1)})$. However, $0 \leq (1 - 2d) < 1$, therefore if m keeps increasing above all bounds, then necessarily $(1 - 2d)^{m-1}$ will be approaching arbitrarily close to 0 and therefore $(M_j^{(m)} - m_j^{(m)})$ must also tend to 0. Now we can see that

$$|p_{ij}^{(m)} - v_j| = |p_{ij}^{(m)} - M_j^{(m)} + M_j^{(m)} - v_j| \leq |M_j^{(m)} - p_{ij}^{(m)}| + |M_j^{(m)} - v_j| \leq (M_j^{(m)} - m_j^{(m)}) + (M_j^{(m)} - v_j).$$

But with increasing m, each of the brackets on the right hand side must approach 0 as we have shown, therefore the expression $|p_{ij}^{(m)} - v_j|$ must also approach 0. In other words, with increasing m all the probabilities $p_{ij}^{(m)}$, $i = 1,2, \ldots, n$ in the j-th column must approach the same number v_j. Therefore, the matrix $P^{(m)}$ must tend to a uniquely determined matrix Q which has all the rows identical, that is,

$$P^{(m)} \text{ approaches } \begin{pmatrix} v_1 & v_2 & \ldots & v_j & \ldots & v_n \\ v_1 & v_2 & \ldots & v_j & \ldots & v_n \\ \vdots & & & \vdots & & \vdots \\ v_1 & v_2 & \ldots & v_j & \ldots & v_n \end{pmatrix} = Q$$

Let v be the row vector of the matrix Q, that is,

$$v = (v_1, v_2, \ldots, v_j, \ldots, v_n).$$

Now we can see that $P^{(m+1)}$ must also approach the same matrix Q. On the other hand, $P^{(m+1)} = P^{(m)}P$ must approach the matrix QP, hence

$$QP = Q$$

or, in other words,

$$vP = v$$

Finally, we know that $p_{1j}^{(m)}$ approaches v_j for each $j = 1,2, \ldots, n$ and therefore $\sum_{j=1}^{n} p_{1j}^{(m)}$ must approach $\sum_{j=1}^{n} v_j$. However, we know that for every m $\sum_{j=1}^{n} p_{1j}^{(m)} = 1$, hence $\sum_{j=1}^{n} v_j = 1$. q.e.d.

→ **THEOREM 6:** Let **P** be a regular transition probability matrix and **p** be an arbitrary vector of initial probabilities. Then with m increasing above all limits, the product $\mathbf{p}\mathbf{P}^{(m)}$ approaches a vector $\mathbf{v} = (v_1, v_2, \ldots, v_n)$, which is uniquely defined by the conditions

$$\mathbf{vP} = \mathbf{v} \qquad \sum_{i=1}^{n} v_i = 1.$$

Proof: We proved the existence of a vector v satisfying both conditions in Theorem 5. Let p be any vector of initial probabilities, and Q be the matrix from Theorem 5. Then the vector $pP^{(m)}$ must approach the vector pQ. Assume that $p = (p_1 p_2 \ldots p_n)$ and let us calculate the j-th component of the vector pQ. It is equal to

$$p_1 v_j + p_2 v_j + \ldots + p_n v_j = v_j \sum_{i=1}^{n} p_i = v_j$$

hence $pP^{(m)}$ must approach v. Let us prove the uniqueness of the vector v. Let $w = (w_1, w_2, \ldots, w_n)$ be any other vector satisfying

$$wP = w \quad \text{and} \quad \sum_{i=1}^{n} w_i = 1.$$

If we use w as an initial probability vector, then necessarily $wP^{(m)}$ must approach v as shown above. However, for every m,

$$wP^{(m)} = wPP^{(m-1)} = wP^{(m-1)} = wP^{(m-2)} = wP = w$$

hence

$$w = v.$$

q.e.d.

EXAMPLE 18

Following are limit distributions for problems introduced in the beginning of this chapter.
(a) Example 2.

$$(x,y,z)\begin{pmatrix} 0 & 1 & 0 \\ \frac{1}{2} & 0 & \frac{1}{2} \\ \frac{2}{3} & \frac{1}{3} & 0 \end{pmatrix} = (x,y,z)$$

The system of linear equations is

$$\begin{array}{l} \frac{1}{2}y + \frac{2}{3}z = x \\ x \quad + \frac{1}{3}z = y \\ \frac{1}{2}y \quad = z \\ \hline x + y + z = 1 \end{array}$$

The third equation gives $y = 2z$. The second equation gives $x = y - \frac{1}{3}z = 2z - \frac{1}{3}z = \frac{5}{3}z$. Substitution of y and x into the last equation gives

$$\frac{5}{3}z + 2z + z = 1 \quad \text{or} \quad z = \frac{3}{14} = .2143.$$

$$y = 2z = \frac{6}{14} = .4286.$$

$$x = \frac{5}{3}z = \frac{5}{14} = .3571.$$

The limit distribution is therefore

$$v = \left(\frac{5}{14}, \ \frac{6}{14}, \ \frac{3}{14}\right) = (.3571, \ \ .4283, \ \ .2143).$$

Thus we can see that Alice will get the ball approximately 36% of the time, Bob approximately 43% and Clint approximately 21% of the time if they will keep playing for a very long period of time.
(b) Example 4.

$$(x,y,z)\begin{pmatrix} 0 & 1 & 0 \\ \frac{2}{3} & 0 & \frac{1}{3} \\ \frac{2}{3} & \frac{1}{3} & 0 \end{pmatrix} = (x,y,z)$$

The system of linear equations is

$$\begin{array}{l} \frac{2}{3}y + \frac{2}{3}z = x \\ x \quad + \frac{1}{3}z = y \\ \frac{1}{3}y \quad = z \\ \hline x + y + z = 1 \end{array}$$

The third equation gives $y = 3z$. The second equation gives $x = y - \frac{1}{3}z = \frac{8}{3}z$. Substitution of y and x into the last equation gives

$$\frac{8}{3}z + 3z + z = 1 \quad \text{or} \quad z = \frac{3}{20} = .15.$$

$$y = 3z = \frac{9}{20} = .45. \quad x = \frac{8}{20} = .40.$$

The limit distribution is

$$\mathbf{v} = (.40, \quad .45, \quad .15).$$

The cougar will spend 40% of the time in area A, 45% in area B and 15% of the time in area C when observed over a very long period of time.

EXAMPLE 19

Social Mobility. The majority of societies consist of an upper class (U), a middle class (M) and a lower class (L). There is usually some mobility of people from class to class in different generations. Imagine that a society has a very pronounced upward trend of social mobility, expressed by the following matrix:

$$\begin{array}{c} \\ U \\ M \\ L \end{array} \begin{array}{c} \begin{array}{ccc} U & M & L \end{array} \\ \begin{pmatrix} \frac{4}{6} & \frac{2}{6} & 0 \\ \frac{2}{6} & \frac{3}{6} & \frac{1}{6} \\ \frac{2}{6} & \frac{3}{6} & \frac{1}{6} \end{pmatrix} \end{array}$$

For example, the last row says that the children of a father in the lower class will move up to the middle class with probability $\frac{1}{2}$, and to the upper class with probability $\frac{1}{3}$ while they will stay in the father's class with a very small probability $\frac{1}{6}$. We can also see that the middle class has a pronounced tendency to move up, while the majority of the upper class will stay. One could feel that this is a very "generous" society and that the lower class will sooner or later completely disappear. The real "just" society in which we are living is obviously a far cry from the ideal conditions in the above society. Let us investigate the future of the hypothetical people.

$$(x,y,z) \begin{pmatrix} \frac{4}{6} & \frac{2}{6} & 0 \\ \frac{2}{6} & \frac{3}{6} & \frac{1}{6} \\ \frac{2}{6} & \frac{3}{6} & \frac{1}{6} \end{pmatrix} = (x,y,z)$$

The system of linear equations is

$$\frac{4}{6}x + \frac{2}{6}y + \frac{2}{6}z = x$$

$$\frac{2}{6}x + \frac{3}{6}y + \frac{3}{6}z = y$$

$$\frac{1}{6}y + \frac{1}{6}z = z$$

$$x + y + z = 1$$

The third equation gives $y = 5z$. The second equation gives $2x = 6y - 3y - 3z$ or $x = 6z$. Substitution into the last equation gives

$$6z + 5z + z = 1.$$

Therefore $$z = \frac{1}{12} = .08\overline{3}, \quad y = \frac{5}{12} = .41\overline{6}, \quad x = .5.$$

In the long run, half of the people will live in the upper class, over 41% in the middle class and well over 8% in the lower class. Therefore, if a society behaves as the above Markov chain, then the lower class will never disappear, and the only way to change living conditions would be a drastic disruption of the "smooth" run of the society and its laws. However, it remains questionable whether the society as a whole would profit from any such change.

EXAMPLE 20

Gossip Spreading. A young man is living in a community of basically honest and truthful people. If a member of this community is entrusted with some news, he will tell it to his friends truly and without any changes—the probability that he will twist it around and tell the opposite is really negligible, say q. Even though there is always a chance that this might happen, perhaps due to a misunderstanding, hence $q > 0$, it is believed that the chances of such gross misconduct and base misbehavior are less than one in a billion.

Once when the young man was walking his dog, the animal sniffed another dog's remainders on the neighbor's lawn and went to investigate. Unfortunately, just then the neighbor stepped out of his house. The young man decided to face this precarious situation with sterling truth, and told the neighbor precisely and accurately what had happened. However, after a certain time, he started hearing disturbing rumors that his dog was dirtying other people's yards. Can anything like that happen in an honest and truthful community? Is it possible that truthful people will spread such dirty gossip?

SOLUTION. If a member of the community receives news, he will tell it to another member truly and unchanged with probability $p = 1 - q$ (where $p < 1$) and he will twist it and tell its opposite with probability $q = 1 - p$ (where $0 < q < 10^{-9}$). We can imagine the spreading of news as a Markov chain with two states:

T—meaning the true statement about the young man's dog.

N—meaning negation of the true statement about the young man's dog.

The transition probability matrix is $\begin{array}{c} \\ T \\ N \end{array}\!\!\begin{array}{c} \begin{array}{cc} T & N \end{array} \\ \begin{pmatrix} p & q \\ q & p \end{pmatrix} \end{array}$ where, e.g., $p_{11} = p$ is the probability that a person who has been told the true statement will tell the next person the true statement, while $p_{12} = q$ is the probability that the person who has been told the true statement will tell its negation to the next person, and so on. The limit distribution is obtained as follows:

$$(x,y)\begin{pmatrix} p & q \\ q & p \end{pmatrix} = (x,y)$$

The linear equations are

$$\begin{aligned} xp + yq &= x \\ xq + yp &= y \\ x \quad + y \quad &= 1 \end{aligned}$$

From the second equation, we get

$$xq + y(p - 1) = 0$$

or

$$xq - yq \qquad = 0.$$

Because $q \neq 0$, we have $x = y$. After substitution into the last equation, we get

$$x = \frac{1}{2} \qquad y = \frac{1}{2}.$$

Therefore, even though the community consists of highly truthful and sincere people who will twist around a statement which has been told to them and tell its negation less than once in a billion situations, after a sufficiently long time half of the people will be telling the true statement and half of them its negation about the dog's adventure. This clearly shows how gossip is malicious and scandalmongering. Incidentally, it would take in fact an exceedingly long time before this excessive distortion of the truth would actually happen, nevertheless it is possible in very large communities even if the chance of each person twisting the truth is arbitrarily small. A Markov chain is not perhaps a perfect model for spreading a rumor because any gossip usually grows like an avalanche, returns back to people who heard it before, and so on. However, if the community is very large and if the interest in the rumor lingers for a long time, so that people who have heard it will have forgotten before it comes back or will consider it as another piece of gossip, then in a long run the Markov chain might give a reasonable approximation of the real situation.

EXAMPLE 21

Consider the queueing model introduced in Example 8. For the sake of simplicity, assume that the waiting room has capacity for a single customer only. Find the limit distribution of the chain.

SOLUTION. The Markov chain can have three states only, namely 0, 1 and 2, and the transition matrix will be of the type 3×3. The condition for the fixed vector is

$$(x,y,z)\begin{pmatrix} (1-q) & q & 0 \\ p(1-q) & [pq + (1-p)(1-q)] & q(1-p) \\ 0 & p(1-q) & [pq + (1-p)] \end{pmatrix} = (x,y,z)$$

The system of linear equations is

$$\begin{array}{rrrr} (1-q)x + & p(1-q)y & & = x \\ qx + & [pq + (1-p)(1-q)]y + & p(1-q)z & = y \\ & q(1-p)y + & [pq + (1-p)]z & = z \\ x + & y + & z & = 1 \\ \hline \end{array}$$

From the first equation, we get $x = \frac{p(1-q)}{q} y$.

From the third equation, we get $z = \frac{q(1-p)}{p(1-q)} y$.

Substituting into the fourth equation, we get

$$y\left(\frac{p(1-q)}{q} + 1 + \frac{q(1-p)}{p(1-q)}\right) = 1$$

or

$$y = \frac{pq(1-q)}{p^2(1-q)^2 + pq(1-q) + q^2(1-p)}.$$

$$x = \frac{p^2(1-q)^2}{p^2(1-q)^2 + pq(1-q) + q^2(1-p)}.$$

$$z = \frac{q^2(1-p)}{p^2(1-q)^2 + pq(1-q) + q^2(1-p)}.$$

If, for example, the probability p that a customer will finish his service during the interval of time is the same as the probability that a new customer will arrive during this interval of time, and both are equal to $\frac{1}{2}$, $p = q = \frac{1}{2}$, then $x = \frac{1}{5}$, $y = \frac{2}{5}$, $z = \frac{2}{5}$. Therefore, in a very long run, 20% of the time there will be nobody to be served, 40% of the time there will be only one customer served and 40% of the time there will be one customer served and one customer in the waiting room.

EXAMPLE 22

Consider the game with coins introduced in Example 11 and examine its second version. For the sake of simplicity, assume that the players have only two silver dollars among them (the solution is similar but more tedious for a higher number of coins). Find the limit distribution.

SOLUTION. The condition for the fixed vector is

$$(x,y,z)\begin{pmatrix} q & p & 0 \\ q^2 & 2pq & p^2 \\ 0 & q & p \end{pmatrix} = (x,y,z)$$

The system of linear equations is

$$\begin{aligned} qx + q^2y \qquad &= x \\ px + 2pqy + qz &= y \\ p^2y + pz &= z \\ x + \quad y + \quad z &= 1 \end{aligned}$$

From the third equation, we get $z = \frac{p^2}{1-p}y.$

From the first equation, we get $x = \frac{q^2}{1-q}y.$

After substitution into the last equation, we get

$$y\left(\frac{q^2}{1-q} + 1 + \frac{p^2}{1-p}\right) = 1.$$

Hence $y = \frac{(1-q)(1-p)}{q^2(1-p) + (1-q)(1-p) + p^2(1-q)}.$

$$z = \frac{p^2}{1-p}y = \frac{p^2(1-q)}{q^2(1-p) + (1-q)(1-p) + p^2(1-q)}.$$

$$x = \frac{q^2}{1-q}y = \frac{q^2(1-p)}{q^2(1-p) + (1-q)(1-p) + p^2(1-q)}.$$

Obviously, $x + y + z = 1$. In the case of symmetrical fair coins, $p = q = \frac{1}{2}$. Therefore, $x = \frac{1}{4}, y = \frac{1}{2}, z = \frac{1}{4}$. Thus if the players will continue the game over a very long period of time, player A will have no silver dollars 25% of the time, one silver dollar 50% of the time and two silver dollars 25% of the time.

EXERCISES

(Answers in Appendix to exercises marked)*

In Exercises 1 through 4, do the following:

(a) Find the transition probability matrix and the initial vector of probabilities.
(b) Calculate the transition probability matrices in two and three steps.
(c) Calculate the probability distributions after one, two and three steps.
(d) Determine whether the transition probability matrix is regular. If yes, find the limit distribution, and give an appropriate interpretation to it.
(e) Use your results from (a) through (d) to answer any questions accompanying the problems.

1. A lion hunts either zebras or antelope. If he hunts zebras one day, the probability that he will hunt antelope the next day is .4. If he hunts antelope, the probability that he will hunt zebras the next day is .5. In the beginning of the survey, he was observed to kill a zebra.
   (a) What proportion of the time does the lion hunt antelope over a long period of time?

(b) If you were in the antelope's skin on a Friday, would you prefer that the lion had hunted zebras or antelope on the previous Wednesday?

*2. An industrious young man has a hard time getting used to the working hours of his new job. When he is late, he has to listen to the picky comments of his boss during the coffee break, and for this reason he is twice as likely to come on time as to be late on the next day.

However, if he manages to be on time on a certain day, he is so pleased with himself that the next day he is as likely as not to be late. He usually entertains his friends on Sundays and therefore has a particularly difficult time on Mondays. His past record shows that he can make it on time on Monday mornings usually once a month (that is, one in four).

(i) Investigate his behavior during the week (i.e., find $\mathbf{P}^{(m)}$, $m = 1,2,3,4$).

(ii) Imagine that after a rather drastic social reform, all seven days of a week have been declared working days (an alternative not so impossible as seen by the experience of certain countries). Investigate the man's behavior in the new situation over a long period of time.

3. Investigate the behavior of a rat in the Y maze introduced in problem 15 on page 113. What is the meaning of the limit distribution?

4. It has been observed that a major airline always offers the same choice of food on its transatlantic flights. Identically prepared packages contain either a steak, a chicken or a vegetarian dish. A frequent passenger decided to fight this monotony with the help of the following scheme: He would never take the same package twice in a row. However, he prefers beef to chicken and chicken to vegetarian food, hence he is twice as likely to take his preferred package. On the first flight after the holidays, he decides between beef and chicken by flipping a coin.

(a) What is more likely that he will have on his fourth flight: chicken or beef?

(b) On what proportion of the flights does he eat meat?

5. Inuk is a trapper and Akklavik is an artist who carves animal figures from soapstone. Inuk had four pelts and Akklavik had three sculptures when they decided to barter their wares. They agreed to each pick one of their possessions at random, and swap it for the item of the other man (with the possibility that an item may go back and forth between them several times).

*(a) Can you describe Akklavik's property by a Markov chain?

(b) If they continue their trading for a long time, with how many soapstone sculptures would you expect Akklavik to end up (i.e., what is the most likely number of sculptures)?

6. Let $p + q = 1$ and the transition probability matrix be given as

$$\mathbf{P} = \begin{pmatrix} p & q & 0 & 0 \\ q & p & 0 & 0 \\ p & 0 & q & 0 \\ 0 & q & 0 & p \end{pmatrix}$$

Investigate the corresponding Markov chain (find $\mathbf{P}^{(m)}$, $m = 1,2,3$ and the limit distribution).

7. Investigate a queueing system with a waiting room capacity for two customers (see Example 8). Assume that within an interval of time a customer departs from the service counter with probability .3 and another customer comes with probability .4. Find the limit distribution of the queue. Assume now that the waiting room has a capacity for one customer only, and calculate the limit distribution. Compare with the result in Example 21.

8. Consider the coin game introduced in Example 11, and assume that the coins are loaded so that the odds are 2:3 in favor of heads.

(a) Can you find the limit distribution for both versions of the game?

(b) Imagine that the players have only two silver dollars. Find numerically the limit distribution (if it exists) and compare it with the result in Example 22.

In the following exercises, find the limit distribution if it exists.

9. A white rat is released at the entrance of the following maze:

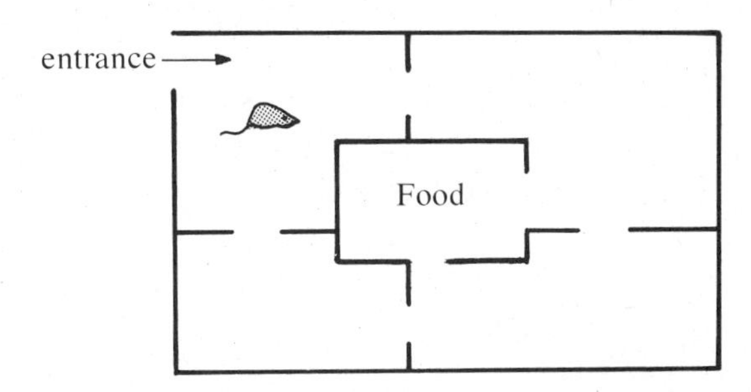

If the rat is in any compartment, then it will be equally likely to enter one of the adjacent compartments. Investigate the rat's behavior over a long period of time if

(a) the door at the main entrance is left open, and the rat can leave the maze (as equally likely as that it will enter the adjacent compartments) and never return.

(b) the main entrance is closed.

(c) the main entrance is closed, and the central compartment contains a large supply of food. If the rat is in any compartment adjacent to the food cell, it will enter the food cell twice as likely as any other compartment.

*10. One urn contains two red and three black balls, while the other urn contains four black and two red balls. One ball is taken at random from each urn and the balls are exchanged. Investigate the number of red balls in the first urn as a Markov chain.

11. Assume that $p > 0$, $q > 0$, $0 < p + q < 1$. Three pirates, Long Tom, Mean Phil and Short Jack are smoking a pipe on the deck of their clipper. Each of them, after having a puff, passes the corncob to the longer of his buddies with probability p, and to the shorter one with probability q (leaving the chance of a second puff for himself). If the pirates spend all night long on their tobacco fiesta, find what part of the night each of them will cherish the pipe.
(a) Find the limit distribution for arbitrary values p and q.
(b) In order to verify (a), select $p = .3$, $q = .5$, calculate numerically the limit distribution and compare with your result in (a).

12. *(a) $\begin{pmatrix} .2 & .8 \\ 1 & 0 \end{pmatrix}$ *(b) $\begin{pmatrix} .2 & .8 \\ 0 & 1 \end{pmatrix}$

(c) $\begin{pmatrix} .2 & .4 & .4 \\ .5 & .2 & .3 \\ .6 & .4 & 0 \end{pmatrix}$

(d) $\begin{pmatrix} .4 & .6 & 0 & 0 \\ .3 & .7 & 0 & 0 \\ .1 & .1 & .3 & .5 \\ 0 & 0 & .6 & .4 \end{pmatrix}$

(e) $\begin{pmatrix} .2 & .6 & .2 & 0 \\ 0 & .2 & .6 & .2 \\ 0 & .5 & 0 & .5 \\ .5 & 0 & .5 & 0 \end{pmatrix}$

13. (a) $\begin{pmatrix} p & q & 0 \\ p(1-p) & q(1-p) & p \\ q(1-q) & p(1-q) & q \end{pmatrix}$

(b) $\begin{pmatrix} p & q & r \\ r^2 & p^2 + 2pq + q^2 & 2(p+q)r \\ q^2 & 2q(p+r) & p^2 + 2pr + r^2 \end{pmatrix}$

THE THEORY OF GAMES

It may seem strange, but the theory of games was not invented either to further gambling adventures in casinos or to entertain. Its origins were laid during the 1920's and 1930's, and its fast development was triggered by publication of the book, *The Theory of Games and Economic Behavior* by *J. von Neumann* and *O. Morgenstern* at the end of the Second World War. The authors suggested a new approach to solving problems in terms of adversaries playing certain types of simple games. For this reason we call the methods we are going to investigate the *theory of games.* We shall also introduce examples derived from card games, children's games, and so on, in order to motivate the theoretical development. Nevertheless, the reader should realize from the beginning that these and similar methods of solving problems are playing ever more important roles in economics, business management, industry, engineering, military science, etc.

Before we start discussing the theory of games, we have to make two fundamental assumptions. First, we shall postulate that the outcome of any conflicting situation involving adversaries of opposite interests can be expressed in the form of some real or artificial "money" paid by one adversary to the other. This money represents loss or gain, depending on the outcome. At the same time, we assume that the amount of money that one adversary wins is paid in full by the other adversary, and that there are no external sources. This approach might seem too materialistic and simplistic, nevertheless it is very useful in many situations. Second, we shall develop the theory of games using the notion of expected value. However, we already know from Chapter XIII on the Law of Large Numbers that this notion has a real practical meaning only if the underlying experiment is repeated many times. Therefore, in order to be able to interpret our results correctly, we shall assume that our games are to be played many times, under identical conditions.

We shall investigate conflicting situations involving two adversaries or *players.* For reasons which will become apparent, we shall call them Mr. Row—or shortly, Mr. *R,* and Mr. Column—or Mr. *C.* We shall assume that each of these players has a choice of two (or more) alternative actions. Each of the two players makes his selection and then the outcome of such a *game* is determined by some rules. We assume that each outcome is represented by a real number that is equal to the amount of "money" one player, say Mr. *R,* will get from the other (Mr. *C*). If Mr. *R* actually gets some money, we shall take this number as positive; if he has to pay Mr. *C,* we shall take it as negative. Such games we shall call *matrix games.*

EXAMPLE 1

Consider the following game of cards. Mr. *R* has Red 5 and Red 8 while Mr. *C* has Red 5, Black 4 and Black 6. They each select a card, and at a given signal they simultaneously expose their cards. If there is a match in color, Mr. *R* pays to Mr. *C* the difference (in dollars) between the numbers of their cards. If there is not a match, Mr. *R* gets the difference (in dollars) from Mr. *C.* The game is to be continued over a fairly long period of time.

It is convenient to express all possibilities in the form of an array.

		Mr. *C*		
		Rd 5	Bl 6	Bl 4
Mr. *R*	Rd 5	0	1	1
	Rd 8	−3	2	4

This array can be considered as a matrix (see Chapter XIV, page 242), and in the theory of games it is called the payoff matrix. In order to play the game, Mr. R selects a row (thereafter, we call him Mr. Row), Mr. C selects a column (therefore, Mr. Column) and the intersection of the selected row and column represents the payoff. A payoff is positive if Mr. R gets the money from Mr. C, and negative if Mr. C gets the money from Mr. R.

For Mr. C, deciding how to play this game is very simple. If he plays Bl 4 he always has to pay more than if he plays Bl 6. For this reason he should not play Bl 4 at all. But for the same reason, he should abandon Bl 6 in favor of Rd 5. Therefore, Mr. C should always play Rd 5, which is obviously his best decision. For Mr. R, the situation is more complicated. He would like to maximize his profit and therefore he hopes for $4 in the right lower corner. For this reason, he should play Rd 8. But he does not know what Mr. C is going to play. Should Mr. C play Rd 5, then Mr. R is in for trouble because he would have to pay $3. But we assume that Mr. R is a *careful* player and decides, therefore, always to play Rd 5. In this case, he has guaranteed that he does not lose *regardless of the behavior* of his adversary. And should Mr. C be unreasonable and play Bl 4 or Bl 6, then Mr. R will actually make some money. Therefore Rd 5 is the best choice for Mr. R. We have found the exact description of how both players should always play. In addition, the suggested method has one great advantage: If a player follows it, he has a certain reasonable outcome guaranteed, regardless of the behavior of his adversary.

DEFINITION 1: A strategy is a set of exact rules telling a player how he should play in every possible situation.

Thus, a strategy for Mr. C can be: "Always play Bl 4" or "Always play Rd 5." A strategy for Mr. R can suggest: "Flip a coin. If you get H, play Rd 5, otherwise play Rd 8." Our discussion above suggests that the strategies for Mr. R: "Always play Rd 5" and for Mr. C: "Always play Rd 5" are in a sense best for both players provided they play carefully.

EXAMPLE 2

Sam and Jim each write a number between 1 and 4, inclusive, on a hidden piece of paper. At a signal, they show their selections. Then Sam pays Jim the amount of dollars equal to the sum of the numbers, while Jim pays Sam the dollars equal to the product. Find the payoff matrix for this game, and list some strategies for Sam and Jim.

SOLUTION. Let us call Sam Mr. C and Jim Mr. R for the payoff matrix. Each has four possible alternatives, namely, to write down 1, 2, 3, or 4. If Sam writes 3 and Jim writes 2, then Sam pays Jim $5 while Jim pays Sam $6, hence Jim ends up losing $1. The payoff matrix entry in the second row and third column is -1, and we can similarly calculate other entries.

		Sam—Mr. C			
	Numbers	1	2	3	4
Jim—Mr. R	1	1	1	1	1
	2	1	0	-1	-2
	3	1	-1	-3	-5
	4	1	-2	-5	-8

Possible strategies for Jim could be: "Always write 1" or "Write 1, 2, 3, 4, 1, 2, 3, 4, 1, 2, etc., in this order for successive games." A strategy for Sam could be: "Toss two coins. If HH appears write 1, if HT appears write 2, if TH appears write 3, otherwise write 4."

EXAMPLE 3

A fast growing city started building three new satellite residential areas *A, B* and *C* of the same size. Two competing banks, the Royal Bank and the Commerce Bank, decided to open one branch each in the new development. It is estimated that if both branches are located within the same area, both will capture 50% of the overall business. However, if only one branch is located within an area, it will get all the business. If no branch is in an area, then everybody will go to the branch in the nearest area. Where should the banks open their branches?

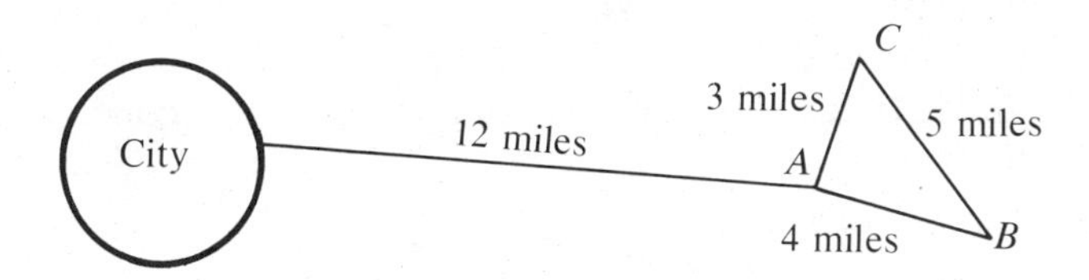

SOLUTION. Let us assume that the business created by each area can be expressed by 2 units. If Royal Bank (call it Mr. *R*) opens its branch in area *A,* while Commerce Bank (call it Mr. *C*) opens its branch in area *B,* then Mr. *R* gets 2 units of business from *A* and 2 units of business from *C.* If both branches are located in the same area, then Mr. *R* gets 1 unit of business from each area or a total of 3 units. Similarly, we can calculate all other entries in the payoff matrix, as follows (each entry is the business volume of Mr. *R* for the given location of branches):

		Mr. *C* Commerce Bank		
	Branch locations	*A*	*B*	*C*
Mr. *R* Royal Bank	*A*	3	4	4
	B	2	3	2
	C	2	4	3

The best outcome for Mr. *R* is to get 4 units of business. From the matrix, we can see that the worst strategy for Mr. *R* is "Open branch in *B.*" "Open branch in *C*" is a better choice. However, the strategy "Open branch in *A*" is best for him. Similarly for Mr. *C,* the strategy "Open branch in *B*" is worst, "Open branch in *C*" is better and "Open branch in *A*" is best. If both players will accept their best strategies, then each will get guaranteed 3 units of business. However, should one of them take another strategy, he will get only 2 units and the other player (keeping to his best strategy) will get 4 units. Therefore 3 units is the guaranteed minimum of business of each player's best strategy, regardless of the behavior of his competitor.

Let us try to formulate in general terms the principles we just used in the above examples. *We assume that both players are very smart, keen and careful people, each anticipating the moves of his adversary and each trying to optimize his situation and minimize his loss.* Let us try to follow the reasoning of each of them.

The reasoning of Mr. *C*: Mr. *R* is very clever. He can select rows, and certainly he tries to maximize his gain. The worst thing that can happen to me is when he selects a row giving him the maximum in a given column. Whichever column I select, I can lose in the *worst case the maximum columnwise.* Therefore, I have to select the column where this maximum is *minimal.*

This reasoning gives Mr. C a very simple rule for analyzing the game. He has to find, first, the maximum of each column, and second, he has to select the column which gives the minimal value.

Mr. C ↓ first maximum

second minimum →

This method will be called *minmax.*

The reasoning of Mr. R: Mr. C is very clever. He can select columns, and certainly he tries to minimize his loss. The worst thing that can happen to me is when he selects a column giving him the minimum in a given row. Whichever row I select, I can make in the *worst case the minimum rowwise.* Therefore, I have to select the row where this minimum is *maximal.*

This reasoning gives Mr. R a very simple rule for analyzing the game. He has to find, first, the minimum of each row, and second, he has to select the column which gives the maximal value.

Mr. R first minimum →

↓ second maximum

This method will be called *maxmin.* Let us apply this approach to our examples.

EXAMPLE 1 (continued)

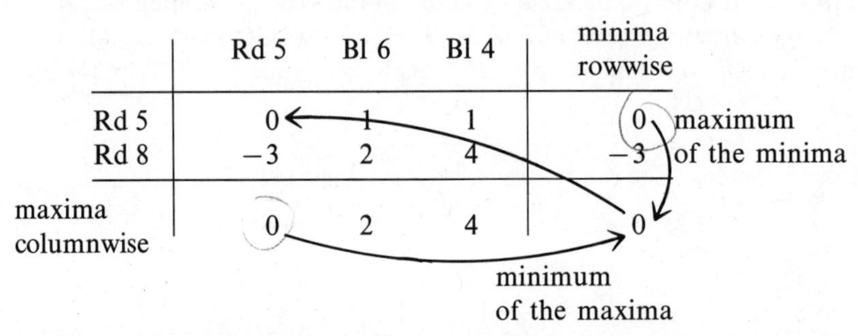

	Rd 5	Bl 6	Bl 4	minima rowwise
Rd 5	0	1	1	0
Rd 8	−3	2	4	−3
maxima columnwise	0	2	4	0

In this case both maxmin and minmax give the same value, namely 0. It can be seen that this value is simultaneously the minimum of the row in which it occurs, and the maximum of the column in which it occurs. This is true always when minmax and maxmin give the same value, say v. Then clearly the best strategies are as follows:

For player R: Always play the row containing v.
For player C: Always play the column containing v.

EXAMPLE 2 (continued)

	Numbers	Mr. C 1	2	3	4	minima rowwise
Mr. R	1	1	1	1	1	1
	2	1	0	−1	−2	−2
	3	1	−1	−3	−5	−5
	4	1	−2	−5	−8	−8
maxima columnwise		1	1	1	1	1

maximum of the minima

minimum of the maxima

Best strategies: Mr. R—"Write down 1."
Mr. C—"Write down 1."

EXAMPLE 3 (continued)

		Mr. C			
Branch locations		*A*	*B*	*C*	minima rowwise
Mr. *R*	*A*	3	4	4	3
	B	2	3	2	2
	C	2	4	3	2
maxima columnwise		3	4	4	3

maximum of the minima: 3; minimum of the maxima: 3

Best strategies: Mr. *R*—"Open branch in *A*."
Mr. *C*—"Open branch in *A*."

DEFINITION 2: An entry in a payoff matrix which is simultaneously the minimum of the row in which it occurs, and the maximum of the column in which it occurs, is called a *saddle point*.

The term saddle point is derived from the analogy to the surface of a saddle. It can be easily seen that if minmax and maxmin yield the same value v, then it is a saddle point of a game. Thus in Example 1 the saddle point is 0, in Example 2 it is 1 and in Example 3 it is 3.

DEFINITION 3: We shall say that a matrix game is strictly determined if its payoff matrix has a saddle point v. The number v is called the value of the game, and the game is fair if $v = 0$. Optimal strategies are as follows: For Mr. *R*—"Always play the row containing v"; For Mr. *C*—"Always play the column containing v."

EXAMPLE 4

Analyze the following game.

	*C*1	*C*2	*C*3	*C*4	minima
*R*1	0	−1	2	3	−1
*R*2	1	0	3	4	0
*R*3	2	−1	1	−3	−3
*R*4	−2	−2	3	0	−2
maxima	2	0	3	4	0 = minmax = maxmin

The game is strictly determined, and its value $v = 0$. Therefore the game is fair. The optimal strategies are $R2$ for Mr. R, and $C2$ for Mr. C.

EXAMPLE 5

Analyze the following game.

	$C1$	$C2$	$C3$	$C4$	minima
$R1$	-1	-3	0	1	-3
$R2$	2	3	1	3	1
$R3$	0	2	-1	0	-1
maxima	2	3	1	3	1 = minmax = maxmin

The game is strictly determined and its value $v = 1$, hence it is an unfair game. The optimal strategies are $C3$ for Mr. C, and $R2$ for Mr. R.

In this game we can notice one peculiar feature. If Mr. C plays $C4$, he always has to pay more than if he plays $C3$. Therefore $C3$ is certainly preferable for him or, as we say, $C3$ *dominates* $C4$. Obviously, Mr. C will never play $C4$, hence we can completely omit $C4$. Similarly, $R2$ is always better for Mr. R than $R3$; we say that $R2$ *dominates* $R3$. We can therefore *omit all dominated rows and columns.* A game without dominated rows and columns is called a game in *reduced form.* The solutions of the original game and its reduced form must be obviously equal.

Reduced form of a game:

	$C1$	$C2$	$C3$	minima
$R1$	-1	-3	0	-3
$R2$	2	3	1	1
maxima	2	3	1	1

EXAMPLE 6

	$C1$	$C2$	$C3$	$C4$
$R1$	1	0	1	2
$R2$	2	2	0	0
$R3$	1	1	-1	-1

In this case, $C3$ dominates $C4$, $C2$ dominates $C1$ and $R2$ dominates $R3$. The reduced form of this game is

	$C2$	$C3$	minima
$R1$	0	1	0
$R2$	2	0	0
maxima	2	1	maxmin = 0 minmax = 1

In this case, minmax $\neq$ maxmin and the game is not strictly determined. Even in this simple example, we cannot find the optimal strategies.

EXAMPLE 7

Mr. Long conceals a $10 bill or a $50 bill in his hand. Mr. Short guesses 10 or 50 and wins the bill if he is right.

		Mr. Short 10	50	minima
Mr. Long	10	-10	0	-10
	50	0	-50	-50
maxima		0	0	maxmin $= -10$ minmax $= 0$

The game is not strictly determined, and we are not able to find optimal strategies for Mr. Long and Mr. Short.

It is obvious that even very simple games with two columns and two rows need not have optimal strategies as defined before. We shall analyze such games, and try to find how to play them. We shall call such games 2×2 matrix games, and denote them, in general:

$$G = \begin{pmatrix} a & b \\ c & d \end{pmatrix}$$

THEOREM 1: A 2×2 matrix game G is not strictly determined if one of the two following conditions is satisfied: ←
(a) $a < b, \quad a < c, \quad d < b, \quad d < c.$
(b) $a > b, \quad a > c, \quad d > b, \quad d > c.$

Proof: Case (a)

			minima
	a	b	a
	c	d	d
maxima	c	b	

But $a < c$, $a < b$, $d < c$, $d < b$, hence max $(a,d) <$ min (c,b).

Case (b)

			minima
	a	b	b
	c	d	c
maxima	a	d	

But $b < a$, $b < d$, $c < a$, $c < d$, hence max $(b,c) <$ min (a,d). q.e.d.

EXAMPLE 8

Let us consider another variation of the game of cards introduced in this chapter (see Example 1). Mr. R has Red 5 and Black 7, Mr. C has Red 6 and Black 6. They select a card and disclose it simultaneously. If there is a match in color, Mr. R pays the difference, otherwise Mr. C pays it.

		Mr. C Rd 6	Bl 6	minima
Mr. R	Rd 5	-1	1	-1
	Bl 7	1	-1	-1
maxima		1	1	maxmin $= -1$ minmax $= 1$

The game is not strictly determined. What does this mean practically? In the beginning, we stipulated that each game has to be played repeatedly over a fairly long period of time. Now imagine that Mr. C will think that Rd 6 is his optimal strategy, and plays it all the time. Sooner or later, Mr. R will discover this and will always play Bl 7, thus winning $1 in each game. Obviously, Mr. C can do better than lose $1 in each game—it is very unreasonable to play Rd 6 systematically. For the same reason, he cannot play Bl 6 systematically because in this case Mr. R would always play Rd 5. Mr. C must alternate somehow between Rd 6 and Bl 6. Unfortunately, if he *determines in advance* any fixed pattern, say Rd 6, Rd 6, Bl 6, Rd 6, Rd 6, Bl 6, etc., Mr. R can always detect it after a certain period of time and beat him—in the example given, simply by playing Bl 7, Bl 7, Rd 5, Bl 7, Bl 7, Rd 5, etc. The only solution for Mr. C is that *he must abandon any strictly determined pattern and play at random,* selecting Rd 6 and Bl 6 with some *probabilities.* Let us denote the probability of selecting the first column by q_1 and the probability of selecting the second column by q_2. Then necessarily

$$q_1 + q_2 = 1.$$

In our example, we intuitively feel that he should select his cards with equal probabilities, that is, $q_1 = q_2 = \frac{1}{2}$.

We can repeat this reasoning for Mr. R. If he always plays Rd 5, Mr. C will discover it and play Rd 6, thus getting $1 in each game. For the same reason, no determined pattern is acceptable for Mr. R. He has to select his rows with certain probabilities. Let us denote the probability of selecting the first row by p_1 and the probability of selecting the second row by p_2. Then necessarily

$$p_1 + p_2 = 1.$$

In our example, we intuitively feel that $p_1 = p_2 = \frac{1}{2}$. We shall derive general formulas for "best probabilities" of each player. Any selection of probabilities (p_1,p_2) can be considered as an exact rule for Mr. R on how to play the game, or as a strategy. He will always select his rows according to the probabilities (p_1,p_2). In the above example, he can toss a coin; if he gets H he plays Rd 5, otherwise he plays Bl 7.

DEFINITION 4: Any rule telling Mr. R "Always play a certain row" or telling Mr. C "Always play a certain column" is called a pure strategy. Any choice of probabilities (p_1, p_2) or (q_1, q_2), respectively, is called a mixed strategy for Mr. R or Mr. C, respectively.

COMMENT. The notion of a mixed strategy can be generalized to larger than 2×2 games in an obvious way. Let us consider the game

$$G = \begin{pmatrix} a & b \\ c & d \end{pmatrix}$$

and let us assume that Mr. R plays according to the mixed strategy (p_1,p_2) while Mr. C plays according to the mixed strategy (q_1,q_2). We assume that their decisions are *independent* of each other and therefore we can use the following probabilistic model.

$$P\{\text{Mr. } R \text{ selects the 1st row and Mr. } C \text{ selects the 1st column}\} = p_1q_1$$
$$P\{\text{Mr. } R \text{ selects the 2nd row and Mr. } C \text{ selects the 1st column}\} = p_2q_1$$

$$P\{\text{Mr. } R \text{ selects the 1st row and Mr. } C \text{ selects the 2nd column}\} = p_1q_2$$
$$P\{\text{Mr. } R \text{ selects the 2nd row and Mr. } C \text{ selects the 2nd column}\} = p_2q_2.$$

In the matrix form, we get

Payoffs $G =$

a	b
c	d

Probabilities

	q_1	q_2
p_1	p_1q_1	p_1q_2
p_2	p_2q_1	p_2q_2

Let the random variable X denote the *gain* of Mr. R in the game. Then the distribution of X is as follows:

$$P(\{X = a\}) = p_1q_1, \quad P(\{X = b\}) = p_1q_2, \quad P(\{X = c\}) = p_2q_1, \quad P(\{X = d\}) = p_2q_2.$$

(This is true if all the values a,b,c,d are distinct. Otherwise we have to make obvious modifications.) The expected value of X is equal to

$$E(X) = ap_1q_1 + bp_1q_2 + cp_2q_1 + dp_2q_2.$$

In accordance with our basic principles, we can give the following definition in the case of a game which is not strictly determined.

DEFINITION 5: The probabilities (p_1^0, p_2^0) and (q_1^0, q_2^0) are optimal mixed strategies for Mr. R and Mr. C, respectively, and a number v is the value of the game if Mr. R can assure for himself an expected payoff v or better by playing (p_1^0, p_2^0), regardless of the behavior of Mr. C. At the same time, Mr. C can assure for himself an expected payoff v or better by playing (q_1^0, q_2^0), regardless of the behavior of Mr. R.

It is not clear at all whether such optimal strategies and the value of the game exist, or whether our definition is void. In Theorem 3 we shall prove for 2×2 games that optimal strategies and the value of the game do exist. Incidentally, from our definition it immediately follows that the value must be unique. If there are two distinct values, say v_1 and v_2, then one, say v_1, will be better for Mr. R and he can force his expected payoff v_1. The other one, say v_2, will be better for Mr. C and he can force his expected payoff v_2. But this is a contradiction to the assumption that one player wins what the other player loses.

THEOREM 2: If the game $G = \begin{pmatrix} a & b \\ c & d \end{pmatrix}$ is not strictly determined, then the number v is its value and (p_1^0, p_2^0), (q_1^0, q_2^0) are optimal strategies for Mr. R and Mr. C, respectively, if the following inequalities are satisfied:

$$p_1^0a + p_2^0c \geq v$$
$$p_1^0b + p_2^0d \geq v$$
$$q_1^0a + q_2^0b \leq v$$
$$q_1^0c + q_2^0d \leq v$$

Then $$v = ap_1^0q_1^0 + bp_1^0q_2^0 + cp_2^0q_1^0 + dp_2^0q_2^0.$$

Proof: (a) In order to prove that $(p_1^0\, p_2^0)$ is an optimal strategy for Mr. R, we shall prove that the expected gain of Mr. R if he plays (p_1^0, p_2^0) is always *at least v regardless of the behavior of Mr. C.* Let Mr. C play the strategy (q_1, q_2), which need not be his optimal strategy. Then the expected gain of Mr. R is

$$p_1^0q_1a + p_1^0q_2b + p_2^0q_1c + p_2^0q_2d = q_1(p_1^0a + p_2^0c) + q_2(p_2^0b + p_2^0d) \geq q_1v + q_2v = (q_1 + q_2)v = v$$

as follows from the given inequalities and from $q_1 + q_2 = 1$.

(b) In order to prove that (q_1^0, q_2^0) is an optimal strategy for Mr. C, we shall prove that the expected loss of Mr. C if he plays (q_1^0, q_2^0) is always *at most v regardless of the behavior of Mr. R.* Let Mr. R play the mixed strategy (p_1, p_2). Then the expected loss of Mr. C is

$$p_1q_1^0a + p_1q_2^0b + p_2q_1^0c + p_2q_2^0d = p_1(q_1^0a + q_2^0b) + p_2(q_1^0c + q_2^0d) \leq p_1v + p_2v = (p_1 + p_2)v = v$$

as follows from the given inequalities and from the fact that $p_1 + p_2 = 1$.

(c) We shall prove that if (p_1^0, p_2^0), (q_1^0, q_2^0) are the optimal strategies, then any number v satisfying the above inequalities must satisfy the equality

$$v = ap_1^0q_1^0 + bp_1^0q_2^0 + cp_2^0q_1^0 + dp_2^0q_2^0.$$

Let us multiply the first inequality by q_1, the second inequality by q_2, the third one by p_1 and the fourth one by p_2. We get

$$ap_1^0q_1^0 + cp_2^0q_1^0 \geq q_1^0v$$
$$bp_1^0q_2^0 + dp_2^0q_2^0 \geq q_2^0v$$
$$ap_1^0q_1^0 + bp_1^0q_2^0 \leq p_1^0v$$
$$cp_2^0q_1^0 + dp_2^0q_2^0 \leq p_2^0v$$

Let us add the first two and the second two inequalities,

$$ap_1^0q_1^0 + cp_2^0q_1^0 + bp_1^0q_2^0 + dp_2^0q_2^0 \geq q_1^0v + q_2^0v = (q_1^0 + q_2^0)v = v.$$
$$ap_1^0q_1^0 + bp_1^0q_2^0 + cp_2^0q_1^0 + dp_2^0q_2^0 \leq p_1^0v + p_2^0v = (p_1^0 + p_2^0)v = v.$$

The only way to satisfy both inequalities is

$$v = ap_1^0q_1^0 + bp_1^0q_2^0 + cp_2^0q_1^0 + dp_2^0q_2^0$$

q.e.d.

THEOREM 3: If $G = \begin{pmatrix} a & b \\ c & d \end{pmatrix}$ is not a strictly determined game, then optimal strategies (p_1^0, p_2^0) for Mr. R and (q_1^0, q_2^0) for Mr. C are given as follows.

$$p_1^0 = \frac{d - c}{a + d - b - c}$$

$$p_2^0 = \frac{a - b}{a + d - b - c}$$

$$q_1^0 = \frac{d - b}{a + d - b - c}$$

$$q_2^0 = \frac{a - c}{a + d - b - c}$$

The value of the game v is $$v = \frac{ad - bc}{a + d - b - c}.$$

COMMENT. An interested reader can show that if a game is not strictly determined, then always $a + d - b - c \neq 0$, hence the optimal strategies and the value can be calculated.

Proof: Let us write $$x = a + d - b - c.$$

In order to prove this theorem, we shall show that $(p_1{}^0, p_2{}^0)$, $(q_1{}^0, q_2{}^0)$ and v satisfy the conditions of Theorem 2.

$$p_1^0 a + p_2^0 c = a\frac{d-c}{x} + c\frac{a-b}{x} = \frac{ad - ac + ac - cb}{x} = \frac{ad - bc}{a + d - b - c} = v.$$

The first inequality of Theorem 2 is satisfied with equality. In exactly the same way we shall also verify the other inequalities.

$$p_1^0 b + p_2^0 d = b\frac{d-c}{x} + d\frac{a-b}{x} = \frac{bd - bc + ad - db}{x} = v.$$

$$q_1^0 a + q_2^0 b = a\frac{d-b}{x} + b\frac{a-c}{x} = \frac{ad - ab + ab - bc}{x} = v.$$

$$q_1^0 c + q_2^0 d = c\frac{d-b}{x} + d\frac{a-c}{x} = \frac{cd - cb + ad - dc}{x} = v.$$

q.e.d.

EXAMPLE 9

We can now solve the game of cards introduced in Example 8.

		Mr. c	
		Rd 6	Bl 6
Mr. R	Rd 5	-1	1
	Bl 7	1	-1

The optimal strategies are

$$p_1^0 = \frac{d-c}{a+d-b-c} = \frac{-1-1}{-1-1-1-1} = \frac{1}{2}.$$

We do not have to calculate $p_2{}^0$ because

$$p_2{}^0 = 1 - p_1{}^0 = 1 - \frac{1}{2} = \frac{1}{2}.$$

For Mr. C, we have $q_1^0 = \dfrac{d-b}{a+d-b-c} = \dfrac{-1-1}{-1-1-1-1} = \dfrac{1}{2}.$

$$q_2^0 = 1 - q_1^0 = \frac{1}{2}.$$

$$v = \frac{ad - bc}{a+d-b-c} = \frac{(-1)\cdot(-1) - 1\cdot 1}{-1-1-1-1} = 0.$$

The optimal strategies for both players are quite simple, and can be expressed as follows:

Mr. *R:* Flip a fair coin. If you get *H* play Rd 5, otherwise play Bl 7.
Mr. *C:* Flip a fair coin. If you get *H* play Rd 6, otherwise play Bl 6.

The value of the game is 0, and the game is fair. Practically, it means that if both players stick to their optimal strategies, then in a long run the average gain of Mr. *R* will very likely be close to 0 and the average loss of Mr. *C* will very likely be close to 0 as well. This is a much better outcome for each of them than if they decide to play using a determined pattern. It is also obvious that neither player can do any better—unless his adversary is dumb.

EXAMPLE 10

Two-Finger Morra. Each player holds up one or two fingers and they simultaneously display their selections. If the number of fingers is even, then Mr. R gets the amount equal to the number of fingers displayed. If the number of fingers is odd, then Mr. C gets the corresponding amount.

Mr. C

Mr. R		1 finger	2 fingers	minima
	1 finger	2	-3	-3
	2 fingers	-3	4	-3
	maxima	2	4	minmax $= 2$ maxmin $= -3$

The game is not strictly determined, and we have to find the optimal mixed strategy.

$$p_1^0 = \frac{d-c}{a+d-b-c} = \frac{4-(-3)}{2+4-(-3)-(-3)} = \frac{7}{12}.$$

$$p_2^0 = 1 - p_1^0 = \frac{5}{12}.$$

$$q_1^0 = \frac{d-b}{a+d-b-c} = \frac{4-(-3)}{2+4-(-3)-(-3)} = \frac{7}{12}.$$

$$q_2^0 = 1 - q_1^0 = \frac{5}{12}.$$

$$v = \frac{2\cdot 4-(-3)\cdot(-3)}{2+4-(-3)-(-3)} = \frac{-1}{12}.$$

The optimal strategies are similar for Mr. R and Mr. C. They can be realized as follows. Toss a coin and a die. If you get H or T and 1, play one finger, otherwise play two fingers. If both players play their optimal strategies, in a long run Mr. R will very likely lose on the average about $\frac{1}{12}$ of a dollar or about 8¢ per game, and he cannot do any better. The game is unfair. However, it is quite difficult to discover this without Theorem 3, and one can usually find people willing to play the game in the role of Mr. R who will, therefore, lose money.

EXAMPLE 11

Attacking bombers attempt to bomb a city which is protected by a small wing of fighters. The formation of bombers comes either "high" or "low" on a regular basis. It is known that a fairly small wing of fighters cannot be further split without losing its effectiveness. The fighters, therefore, try to intercept the bombers either in the low altitude or the high altitude. If the bombers are intercepted by the fighters in high altitude, then they are chased away and cannot do any damage. If the bombers are intercepted by the fighters in low altitude, they are chased away but still they can do some damage—credit the bombers with 1 point. If the fighters are waiting low and the bombers come high, then they can do a lot of damage before the fighters can climb up—credit the bombers with 2 points. If the fighters are waiting high and the bombers attack low, this is disaster for the city—credit the bombers with 3 points. What is the optimal strategy of the commander in chief of the wing of fighters?

SOLUTION.

		Fighters		
		Hi	Lo	minima
Bombers	Hi	0	2	0
	Lo	3	1	1
	maxima	3	2	minmax = 2, maxmin = 1

The game is not strictly determined. This means that if the commander in chief decides on any predetermined pattern, it can be detected, which would be disastrous to the city. The optimal mixed strategies are

$$q_1^0 = \frac{d-b}{a+d-b-c} = \frac{1-2}{0+1-2-3} = \frac{1}{4}.$$

$$q_2^0 = 1 - \frac{1}{4} = \frac{3}{4}.$$

$$p_1^0 = \frac{d-c}{a+d-b-c} = \frac{1-3}{0+1-2-3} = \frac{1}{2}.$$

$$p_2^0 = 1 - p = \frac{1}{2}.$$

$$v = \frac{0 \cdot 1 - 2 \cdot 3}{0+1-2-3} = \frac{-6}{-4} = 1\frac{1}{2}.$$

The optimal strategy of the commander in chief of the wing of fighters is very simple. He should flip a dime and a nickel. If he gets two heads, he should send his fighters high, otherwise he should send them low. Such a decision might seem irresponsible when we consider the material value and human lives at stake. However, if the commander does not have more complete information (and this is often the case), then tossing the coins is the best he can do.

EXAMPLE 12

Solve the following game:

	$C1$	$C2$	$C3$
$R1$	1	4	3
$R2$	3	2	1

SOLUTION. Minmax = 3, maxmin = 1 and the game is not strictly determined. We cannot use Theorem 3 to get a solution. However, column $C3$ dominates column $C2$, hence Mr. C would never use $C2$. Thus the original game is equivalent to the following game in reduced form.

	$C1$	$C3$
$R1$	1	3
$R2$	3	1

This game is not strictly determined, and by Theorem 3 optimal strategies are

$$P(R_1) = \frac{1-3}{1+1-3-3} = \frac{1}{2} \qquad P(R_2) = \frac{1}{2}$$

$$P(C_1) = \frac{1-3}{1+1-3-3} = \frac{1}{2} \qquad P(C_2) = \frac{1}{2}.$$

The value of the game is

$$v = \frac{1 \cdot 1 - 3 \cdot 3}{1+1-3-3} = 2.$$

We get the optimal strategy for the original complete game if we put

$$P(R_1) = \frac{1}{2} \qquad P(R_2) = \frac{1}{2}$$

$$P(C_1) = \frac{1}{2} \qquad P(C_2) = 0 \qquad P(C_3) = \frac{1}{2}.$$

Thus sometimes we can solve large games by eliminating dominated strategies, and reducing the game to the 2×2 type.

EXAMPLE 13

The refreshment stand at a race track offers ice cream and popcorn, but the actual sales depend very much on the weather. The manager estimates that on cold days he can sell 1000 bags of popcorn but no ice cream. On hot days, he can sell 200 bags of popcorn and 2000 ice cream cones. He buys goods worth \$500 each time. A bag of popcorn costs him 50¢ and sells for \$1, while an ice cream cone costs him 20¢ and sells for 50¢. Can you give any expert advice to the manager on how to buy the goods? (Assume that everything that is not sold is a complete loss.)

SOLUTION. If he buys for a cold day, he spends the \$500 on popcorn, getting 1000 bags. If he buys for a hot day, he gets \$400 worth of ice cream (2000 cones) and \$100 worth of popcorn (200 bags). His payoff matrix is as follows:

		Actual weather	
		Cold	Hot
Manager buying for	Cold day	500	−300
	Hot day	−300	700

minmax = 350
maxmin = −300

The game is not strictly determined. The optimal strategy for the manager is

$$p_1{}^0 = \frac{700-(-300)}{500+700-(-300)-(-300)} = \frac{5}{9} \qquad p_2{}^0 = \frac{4}{9}.$$

The value of the game is $v = \dfrac{500 \cdot 700 - (-300)}{1800} = 144.44$ dollars. In five cases out of nine, the manager should buy for a cold day; in the remaining four cases he should buy for a hot day. However, it is better to invest $\frac{5}{9}$ of his capital in "cold day goods" and $\frac{4}{9}$ of his capital in "hot day goods," thus buying $322.22 worth of popcorn and $177.78 worth of ice cream (that is, he should buy 644 bags of popcorn and 889 ice cream cones). Then he will always make $144 profit, regardless of the weather.

EXAMPLE 14

The contract which Syrinxo Oil Company has with the government of Kuwadi is the typical contract the company has with many governments in that part of the world: it allows extraction of the amount of oil possible by the capacity of the present equipment, but further expansions are not permitted. Due to the unstable political situation in Kuwadi, there could be nationalization of the economy next year. However, it is estimated that the annual net profit of the company is equal to the actual value of the equipment. In this situation, the company has three possible plans. If it keeps the contract, in the case of nationalization the equipment will be repaid at its actual value (without profit). If the company cancels the contract and removes the equipment from the country, then in the case of nationalization it can sell the equipment back to the government with 100% profit, but if there is no nationalization, there is no profit. If the company gambles and secretly doubles its investment, it can double its annual profit but if there is nationalization then it will forfeit all its investment due to the disclosed breach of contract. What should the company do?

SOLUTION. Let us denote the present value of the equipment and the annual profit it is producing by 1 unit. The payoff matrix for different plans is:

	Nationalization Yes	Nationalization No	Minima
Keep contract	0	1	0
Remove equipment	1	0	0
Double equipment	−2	2	−2
Maxima	1	2	Minmax = 1 Maxmin = 0

The game is not strictly determined, and we cannot solve it using Theorem 3. However, we can consider reduced 2 × 2 games if we omit one of the company plans.

Game 1:

	Nationalization Yes	Nationalization No	Minima
Keep contract	0	1	0
Remove equipment	1	0	0
Maxima	1	1	Minmax = 1 Maxmin = 0

The game is not strictly determined. The optimal strategy for the company is

$$P(\text{keep contract}) = \frac{0-1}{0+0-1-1} = \frac{1}{2}.$$

$$P(\text{remove equipment}) = \frac{1}{2}.$$

The value of this game is $v_1 = \frac{-1}{-2} = \frac{1}{2}$.

Game 2:

Nationalization

	Yes	No	Minima
Keep contract	0	1	0
Double equipment	−2	2	−2
Maxima	0	2	Minmax = 0 Maxmin = 0

The game is strictly determined; the company should keep the contract. The value of this game is $v_2 = 0$.

Game 3:

Nationalization

	Yes	No	Minima
Remove equipment	1	0	0
Double equipment	−2	2	−2
Maxima	1	2	Minmax = 1 Maxmin = 0

The game is not strictly determined. The optimal strategy for the company is

$$P(\text{remove equipment}) = \frac{2-(-2)}{1+2-0-(-2)} = \frac{4}{5}.$$

$$P(\text{double equipment}) = \frac{1}{5}.$$

The value of this game is $$v_3 = \frac{2}{5}.$$

Thus the company has three "subgames" with three values, of which $v_1 = \frac{1}{2}$ is highest and best. Let us investigate what will happen if the company accepts the following mixed strategy for the original complete game (cf. Game 1):

$$p_1 = P(\text{keep contract}) = \frac{1}{2}.$$

$$p_2 = P(\text{remove equipment}) = \frac{1}{2}.$$

$$p_3 = P(\text{double equipment}) = 0.$$

Then if nationalization will be done with probability q_1 and the status quo will be kept by the government with probability $q_2 = 1 - q_1$, the expected payoff of the company will be

$$\frac{1}{2}\cdot q_1\cdot 0 + \frac{1}{2}\cdot q_2\cdot 1 + \frac{1}{2}\cdot q_1\cdot 1 + \frac{1}{2}\cdot q_2\cdot 0 + 0\cdot q_1\cdot(-2) + 0\cdot q_2\cdot 2 = \frac{1}{2}q_1 + \frac{1}{2}q_2 = \frac{1}{2}.$$

Regardless of the government's behavior, the company will have a guaranteed profit $v = \frac{1}{2}$. Therefore, it follows from Definition 5 (modified for larger than 2×2 games) that $p_1 = \frac{1}{2}$, $p_2 = \frac{1}{2}$, $p_3 = 0$ is optimal strategy for the company. Any time the company faces the threat of nationalization, it should either keep the contract or remove the equipment with probability $\frac{1}{2}$. And because the company has contracts with many governments, this policy is the best that can be done under these conditions.

Example 14 shows how we can sometimes solve games when one player has two pure strategies while the second player has more than two. We consider all possible 2×2 games, obtained by deleting some of the strategies of the second player. We solve all these games, then we check whether one of these solutions could be considered as optimal strategy for the original game.

And finally, we mention one more complicated game, the solution of which we cannot calculate, but we can guess it and prove that it is optimal.

EXAMPLE 15

Stone–Scissors–Paper is a well known favorite children's game. Let us specify payments to make it more interesting. Each of the two players shows by a hand sign one of the three objects: stone, scissors or paper. Sone breaks scissors, scissors cuts paper and paper covers stone. The players show their signs simultaneously, and the owner of the weaker object has to pay according to these rules:

(a) If there is a match, then nobody pays.
(b) If "stone breaks scissors," the payment is \$1.
(c) If "scissors cut paper," the payment is \$2.
(d) If "paper covers stone," the payment is \$3.

	Stone	Scissors	Paper	Minima
Stone	0	1	−3	−3
Scissors	−1	0	2	−1
Paper	3	−2	0	−2
Maxima	3	1	2	Maxmin = −1 Minmax = 1

The game is not strictly determined. All our reasoning is still applicable, but we cannot give an exact numerical answer because we cannot use Theorem 3.

The analysis of more complicated games such as this one does not require any conceptual innovations. We only need an appropriate numerical method to find the optimal mixed strategies for both players. These methods are of a rather involved technical character, and we cannot discuss them here. Nevertheless it can be shown that in this example, the optimal strategies for both players are

$$P(\text{stone}) = \frac{1}{3} \qquad P(\text{scissors}) = \frac{1}{2} \qquad P(\text{paper}) = \frac{1}{6}$$

and the value of the game is 0. Indeed, we can generalize Definition 5 to 3×3 games just by considering mixed strategies of the type (p_1, p_2, p_3), where $p_1 \geq 0$, $p_2 \geq 0$, $p_3 \geq 0$ and $p_1 + p_2 + p_3 = 1$. We can easily verify that the conditions of this definition are satisfied.

If one child plays the strategy $p_1{}^0 = \frac{1}{3}, p_2{}^0 = \frac{1}{2}, p_3{}^0 = \frac{1}{6}$ and the other child plays any mixed strategy (q_1,q_2,q_3), then the expected payoff is

$$q_1 \cdot \frac{1}{3} \cdot 0 + q_2 \cdot \frac{1}{3} \cdot 1 + q_3 \cdot \frac{1}{3} \cdot (-3) + q_1 \cdot \frac{1}{2} \cdot (-1) + q_2 \cdot \frac{1}{2} \cdot 0 + q_3 \cdot \frac{1}{2} \cdot 2$$

$$+ q_1 \cdot \frac{1}{6} \cdot 3 + q_2 \cdot \frac{1}{6} \cdot (-2) + q_3 \cdot \frac{1}{6} \cdot 0 = q_1\left(\frac{1}{2} - \frac{1}{2}\right) + q_2\left(\frac{1}{3} - \frac{1}{3}\right) + q_3(1 - 1) = 0 = v.$$

Hence $p_1{}^0 = \frac{1}{3}, p_2{}^0 = \frac{1}{2}, p_3{}^0 = \frac{1}{6}$ is an optimal strategy, and the value is 0.

EXERCISES

(Answers in Appendix to exercises marked)*

1. In the following examples, apply the minmax and maxmin methods to find whether the game is strictly determined or not. Also find pairs of dominated-dominating strategies.

*(a)

	$C1$	$C2$
$R1$	2	3
$R2$	1	7

*(b)

	$C1$	$C2$
$R1$	2	5
$R2$	7	1

*(c)

	$C1$	$C2$	$C3$
$R1$	2	1	3
$R2$	2	0	2

*(d)

	$C1$	$C2$	$C3$
$R1$	2	5	7
$R2$	3	9	14

(e)

	$C1$	$C2$	$C3$
$R1$	−1	1	−3
$R2$	5	2	7
$R3$	9	0	5

(f)

	$C1$	$C2$	$C3$
$R1$	2	2	5
$R2$	1	2	1
$R3$	6	7	9

(g)

	$C1$	$C2$	$C3$	$C4$	$C5$	$C6$
$R1$	6	5	1	−1	2	4
$R2$	3	5	−1	0	0	−3
$R3$	8	9	5	4	7	6
$R4$	2	8	3	2	−1	−2
$R5$	−3	−4	2	1	−5	6

2. Mr. R has Rd 3 and Bl 4 cards while Mr. C has Bl 2, Rd 5, Rd 7 and Rd 8 cards. They each select a card, and at a given signal simultaneously disclose it. If there is a match in color, Mr. C gets the difference in denominations from Mr. R, while if there is a mismatch, Mr. R gets the difference.
   (a) Set up the payoff matrix.
   (b) Is the game strictly determined?

3. Show that if in a 2×2 game both entries in a column are equal, then the game is strictly determined.

4. Show that if a 2×2 or 2×3 game has a saddle point, then either one row dominates another or one column dominates another.

5. For what values of x are the following games strictly determined?

*(a)

	$C1$	$C2$	$C3$
$R1$	2	1	7
$R2$	3	x	5
$R3$	−1	0	8

*(b)

	$C1$	$C2$	$C3$
$R1$	x	2	3
$R2$	5	x	6
$R3$	2	3	7

(c)

	$C1$	$C2$	
$R1$	x	5	2
$R2$	−1	x	−6
$R3$	−2	3	x

6. Find the solutions of the following 2 × 2 games.

*(a)

	C1	C2
R1	1	0
R2	3	4

*(b)

	C1	C2
R1	6	5
R2	1	9

*(c)

	C1	C2
R1	1	2
R2	3	1

(d)

	C1	C2
R1	−5	3
R2	4	0

(e)

	C1	C2
R1	2	3
R2	2	3

(f)

	C1	C2
R1	0	5
R2	3	1

7. Solve the following games.

*(a)

	C1	C2	C3	C4
R1	5	8	3	1
R2	−1	2	9	7

(b)

	C1	C2
R1	−1	0
R2	0	2
R3	3	1
R4	2	−1

*(c)

	C1	C2	C3
R1	0	3	2
R2	4	2	3

(d)

	C1	C2	C3
R1	1	6	4
R2	8	4	5

8. Another version of the game *Two-Finger Morra* (cf. Example 10), which has been popular in Italy since antiquity, is played as follows: each of two players raises one or two fingers and simultaneously calls his guess of the number of fingers his opponent will show. If one player guesses correctly, he wins an amount equal to the sum of the fingers shown, otherwise the game is a tie.
   (a) Denote the possibilities of each player by a pair (x,y), where x is the number of fingers shown and y is the number guessed by him. Set up the payoff matrix.
   (b) Using maxmin and minmax, analyze the game. Is there a saddle point?

*9. *Three-Finger Morra* has the same rules as the game in the previous exercise, except that the players use three fingers instead of two. Analyze the game.

10. Let us have two games G_1 and G_2 with payoff matrices given as follows,

$$G_1 = \begin{pmatrix} a & b \\ c & d \end{pmatrix} \qquad G_2 = \begin{pmatrix} ka & kb \\ kc & kd \end{pmatrix}$$

where k is a positive constant. Show that both games have the same optimal strategies, and if v_1 is the value of G_1 and v_2 is the value of G_2, then $v_2 = kv_1$.

11. Calf Arms Company buys firing pins for its top of the line revolvers. A regular pin costs \$1, but if it is defective after being installed it has to be replaced at a total cost of \$10 per gun. The company also has a choice of two brands of better quality pins. The first type costs \$6 and if it fails, a new pin is installed at no cost. The other type costs \$10, but if it is defective, Calf Arms Company gets a \$10 refund and a new good pin installed at no cost.
   (a) Set up the payoff matrix.
   (b) Is the game strictly determined?
   (c) Solve the game (cf. Example 14).

*12. Solve the following game.

	C_1	C_2	C_3
R_1	0	2	3
R_2	2	0	2
R_3	−4	4	5

SOME PRACTICAL APPLICATIONS OF FINITE SPACES

In this chapter, we shall investigate some rather sophisticated and complex applications of the basic methods of the theory of probability. All these examples are of a quite practical nature, and even though they are rather complicated and involved, the elementary probabilistic methods of the previous chapters will suffice. However, we always need some knowledge of the particular areas of application: we will give this background information in a very concise form for the sake of brevity.

BLOOD GROUPS AND BLOOD TRANSFUSIONS

Blood transfusions have been used in practical medicine for several centuries. However, the theory of blood groups (blood phenotypes) and the conditions of safe transfusion were unveiled only several decades ago. Nevertheless, this ignorance did not preclude frequent administrations of transfusions—sometimes on a rather extensive scale. For example, Napoleon's army had a large staff of field surgeons who gave transfusions indiscriminately to hundreds of wounded soldiers suffering blood loss. Even though blood groups were unknown, and therefore the recipient and the donor were matched more or less at random (though not completely), it was empirically discovered that slightly over 50% of the soldiers recovered. Let us investigate the theoretical probability of this phenomenon in the light of modern medicine and probability theory.

A blood group is specified by the presence or absence of certain bodies, called antigens, on red blood cells. There are two main blood group systems that are important in blood transfusions: the ABO system with two antigens called A and B and the Rh system with several antigens called C, D, E, etc. According to the ABO system, blood is divided into four groups as follows:

	Antigen on red cells	
Blood group	A	B
A	Yes	No
B	No	Yes
AB	Yes	Yes
O	No	No

Blood can be further classified as Rh plus (notation A^+, B^+, AB^+, O^+) or Rh minus (notation A^-, B^-, AB^-, O^-), according to the presence or absence of certain antigens of the Rh system. We have therefore eight main blood groups, and let the letters A^+, B^+, AB^+, O^+, A^-, B^-,

AB^-, O^- denote the sets of people having the corresponding blood groups. Let us further define:

A: the set of people having A antigen on their red blood cells,
B: the set of people having B antigen on their red blood cells,
Rh: the set of people having corresponding Rh antigens on their red blood cells.

Then we can see the following relations.

$$A^+ = A \cap B' \cap Rh \qquad A^- = A \cap B' \cap Rh'$$
$$B^+ = A' \cap B \cap Rh \qquad B^- = A' \cap B \cap Rh'$$
$$AB^+ = A \cap B \cap Rh \qquad AB^- = A \cap B \cap Rh'$$
$$O^+ = A' \cap B' \cap Rh \qquad O^- = A' \cap B' \cap Rh'$$

These eight sets form a partition of Ω(the set of all people). Let us consider a blood transfusion. If D and R are the sets of types of antigens in the donor's and the recipient's blood, respectively, then generally we can expect the transfusion to be safe if

$$D \subset R.$$

Let us denote UD and UR the sets of people who are universal donors and universal recipients. (UD are those who can safely donate to and UR are those who can safely receive blood from anybody.)

$$UD = A' \cap B' \cap Rh'. \qquad UR = A \cap B \cap Rh.$$

Following is a table of the different blood groups of a recipient, and corresponding sets of safe donors:

Recipient	Safe donors
A^+	B'
A^-	$B' \cap Rh'$
B^+	A'
B^-	$A' \cap Rh'$
AB^+	Ω
AB^-	Rh'
O^+	$A' \cap B'$
O^-	$A' \cap B' \cap Rh'$

It has been found that the blood group systems ABO and Rh are independent of each other. The percentages of people belonging to different blood groups are specific and typical for each population (for example, American Indians differ markedly from Europeans). However, we can assume that the average distribution in western Europe is given by the following figures:

System ABO

Blood group	Percentage
A	41%
B	10%
AB	4%
O	45%

System Rh

Blood group	Percentage
Rh positive	85%
Rh negative	15%

Given these figures, we can calculate the distribution of the eight phenotypes from the fact that the systems ABO and Rh are independent. For example,

$$P(A^+) = .41 \times .85 = .3485.$$

Blood group X	A^+	A^-	B^+	B^-	AB^+	AB^-	O^+	O^-
Probability of occurrence $P(X)$	.3485	.0615	.0850	.0150	.0340	.0060	.3825	.0675

Let us denote the event of a safe and successful transfusion by S. Then we can calculate the conditional probabilities of S, given the blood group of a recipient. (Use the notation of the table of safe donors.)

Recipient's blood group X	A^+	A^-	B^+	B^-	AB^+	AB^-	O^+	O^-
Probability of safe transfusion given $X, P(S \mid X)$	.86	.129	.55	.0825	1	.15	.45	.0675

For example,

$$P(S/A^+) = P(B') = 1 - P(B) = 1 - (.1 + .04) = .86.$$

$$P(S/O^+) = P(A' \cap B') = .45.$$

$$P(S/O^-) = P(A' \cap B' \cap Rh') = P(A' \cap B')P(Rh') = .45 \times .15 = .0675.$$

Using the formula of total probability, we can get the probability of a successful transfusion if the donor and recipient are chosen at random:

$$P(S) = \Sigma P(X)P(S/X) = .5672.$$

Thus we can theoretically confirm the practical results of Napoleon's surgeons; we can really expect that slightly over one half of transfusions will be successful.

Using Bayes' formula, we can calculate the conditional probability that the recipient has blood group A^+, given that the transfusion was safe,

$$P(A^+/S) = \frac{P(A^+)P(S/A^+)}{P(S)} = \frac{.3485 \times .86}{.5672} = .5284.$$

Similarly, we can calculate the conditional probabilities $P(X/S)$ for all blood groups X.

Blood group X	A^+	A^-	B^+	B^-	AB^+	AB^-	O^+	O^-
$P(X/S)$	.5284	.0140	.0824	.0022	.0599	.0016	.3035	.0080

POPULATION GENETICS AND CASES OF DISPUTED PATERNITY

It is well known that the theory of the inheritance of blood phenotypes can often be used by courts to decide cases of disputed paternity. Of two possible fathers of a child, one can sometimes be completely excluded just from the relation of the blood groups of the child, the mother and the men. However in some cases, even after the blood analysis all the men involved are possible fathers. We will derive methods for finding who is *most likely* to be the father. This analysis requires a much more detailed knowledge of blood phenotypes than the previous discussion of safe transfusions.

The investigation of the inheritance of blood phenotypes is based on laws of population genetics. The phenotype of a person is uniquely determined by his or her genotype, which in turn is determined by a combination of so-called alleles. There are three alleles of the ABO blood system present in the human population: I^A producing A antigen, I^B producing B antigen and i which does not produce any antigen. The genotype of a person is derived from a combination of any two out of the three alleles. We have therefore $\binom{4}{2} = 6$ different genotypes (see combinations with repetitions, page 58), I^AI^A, I^Ai, I^BI^B, I^Bi, I^AI^B, ii. Let us assume that the probability (proportion) of I^A alleles in a population is p, the probability of I^B is q and the probability of i is r where $p + q + r = 1$, and that these values remain constant from generation to generation. We can imagine that the alleles are chosen and paired at random and independently to create genotypes. Under these assumptions, the distribution of genotypes (unordered pairs of alleles) is multinomial. This fact is known as the Hardy-Weinberg law of equilibrium for randomly mating (panmictic) populations.

Genotype	Antigen produced		Phenotype	Multinomial probability of occurrence
	A	B		
I^AI^A	Yes	No	A	p^2
I^Ai	Yes	No	A	$2pr$
I^BI^B	No	Yes	B	q^2
I^Bi	No	Yes	B	$2qr$
I^AI^B	Yes	Yes	AB	$2pq$
ii	No	No	O	r^2

Let us consider now the genotypes of offspring. Given the genotypes of the parents (that is, given two pairs of alleles of the parents), the genotype of a child is determined by a random selection of a single allele from each of the parents. For example, if one parent is of I^AI^B genotype and the other one is of I^Ai genotype, then the children can be of I^AI^A, I^Ai, I^AI^B, I^Bi genotypes.

Unfortunately, in the majority of situations we do not know the genotypes of the people involved; we can only find their phenotypes. Nevertheless, given the phenotypes of the parents, we can find the probability distribution of the possible phenotypes of the offspring from the distribution of the genotypes in the population. Let us say that one of the parents is of O phenotype, while the other is of A phenotype.

Phenotype	Possible genotypes	Probability of occurrence
O	ii	r^2
A	I^Ai	$2pr$
	I^AI^A	p^2

Now if the specified A parent is of the I^Ai genotype, then there are two possibilities for the child, namely I^Ai and ii with equal probabilities $\frac{2pr\,r^2}{2}$. In the case of the I^AI^A genotype, there is just one possibility for the child, namely I^Ai with the probability p^2r^2. However, we have to multiply all these probabilities by 2 because $A \times O$ type mating can occur in two different ways: either A is the father and O is the mother, or vice versa. Therefore, $A \times O$ type mating can occur with the total probability $2(p^2 + 2pr)r^2 = 2pr^2(p + 2r)$. (As a matter of fact, this probability of mating is again a multinomial distribution.) The possible genotypes of the child are I^Ai with the probability $(2pr^3 + 2p^2r^2) = 2pr^2(r + p)$ and ii with the probability $2pr^3$. In the first case, the child is of A phenotype, in the second case, the child is of O phenotype. Similarly, we can calculate the probability distribution of phenotypes of the offspring for any combination of phenotypes of parents. For example, in the case of $A \times B$ mating the A parent can be either I^AI^A or I^Ai with probabilities p^2 and $2pr$, whereas the B parent can be I^Bi^B or I^Bi with probabilities q^2 and $2qr$, respectively. Corresponding probabilities of genotypes of children must be multiplied by 8 because there are eight ways of assigning four genotypes to two parents. We therefore get the following data for the child:

Genotype I^Ai representing phenotype A with probability $8\left(\frac{2p^22qr}{16} + \frac{2pr2qr}{16}\right)$.

Genotype I^Bi representing phenotype B with probability $8\left(\frac{2q^22pr}{16} + \frac{2pr2qr}{16}\right)$.

Genotype I^AI^B representing phenotype AB with probability

$$8\left(\frac{4p^2q^2}{16} + \frac{2p^22qr}{16} + \frac{2q^22pr}{16} + \frac{2pr2qr}{16}\right).$$

Genotype ii representing phenotype O with probability $8\frac{2pr2qr}{16}$.

We can summarize the probabilities of all the types of mating and the probabilities of offspring in the accompanying table.

		Offspring			
Type of mating	Probability of mating	O	A	B	AB
$O \times O$	r^4	r^4	....	....	....
$O \times A$	$2pr^2(p + 2r)$	$2pr^3$	$2pr^2(p + r)$	....	....
$O \times B$	$2qr^2(q + 2r)$	$2qr^3$	....	$2qr^2(q + r)$	....
$O \times AB$......	$4pqr^2$	....	$2pqr^2$	$2pqr^2$	....
$A \times A$	$p^2(p + 2r)^2$	p^2r^2	$p^2(p + r)(p + 3r)$	....	....
$A \times B$	$2pq(p + 2r)(q + 2r)$	$2pqr^2$	$2pqr(p + r)$	$2pqr(q + r)$	$2pq(p + r)(q + r)$
$B \times B$	$q^2(q + 2r)^2$	q^2r^2	....	$q^2(q + r)(q + 3r)$	....
$A \times AB$......	$4p^2q(p + 2r)$	....	$2p^2q(p + 2r)$	$2p^2qr$	$2p^2q(p + r)$
$B \times AB$......	$4pq^2(q + 2r)$	....	$2pq^2r$	$2pq^2(q + 2r)$	$2pq^2(q + r)$
$AB \times AB$.....	$4p^2q^2$	....	p^2q^2	p^2q^2	$2p^2q^2$
TOTAL	1.00	r^2	$p^2 + 2pr$	$q^2 + 2qr$	$2pq$

This table gives the joint probability distribution of the type of mating (for example $O \times AB$ means that one parent has O phenotype and the other has AB phenotype) and the phenotype of the offspring (see page 286). The second column gives the marginal probabilities of occurrence for all 10 types of mating. If we now want to get the conditional probability of a child's phenotype given the phenotypes of the father and mother, we have to divide the corresponding joint probability by the corresponding marginal probability. For example,

$$P(\text{child } A/\text{father } B \text{ mother } AB) = \frac{2pq^2r}{4pq^2(q+2r)} = \frac{r}{2(q+2r)}.$$

$$P(\text{child } A/\text{father } O \text{ mother } AB) = \frac{2pqr^2}{4pqr^2} = \frac{1}{2}.$$

$$P(\text{child } A/\text{father } AB \text{ mother } AB) = \frac{p^2q^2}{4p^2q^2} = \frac{1}{4}.$$

If we know the actual values of p, q and r, we can easily calculate all the conditional probabilities numerically. However, the numbers p, q and r are specific for each ethnic group or for any population in equilibrium. For example, in the case of the population which was considered in the previous section on blood transfusion, we have approximately $p = .26$, $q = .07$ and $r = .67$.

All these facts can be used in a case of disputed paternity. Consider for example the case of Mary Ann. She is of an independent mind and dates three men freely and regularly. She likes Ed, who is a good friend, but she spends on the average twice as much time with athletic Joe. However, she divides her favors equally between Joe and intellectually minded Ben. After some time she finds herself pregnant and gives birth to a girl. Blood tests reveal that Mary Ann has AB blood group while her daughter has A blood group. She casually questions the men and finds that Ed has AB, Joe has O and Ben has B blood group. Who is most likely to be the father?

Given the frequency of relations, we can find the à priori probabilities of the men being fathers:

$$P(\text{Ed}) = .2 \qquad P(\text{Joe}) = .4 \qquad P(\text{Ben}) = .4.$$

Using the previously mentioned formulas for the conditional probabilities of the blood group of a child, given phenotypes of the parents and corresponding values of p, q, r, we get

$$P(\text{child } A/\text{father Ed}) = .25$$
$$P(\text{child } A/\text{father Joe}) = .5$$
$$P(\text{child } A/\text{father Ben}) = \frac{r}{2(q+2r)} = .24$$

Now, using Bayes' theorem (see page 85), we can calculate the à posteriori probabilities of the paternity of the men, given the fact that the little girl has A blood group:

$$P(\text{father Ed/child } A) = \frac{.2 \times .25}{.2 \times .25 + .4 \times .5 + .4 \times .24} = .14$$

$$P(\text{father Joe/child } A) = \frac{.4 \times .5}{.2 \times .25 + .4 \times .5 + .4 \times .24} = .58$$

$$P(\text{father Ben/child } A) = \frac{.4 \times .24}{.2 \times .25 + .4 \times .5 + .4 \times .24} = .28$$

Therefore Joe is most likely to be the father, given the facts about the phenotypes of all the people involved. As a matter of fact, he is more than twice as likely as Ben, and more than four times as likely as Ed. This, of course, is not positive proof uniquely deciding the disputed paternity, however, it is the best Mary Ann can do if she is lacking more information.

DECISION ANALYSIS

Modern man faces a continuous stream of *decisions* in his everyday life. Some of them are quite simple—for example, the selection of new shoes, of lunch in the cafeteria, of the color of roses for a girl friend, and so on. Other decisions are quite complicated and involved, and require a lot of mental energy—for example, enrollment at a university, whether to get married, the selection of a career, and so on. *Alvin Toffler,* in his brilliant book *Future Shock,* writes: "The number and type of decisions demanded of us are not under our autonomous control. It is the society that basically determines the mix of decisions we must make and the pace at which we must make them. Today there is a hidden conflict in our lives between the pressures of acceleration and those of novelty. One forces us to make faster decisions while the other compels us to make the hardest, most time-consuming type of decisions." A possible solution of this stressful problem is to devise and develop fast and effective methods for *decision making.* Many mathematical models have been developed for this purpose (the theory of decision functions, utility theory, PERT, CPM, etc.), and decision analysis can be considered as one of them.

Two specific features arise in the majority of decisions. There is usually some risk involved, and different possible decisions are then evaluated on a penalty-reward basis. We shall assume that all such preferences can be expressed in a numerical form as some artificial (or real) "money." We shall also see that in the majority of the situations requiring a decision, one cannot avoid the influence of chance or of some random factors. Our final decision among different alternatives will therefore be based on the notions of *expected gain* or *expected loss.* This is one of the reasons why we had to study in detail the notion of a random variable and its expected value. A model based on these assumptions will have practical meaning and importance in situations where a decision must be made many times under the same conditions (for example in industry, business, gambling, etc.). Each time, the decision is followed by a gain or a loss, which can be considered as a random variable. From the Law of Large Numbers (see Chapter XIII), we know that the average gain per decision will converge in some sense to the expected gain, and from this point of view the expected gain is of top importance for the evaluation of different decisions. We shall tacitly accept a rule claiming that the decision with the higher expected gain is better. However, there are many situations in life when the decision to be taken is unique and can never be repeated—for example, loss of virginity for a girl or vasectomy for a man—our theory is then completely inadequate to say the least.

We shall not try to develop any general theory of decision analysis, but we shall demonstrate some basic ideas in simple examples. Imagine the following situation: after paying an admission charge, a gambler can enjoy himself at a peculiar game. There are 10 identical indistinguishable urns: two of them, called θ_1, contain one silver and one gold coin each, three of them called θ_2 contain one silver and two gold coins each, and five of them, called θ_3, contain four silver and one gold coin each.

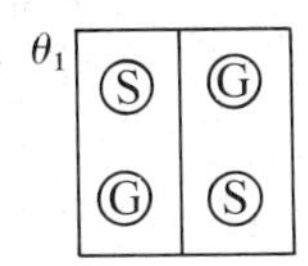

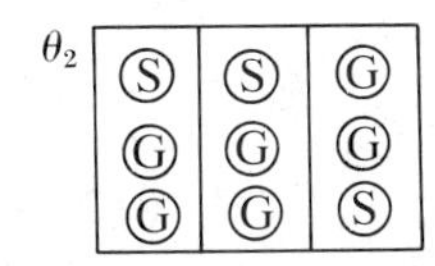

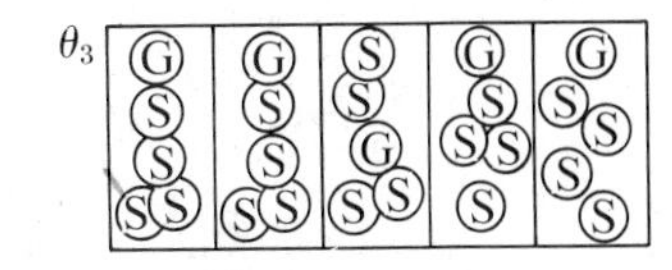

The dealer selects one of these urns at random. Then the gambler has to decide (before seeing the contents) on one of the following *actions* or *strategies:* S_1—to keep all the silver coins from the chosen urn; S_2—to keep all the gold coins from the chosen urn; S_3—to guess that it is a type θ_1 urn and get $60 if he is correct, otherwise to get nothing; S_4—to guess that it is a type θ_2 urn and get $40 if he is correct and nothing otherwise; S_5—to guess that it is a type θ_3 urn and get $20 if he is right, otherwise nothing. The gambler knows that each silver coin is worth $5 and each gold coin is worth $10. What strategy should he play?

This seemingly rather artificial situation is analogous to problems arising frequently in life. A businessman knows that the market demand might be for a special type of large, medium or small car, but he does not know ahead of time what the *actual or true state* of the demand is. However, he has to decide on a certain action or strategy specifying which types of commodities to stock and in what amounts. If he does not "hit" it right, he will lose money. Or imagine a factory producing its wares in runs or batches. Each run has a certain percentage of defectives caused by the alignment of the machines. Every item has to be completely finished, including its casing, before the next item is produced. The casing itself cannot influence the defectiveness of the run. However, good quality runs can be made very attractive and profitable with an expensive casing while the expensive casing on low quality runs might drastically reduce the profit due to frequent replacements under the warranty. The quality of the run is not known ahead of time but the production engineer has to decide what type of casing to use: cheap, medium priced or expensive. The writing of this book involved the same kind of problem: the present tendency of the educational system might be towards more theoretical or more applied treatment of the material. The author's emphasis on a theoretical or an applied presentation or their balance, if he knew the actual tendency, might influence to a certain extent the success of his textbook.

We could easily mention many other practical examples. However, let us try to generalize the common features of these situations. There are usually several *parameters* or *states of nature* (denoted by $\theta_1, \theta_2, \theta_3$, etc.), which influence substantially the whole situation (the types of urn selected by the dealer, the type of market demand, the quality of the production run given by the proportion of defectives, the trend of the educational system, etc.). Unfortunately the true or actual state of nature is not known when the decision is to be made. One knows only the *à priori* (probability) *distribution* of the states or parameters. This distribution is usually denoted by $P(\theta)$ $\left(\text{for example, in the case of the gambler the first type of urn can be chosen as 2 out of 10, hence } P(\theta_1) = \frac{2}{10}, P(\theta_2) = \frac{3}{10}, P(\theta_3) = \frac{5}{10}\right)$. The person making the decision has to choose one action or strategy (denoted by S_1, S_2, S_3, etc.) among several alternatives. For example, the gambler may decide to keep the gold coins, the businessman may decide to stock only small and medium cars in equal amounts, the production engineer may decide to use the medium priced casing and the textbook writer may decide to write an abstract theoretical book boosting his ego and disregarding the students. Given the true (but unknown) state of nature θ_i and the selected strategy S_j of the decision maker, one can usually evaluate the outcome of the situation by a certain number X_{ij}. For example, if the dealer selects urn θ_1 (containing one silver and one gold coin), and the gambler selects the strategy S_2 (keep the gold coins) then he gets $X_{12} = \$10$ (each gold coin is worth \$10). Or if the actual demand is for big cars and the dealer stocks only small and medium cars, then he will lose a certain amount of money. We can usually express the value X_{ij} in the form of real or artificial money, and for this reason it is called a *payoff*. Let us return to the gambling problem and summarize all the facts in the form of a table.

θ	$P(\theta)$	S_1	S_2	S_3	S_4	S_5
θ_1	$\frac{2}{10}$	5	10	60	0	0
θ_2	$\frac{3}{10}$	5	20	0	40	0
θ_3	$\frac{5}{10}$	20	10	0	0	20
Expectation $E(S_j)$		12.5	13	12	12	10

Column S_1 corresponds to the first strategy of the gambler, i.e., to keep all the silver coins. There are one silver coin worth \$5 in the urns of type θ_1, one silver coin worth \$5 in the urns of type θ_2 and four silver coins worth \$20 in the urns of type θ_3. The second column labeled $P(\theta)$ gives the probabilities of particular types of urns being selected. Imagine that the gambler selects the first strategy and sticks to it consistently. Then the expected payoff is

$$E(S_1) = \frac{2}{10} \cdot 5 + \frac{3}{10} \cdot 5 + \frac{5}{10} \cdot 20 = 12.5.$$

This is the value at the bottom of column S_1. The last row of the table gives the expected payoffs for all the strategies. We immediately see that the strategy S_2 is best because it guarantees the highest payoff. We shall call this strategy *optimal,* that is,

$$S_2 = Sopt \text{ and } E(Sopt) = \$13.$$

Therefore, with no information about the selection of urns by the dealer, the gambler can expect on the average \$13 per game if he plays for a long time.

Imagine now that the dealer is a grafter and makes a deal with the gambler: He will let the gambler know what type of urn has been chosen before the selection of the strategy. In this situation, the gambler has *perfect information* about the outcome. There is no uncertainty, and he can follow the *perfect information strategy,* namely, for each type of θ, he can select the strategy giving him the highest profit: for θ_1 it is S_3, giving him \$60, for θ_2 it is S_4, giving him \$40 and for θ_3 it is either S_5 or S_1, each giving him \$20. In this ideal case the gambler may expect to win, using the perfect information strategy, the following amount:

$$E(PIS) = \frac{2}{10} \cdot 60 + \frac{3}{10} \cdot 40 + \frac{5}{10} \cdot 20 = \$34.$$

This is the highest possible amount the gambler can ever expect to get unless he robs the casino. As a matter of fact, without the grafter's help he could get $E(Sopt) = \$13$ while with the help of his accomplice he raises his gain to $E(PIS) = \$34$. This increment is actually the *expected value of perfect information* about the game.

$$E(PI) = E(PIS) - E(Sopt) = 34 - 13 = \$21.$$

Thus the gambler should be willing to pay up to \$21 for this kind of information.

However, not all dealers are swindlers, and perfect information is practically never available. It serves only as an upper bound of gambler's expectations under the most favorable conditions; under practical conditions the gambler must expect less than $E(PIS)$. However, the gambler may be offered another deal. Immediately after the urn has been selected but before the gambler makes a decision about his strategies, he is allowed to *sample* the contents of the urn by selecting one coin at random from it to see whether it is gold or silver. The gambler is obviously getting very valuable *information* about the type of urn. Imagine that he draws a gold coin. This is corroborating evidence in favor of an urn of type θ_2 containing the highest proportion of gold coins. Actually, the gambler can find a posteriori probabilities (after sampling the urn that has been selected) using Bayes's theorem. We can summarize the situation after the selection of a *gold coin* as follows:

θ	$P(\theta)$	$P(G/\theta)$	$P(G \cap \theta)$	$P(\theta/G)$	$S1$	S_2	S_3	S_4	S_5
θ_1	$\frac{2}{10}$	$\frac{1}{2}$	$\frac{1}{10}$	$\frac{1}{4}$	5	10	60	0	0
θ_2	$\frac{3}{10}$	$\frac{2}{3}$	$\frac{2}{10}$	$\frac{2}{4}$	5	20	0	40	0
θ_3	$\frac{5}{10}$	$\frac{1}{5}$	$\frac{1}{10}$	$\frac{1}{4}$	20	10	0	0	20
			Expectation $E(S_j/G)$		$8\frac{3}{4}$	15	15	20	5

The third column contains the probabilities $P(G/\theta)$ of selecting a gold coin from an urn of type θ. Say for θ_1 the urn contains one gold and one silver coin, hence $P(G/\theta_1) = \frac{1}{2}$. The third column contains the probabilities of the simultaneous occurrence of an urn of type θ and getting a gold coin. Obviously

$$P(G \cap \theta_1) = P(\theta_1) \cdot P(G/\theta_1) = \frac{2}{10} \cdot \frac{1}{2} = \frac{1}{10}.$$

(See the comment in Chapter VIII, page 78.) These probabilities are very important for the calculations in Bayes's theorem. The fourth column entry is obtained as a product of entries in the second and third columns. Incidentally, the sum of the fourth column will give the probability of getting a gold coin (see the formula of total probability, Chapter VIII, page 84).

$$P(G) = \sum_{i=1}^{n} P(G \cap \theta_i) = \frac{1}{10} + \frac{2}{10} + \frac{1}{10} = \frac{4}{10}.$$

The fifth column contains the à posteriori probabilities $P(\theta/G)$ of an urn of type θ, given the fact that the selected coin is gold. Using Bayes' theorem, we get (see Chapter VIII, page 85):

$$P(\theta_1/G) = \frac{P(G \cap \theta_1)}{P(G)} = \frac{\frac{1}{10}}{\frac{4}{10}} = \frac{1}{4}.$$

The fifth column entries are obtained from the entries of the fourth column by dividing by $P(G) = \frac{4}{10}$. Therefore the à posteriori probabilities represent new probabilities of urns of different types after obtaining the *sample information.* We can again calculate the expected payoffs for different strategies, using $P(\theta/G)$. For example, if the gambler now selects strategy S_1, then he can expect

$$E(S_1/G) = \frac{1}{4} \cdot 5 + \frac{2}{4} \cdot 5 + \frac{1}{4} \cdot 20 = \frac{35}{4} = \$8\frac{3}{4}.$$

The last row contains these expected values, given the condition that the selected coin is gold. In this situation strategy S_4 is best, giving him $20.

$$E(Sopt/G) = E(S_4/G) = \$20.$$

Let us make the same analysis in the event that the gambler draws a silver coin S.

θ	$P(\theta)$	$P(S/\theta)$	$P(S \cap \theta)$	$P(\theta/S)$	S_1	S_2	S_3	S_4	S_5
θ_1	$\frac{2}{10}$	$\frac{1}{2}$	$\frac{1}{10}$	$\frac{1}{6}$	5	10	60	0	0
θ_2	$\frac{3}{10}$	$\frac{1}{3}$	$\frac{1}{10}$	$\frac{1}{6}$	5	20	0	40	0
θ_3	$\frac{5}{10}$	$\frac{4}{5}$	$\frac{4}{10}$	$\frac{4}{6}$	20	10	0	0	20
				Expectations $E(S_j/S)$	15	$11\frac{2}{3}$	10	$6\frac{2}{3}$	$13\frac{1}{3}$

The probability of getting a silver coin is (from the sum of the fourth column)

$$P(S) = \frac{6}{10}.$$

The optimal strategy in this case is S_1, giving the expected payoff

$$E(Sopt/S) = E(S_1/S) = \$15.$$

We can summarize all the previous calculations in this simple advice to the gambler: If you draw a gold coin play S_4—this may happen with probability $\frac{4}{10}$ and you may expect $20 in these situations; if you draw a silver coin play S_1—it may happen with probability $\frac{6}{10}$ and you may expect $15 in these situations. If the gambler follows such an *information strategy,* he may expect to win on the average:

$$E(IS) = P(G) \cdot P(S_4/G) + P(S)P(S_1/S) = \frac{4}{10} \cdot 20 + \frac{6}{10} \cdot 15 = \$17.$$

Remember that the gambler could win $E(Sopt)$ = $13 without the possibility of sampling the urns. The increment between these values represents the *expected value of the sample information:*

$$E(I) = E(IS) - E(Sopt) = 17 - 13 = \$4.$$

Therefore, the gambler should be willing to pay up to $4 for the privilege of sampling the urn before making a decision.

Draw	Play	$E(IS)$	$E(I)$
G	S_4	\$17	\$4
S	S_1		

Decision table for the gambler.

This problem might appear rather trifling and superficial. However, it captures basic ideas which can be used for an analysis of situations of a more practical and complicated nature. GLOBALL Company Ltd. is manufacturing a certain kind of ball bearings. The final marketed product consists of a two-piece circular casing with the inner groove filled with nine steel balls.

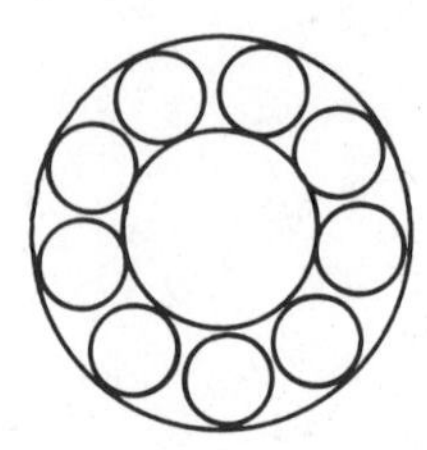

Balls for the bearings are produced in large runs, each of which is quite homogeneous in its physical parameters. However, due to circumstances in the manufacturing process beyond engineering control, there is always some very small difference in the diameters of the balls from run to run. From this point of view, all the production runs are divided into four categories:

θ_1—runs with 20% of the balls having diameters slightly above the recommended standard (or runs with approximately 20% of such balls),
θ_2—runs with 40% (or close to 40%) of the balls having diameters slightly above the recommended standard,
θ_3—runs with 60% (or close to 60%) of the balls having diameters slightly above the recommended standard,
θ_4—runs with 80% (or close to 80%) of the balls having diameters slightly above the recommended standard.

Incidentally, it is a characteristic of *every* manufacturing process that the actual physical dimension of the item produced is a random variable fluctuating around the recommended standard within certain tolerance limits; it is practically impossible to produce items with dimensions exactly equal to the recommended standard. Obviously, we could imagine runs of the θ_1 type as somehow being "small" balls, runs of the θ_4 type as "large" balls, and so on. Even though the diameters of the balls might differ slightly, they are all perfectly useful for the assembly of ball bearings. The company has three types of casing available: C_1 with somewhat "undersized" grooves, C_2 with average grooves and C_3 with somewhat "oversized" grooves. The casings of type C_1 fit perfectly the balls from θ_1 runs, giving high quality bearings which bring a high profit to the company. However, if C_1 casings are used on θ_4 balls, the resulting bearings are still good for use but increased materials tension causes accelerated aging, which might lead to early failures; then the company could incur extra expense in replacing bearings under the warranty. The situation with the C_3 casings is similar because they fit θ_4 perfectly, yielding a high profit, while if used on θ_1 balls they yield a rather poor profit. There are many other combinations of ball types and casing sizes that can be used, and the corresponding profits *per run* can be summarized as follows:

Type of run θ	Type of casing used C_1	C_2	C_3
θ_1	$400	$280	$120
θ_2	$300	$290	$220
θ_3	$200	$310	$280
θ_4	$100	$230	$410

If the company knew exactly what kind of run of balls is being produced, it would be easy to decide what kind of casing to use. Unfortunately, it is impossible to measure all the balls because of prohibitive costs, therefore it is *never known* what kind of run is in production. However, it has been observed from past experience that θ_1 runs represent 20%, θ_2 represent 30%, θ_3 represent 20% and θ_4 represent 30% of the overall production. If the company could ever find an easy and cheap way of measuring all the balls and discovering exactly the type of run in production, the decision for what casings to use would be as follows:

Type of run θ	$P(\theta)$	Use casing	Profit per run
θ_1	.2	C_1	$400
θ_2	.3	C_1	$300
θ_3	.2	C_2	$310
θ_4	.3	C_3	$410

Given *perfect information* about the run, the company could expect the following average profit per run:

$$E(PIS) = .2 \cdot 400 + .3 \cdot 300 + .2 \cdot 310 + .3 \cdot 410 = \$355.$$

If nothing is known about the run, the company should decide on the optimal strategy giving the best expected profit:

θ	$P(\theta)$	Type of casing used C_1	C_2	C_3
θ_1	.2	400	280	120
θ_2	.3	300	290	220
θ_3	.2	200	310	280
θ_4	.3	100	230	410
		240	274	269

The optimal strategy in this case is obviously $C_2 = Sopt$, giving the expected profit of $274. We mentioned that complete and exhaustive measurement of the whole run is prohibitively expensive. However, the production engineer might decide on the following *information strategy:* Take a sample of two balls, and determine how many have a diameter above the recommended standard. Use the *sample information* for a decision about the type of casing to be used. Imagine that the engineer finds that both balls in the sample have a diameter above the standard. Given this fact, he can recalculate the à priori probabilities of different runs as follows: Let us denote the observed fact by $2L$ (getting two "large" balls). If the run in production is θ_1, then the probability of a large ball being produced is .2, hence the probability of exactly two large balls among the two chosen is obtained using the binomial distribution:

$$P(2L/\theta_1) = \binom{2}{2}(.2)^2(.8)^0 = .04.$$

Similarly, for the production runs θ_2, θ_3 and θ_4, we get

$$P(2L/\theta_2) = \binom{2}{2}(.4)^2(.8)^0 = .16.$$

$$P(2L/\theta_3) = .36. \qquad P(2L/\theta_4) = .64.$$

Having these probabilities, we can calculate the probabilities of the simultaneous occurrence of $2L$ and θ_i to be used in Bayes' theorem:

$$P(2L \cap \theta_1) = P(\theta_1)P(2L/\theta_1) = .2 \cdot .04 = .008.$$

$$P(2L \cap \theta_2) = P(\theta_2)P(2L/\theta_2) = .3 \cdot .16 = .048.$$

$$P(2L \cap \theta_3) = .2 \times .36 = .0720. \qquad P(2L \cap \theta_4) = .3 \times .64 = .1920.$$

Using the formula of total probability, we get

$$P(2L) = \sum_{i=1}^{4} P(\theta_i)P(2L/\theta_i) = .008 + .048 + .072 + .1920 = .32.$$

Using Bayes' theorem, we get the probability of run θ_i being in production, given the fact that we got two large balls as

$$P(\theta_1/2L) = \frac{P(2L \cap \theta_1)}{P(2L)} = \frac{.008}{.32} = .0250, \qquad P(\theta_2/2L) = \frac{.048}{.32} = .15,$$

$$P(\theta_3/2L) = \frac{.0720}{.32} = .225, \qquad P(\theta_4/2L) = \frac{.1920}{.32} = .6.$$

We can summarize the obtained information in the following table. The last row contains the expected profit, using the corresponding type of casing.

θ	$P(L)$	$P(\theta)$	$P(2L/\theta)$	$P(2L \cap \theta)$	$P(\theta/2L)$	C_1	C_2	C_3
θ_1	.2	.2	.04	.008	.025	400	280	120
θ_2	.4	.3	.16	.048	.15	300	290	220
θ_3	.6	.2	.36	.0720	.225	200	310	280
θ_4	.8	.3	.64	.1920	.6	100	230	410
			Expectation $E(C_j/2L)$			160	258.25	345

For example, $E(C_3/2L) = .025 \cdot 120 + .15 \cdot 220 + .225 \cdot 280 + .6 \cdot 410 = \$345.$

Obviously, if the engineer gets two large balls in the sample of two, then casing C_3 is the best to use, giving the expected profit:

$$E(Sopt/2L) = E(C_3/2L) = \$345.$$

However, a sample of two can yield one ball with a diameter larger than the recommended standard—denote this event $1L$. Then we can calculate again all the probabilities as follows:

θ	$P(L)$	$P(\theta)$	$P(1L/\theta)$	$P(1L \cap \theta)$	$P(\theta/1L)$	C_1	C_2	C_3
θ_1	.2	.2	.32	.064	.16	400	280	120
θ_2	.4	.3	.48	.144	.36	300	290	220
θ_3	.6	.2	.48	.096	.24	200	310	280
θ_4	.8	.3	.32	.096	.24	100	230	410
			Expectation $E(C_j/1L)$			244	278.8	264

From the fifth column, we find that $P(1L) = .4$ and from the last row we see that in the case of exactly one large ball in the sample of two, it is best to use casing C_2, giving the expected profit:

$$E(Sopt/1L) = E(C_2/1L) = \$278.8.$$

Similarly, we can calculate what happens if there is no large ball in the sample—denote this event $0L$. It happens with probability .28, and the best casing to use is C_1, giving the expected profit:

$$E(Sopt/0L) = E(C_1/0L) = \$325.71.$$

The following table summarizes our analysis of samples of two:

No. of large balls k	$P(k)$	Best casing $Sopt$	Expected profit $E(Sopt/kL)$
0	.28	C_1	\$325.71
1	.4	C_2	\$278.8
2	.32	C_3	\$345

Decision table for $n = 2$.

If the production engineer will always take a sample of two from each run, and decide which casing to use from the above table, with this *information strategy* the company may expect to make on the average the following profit per run:

$$E(IS) = (.28) \cdot (325.71) + (.4) \cdot (278.8) + (.32) \cdot 345 = \$313.12.$$

We remember that without sampling the best profit to be expected is $E(C_2) = \$274$. However, taking a sample of size $n = 2$ and following the appropriate information strategy increases the expected profit to $E(IS) = \$313.12$. The increase is therefore the *expected value of sample information* in a sample of size $n = 2$.

$$E(I_2) = E(IS) - E(C_2) = 313.12 - 274 = \$39.12.$$

Measuring two balls certainly does not cost $39.12, therefore the appropriate information strategy tremendously increases the average profit per run. However, the production engineer will immediately ask the following question: What happens if we take samples of other sizes, say $n = 3, 4, 5$, etc.? Is there any *optimal* sample size that will give us the largest increment of expected profits without costing prohibitively? In order to find the optimal sample size, we have to undertake extensive calculations and find expected values of sample information in samples of size $n = 1, 2, 3, 4, 5$, etc. These calculations are summarized in the accompanying table, which also gives the optimal information strategies. Let us denote the symbols as follows:

n = sample size
k = the number of balls in the sample having a diameter exceeding the recommended standard
$Sopt$ = optimal strategy—the best casing to be used for a given n and k, contained in the row for k and the column for n
$E(IS)$ = expected value of the information strategy for a sample of size n in dollars

n	0	1	2	3	4	5	7	10	15	20	25	30	34
$E(IS)$	274	303.2	313.12	319.42	326.83	328.79	334.72	339.94	344.84	347.63	349.11	350.00	350.62
k													
0	—	1	1	1	1	1	1	1	1	1	1	1	1
1		3	2	1	1	1	1	1	1	1	1	1	1
2			3	3	2	2	1	1	1	1	1	1	1
3				3	3	3	2	1	1	1	1	1	1
4					3	3	2	1	1	1	1	1	1
5						3	3	2	1	1	1	1	1
6							3	3	1	1	1	1	1
7							3	3	2	1	1	1	1
8								3	2	1	1	1	1
9								3	2	2	1	1	1
10								3	3	2	1	1	1
11									3	2	2	1	1
12									3	2	2	1	1
13									3	3	2	2	1
14									3	3	2	2	1
15									3	3	2	2	2
16										3	3	2	2
17										3	3	2	2
18										3	3	2	2
19										3	3	3	2
20										3	3	3	2
21											3	3	2
22											3	3	3
and more													

The accompanying graph clearly shows how the expected value of the sample information increases from $274 for no sampling up to close to $355, the value of the perfect information. Imagine that setting up the measuring equipment costs $11, and precise measurement of each ball costs about $1.50. The straight line represents the cost of testing as a function of

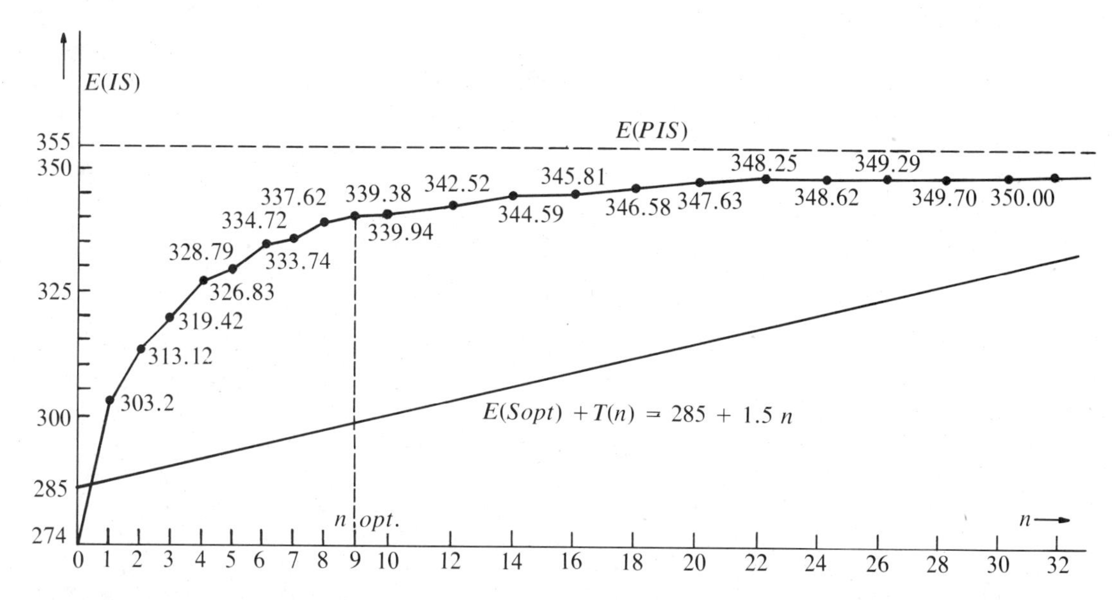

sample size $T(n) = 11 + 1.5n$. From the graph and the calculations, we find that the biggest difference between the expected value of the sample information and the cost of testing is reached for $n = 9$, which represents the *optimal sample size*. The actual expected *sampling profit* using the optimal sample size is

$$E(SP) = E(IS) - E(Sopt) - T(n) = 339.4 - 274 - 24.5 = \$40.9.$$

We can see that a well designed optimal sampling strategy can considerably increase the profit per run. Incidentally, it is interesting to note that the expected value of the information strategy as a function of n is not *concave* as one would expect. The expected value of the sample information is 333.74 for $n = 6$, 334.72 for $n = 7$ and finally, it is 337.62 for $n = 8$. This means that one item expanding the sample from six to seven is "worth" only 98¢, while one item expanding the sample from seven to eight is worth $2.90. This seemingly strange fact is caused by the finite number of values we accept for θ, because in reality there will be runs with many different values of θ.

ENTROPY AND INFORMATION

The notion of *information* is becoming one of the most important categories used in analysis of the natural and social aspects of human life. Some thinkers even classify the second half of the twentieth century as the era of the "information explosion" or "revolution," clearly hinting at a comparison with the feudal and industrial revolutions and their great impact on the evolution of mankind. Even though no one will ever dispute the important role of information in modern society, it is really very difficult to give a concise and logically consistent definition of this significant notion. Just one aspect is clear: one can speak about *information* in relation to a phenomenon only if there is some *uncertainty* or *lack of knowledge* preceding the act of emitting and transmitting the information. From this point of view, any information theory must be closely related to the theory of probability. There is no general information theory available to cover all the multifaceted aspects of this notion. However, there are some probabilistic models describing the creation, transformation and use of information in some special areas of application. The previous section analyzed the notion of information in the realm of decision making. In the present section, we shall investigate this notion from the point of view of transmission of messages through information channels and communications. These aspects are very closely related to electrical engineering, and the first impetus in this direction came from *Claude E. Shannon* in his early work for Bell Laboratories (1948).

— 愛は惜しみなく与う —

Imagine a simple lad from Texas who is given a novel in Japanese by his dainty girl friend from the Land of the Rising Sun. He does not know any Japanese but he examines the sequence of *symbols,* each of which contains some amount of *information.* He knows that each symbol represents an idea, and that a *message* is formed by a sequence of more or less independent symbols,* as opposed to the structure of English where each symbol's occurrence is greatly influenced by the preceding symbol. As a matter of fact, we usually speak about *letters* instead of symbols. Each letter-symbol has a certain probability of occurrence in a message. In the first simplified approximation, we could describe an *information source* by the set of all its letters and by the probabilities of their occurrence. The set of all available letters is usually called the *alphabet,* and we shall assume that it is finite. To begin, we shall investigate a source of information emitting a single letter. An alphabet can consist of *any* set of distinguishable symbols: perhaps $\cdot -$, A, β, Ň and a red flag will do well. Emission of a single letter from the source can be understood as an experiment, with the set of all possible outcomes being equal to the alphabet. Thus the alphabet can be viewed as a sample space. We shall assume that the source of information is uniquely described by its alphabet and the probabilities of occurrence of its letters. From this point of view, *we define the source of information as a (finite) probability measure space, in which the sample space is called the alphabet and the outcomes are called the letters.*

For the sake of uniformity, let us denote the source of information (Ω, P) where $\Omega = \{\omega_1, \omega_2, \ldots, \omega_n\}$ is the corresponding alphabet and P is the probability measure. For each letter ω_i, we have a uniquely defined probability

$$P_i = P(\{\omega_i\}),$$

which represents the probability of emission of the letter ω_i from the source (Ω, P). There is an *uncertainty* about which letter will be sent from the source. If the letter ω is dispatched, then this *uncertainty is eliminated by the information contained in* ω. We would like to be able to measure the *amount* of information contained in this letter. We shall denote this amount by $I(\omega)$. It is expedient to define (assuming $P(\{\omega\}) > 0$):

$$I(\omega) = \log \frac{1}{P(\{\omega\})},$$

where the base of the logarithm is 2. The unit of information is called the *bit* (a contraction of *binary unit* because of the base 2). Imagine that a peculiar source emits symbols H and T as its alphabet. This source is sending the information about the outcome of the toss of a fair coin performed by your friend (as a matter of fact, the source might simply be your friend saying H or T after each toss). How much information is contained in one symbol, say H?

$$I(H) = \log \frac{1}{P(H)} = \log \frac{1}{\frac{1}{2}} = \log 2 = 1.$$

*This is almost true about Japanese written in ancient symbols, kanji, not so much about hiragana or katakana scripts.

Therefore, the amount of information in a message about the outcome of a single toss of a fair coin is one bit. As a matter of fact, we can easily see that *one bit is the amount of information received if one of the two equally likely events is specified.*

It is well known that a computer's memory consists of electronic units, each of which can be in one of two states: either 0 or 1. If the information brought into the memory unit will lead it into one of these two states with equal likelihood, then the unit will store an amount of information equal to exactly one bit. We say that the unit has an *information capacity* of one bit. A certain number of these units combined together form a *word* (frequently 60 bits), and modern computers contain tens of thousands of words. The overall information capacity of the memory is therefore millions of bits.

Let us return to the definition of the amount of information. We have defined it for every letter ω as $I(\omega) = \log \frac{1}{P(\{\omega\})}$. It follows that $I(\omega)$ is a *random variable,* and we can speak about its expectation, which we shall denote by $H(\Omega, P)$, to show the source of information.

$$H(\Omega, P) = E(I(\omega)) = \sum_{i=1}^{n} P_i \log \frac{1}{P_i} = -\sum_{i=1}^{n} P_i \log P_i.$$

If any of the probabilities $P_i = 0$, then we set $P_i \log P_i = 0$. The number $H(\Omega, P)$ *is the expected amount of information received per letter of the source* (Ω, P), *and it is called the entropy of the source.* Imagine that three fair coins are tossed one after another so that all $2^3 = 8$ outcomes are equally likely. You are told the outcome (it may be very important, for example, for the game described in Example 17, page 116). In this case,

$$n = 8, \quad P_1 = P_2 = \cdots = P_8 = \frac{1}{8}$$

therefore the entropy of the source of information

$$H(\Omega, P) = -\sum_{i=1}^{\infty} P_i \log P_i = -8 \cdot \frac{1}{8} \log \frac{1}{8} = \log 8 = \log 2^3 = 3 \text{ bits}.$$

Therefore, if nothing is known about the outcome, then the message specifying which sequence of heads and tails has occurred contains 3 bits of information.

We have already mentioned that each source of information (or probability measure space, which is the same thing) is always associated with an *uncertainty* about the emission of a letter or the realization of an outcome. This uncertainty exists only up to the moment that a letter or a symbol for the outcome is received. At this moment the uncertainty is suppressed and nullified by the information in the message (provided the outcome is fully specified). For this reason, the entropy (the average information per letter) is considered to be a *measure of uncertainty* associated with the source.

Imagine the following two experiments: In the first one, a fair coin is tossed. The sample space has two points and the entropy is

$$H(\Omega_1, P_1) = \frac{1}{2} \log 2 + \frac{1}{2} \log 2 = 1 \text{ bit}.$$

In the second experiment, a fair die is rolled and two possible outcomes observed: either a success, meaning that a 6 occurred or a failure meaning that anything else occurred. The sample space again has two points and the entropy is

$$H(\Omega_2, P_2) = \frac{1}{6}\log 6 + \frac{5}{6}\log\frac{6}{5}.$$

The logarithm used so far has the base 2, but for evaluation it might be better to use the natural logarithm, ln. We know that

$$\log x = \log_2 x = \frac{\ln x}{\ln 2}.$$

$$\log 6 = \frac{1.791759}{.693147} = 2.584962.$$

$$\log\frac{6}{5} = \frac{.182322}{.693147} = .263034.$$

Hence $$H(\Omega_2, P_2) = .650022 \text{ bit.}$$

Therefore, there is much more uncertainty in the first experiment than in the second. As a matter of fact, it is completely obvious intuitively. Both experiments have two outcomes: In the first one, both outcomes are equally likely and nothing can be said ahead of time about which one will occur. However, in the second experiment we know that the first outcome is much less likely to occur than the second; we can predict that on the average in five out of six repetitions the second outcome will occur. In this sense, there is much less uncertainty in the second experiment than in the first one. As a matter of fact, if the experiment has n possible outcomes then it is very plausible to expect that *the highest uncertainty is achieved when all the outcomes are equally likely.* Using the notion of entropy we should have the following theorem.

→ **THEOREM.** $$H(\Omega,P) \leq \sum_{i=1}^{n} \frac{1}{n}\log n = \log n$$

for every probability measure space with n possible outcomes. The equality is reached only for the equally likely space.

Proof: Let us investigate the behavior of the functions $y = x - 1$ and $y = \ln x$.

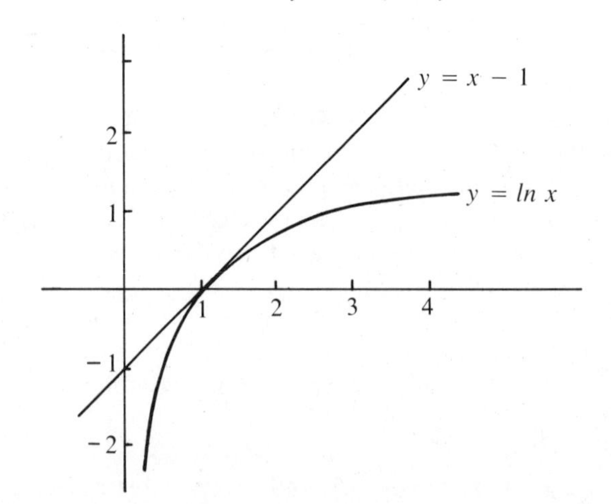

It can be easily seen that

$$\ln x \leq x - 1$$

for all x, and the equality is reached only for $x = 1$. Further we get (after multiplying by (-1))

$$\ln \frac{1}{x} \geq 1 - x.$$

Let us investigate the entropy $H(\Omega,P)$.

$$\log n - H(\Omega,P) = \sum_{i=1}^{n} P_i \log n - \sum_{i=1}^{n} P_i \log \frac{1}{P_i} = \sum_{i=1}^{n} P_i \log nP_i = \frac{1}{\ln 2} \sum_{i=1}^{n} P_i \ln nP_i.$$

If we put $x = \frac{1}{nP_i}$ into the above inequality, we get

$$\ln nP_i \geq \left(1 - \frac{1}{nP_i}\right)$$

hence

$$\log n - H(\Omega,P) \geq \frac{1}{\ln 2} \sum_{i=1}^{n} P_i \left(1 - \frac{1}{nP_i}\right) = \frac{1}{\ln 2}\left(\sum_{i=1}^{n} P_i - \sum_{i=1}^{n} \frac{1}{n}\right) = 0.$$

Therefore we have $$H(\Omega,P) \leq \log n.$$ q.e.d.

The above theorem can be interpreted practically as follows: If we observe a source of information that can emit n different letters (symbols, signals, etc.), then the largest information per letter is obtained if all the letters-symbols have the same probability.

Assume now that a new source, which we denote $(\Omega^m, P^{(m)})$, emits a message consisting of m *independently* produced letters from the original source (Ω,P). If the original source is thought of as a probability measure space associated with an experiment, then this new *extended* source (transmitting messages of the length m) can be considered as m independent repetitions of the original experiment. We have studied models of independent repeated trials in Chapters XI (binomial distribution) and XIII (relating to the Law of Large Numbers), and we know that the corresponding "new" sample space Ω^m consists of all sequences of the length m:

$$\underline{\omega} = \underbrace{(\omega_{i_1}, \omega_{i_2}, \ldots, \omega_{i_m})}_{m \text{ times}}$$

$$\Omega^m = \underbrace{\Omega \times \Omega \times \ldots \times \Omega}_{m \text{ times}}$$

$\omega_{i_1}, \omega_{i_2}, \ldots, \omega_{i_m} \in \Omega$ (where Ω is the original alphabet). The probability of each message is equal to

$$P^{(m)}(\{\underline{\omega}\}) = P_{i_1} P_{i_2} P_{i_3} \ldots P_{i_m}$$

because of independence. Let us calculate the entropy of such an extended source, which emits messages of the length m.

$$H(\Omega^m, P^{(m)}) = \sum_{\underline{\omega} \in \Omega^m} P^{(m)}(\{\underline{\omega}\}) \log \frac{1}{P^{(m)}(\{\underline{\omega}\})}$$

$$= \sum_{\underline{\omega}\in\Omega^m} P_{i_1} P_{i_2} P_{i_3} \ldots P_{i_m} \log \frac{1}{P_{i_1} P_{i_2} P_{i_3} \ldots P_{i_m}}$$

$$= \sum_{\underline{\omega}\in\Omega^m} P_{i_1} P_{i_2} P_{i_3} \ldots P_{i_m} \left[\log \frac{1}{P_{i_1}} + \log \frac{1}{P_{i_2}} + \log \frac{1}{P_{i_3}} \cdots + \log \frac{1}{P_{i_m}}\right]$$

$$= \sum_{i_1 i_2 \ldots i_m} P_{i_1} P_{i_2} P_{i_3} \ldots P_{i_m} \log \frac{1}{P_{i_1}} + \sum_{i_1 i_2 \ldots i_m} P_{i_1} P_{i_2} P_{i_3} \ldots P_{i_m} \log \frac{1}{P_{i_2}}$$

$$+ \sum_{i_1 i_2 \ldots i_m} P_{i_1} P_{i_2} P_{i_3} \ldots P_{i_m} \log \frac{1}{P_{i_3}} + \cdots$$

$$+ \sum_{i_1 i_2 \ldots i_m} P_{i_1} P_{i_2} P_{i_3} \ldots P_{i_m} \log \frac{1}{P_{i_m}}$$

$$= m \sum_{i=1}^{n} P_i \log \frac{1}{P_i} = m\, H(\Omega, P).$$

Therefore, we can conclude that if the new extended source emits messages of the length m consisting of independently produced letters of the original source, its entropy is m times larger, namely,

$$H(\Omega^m, P^{(m)}) = m\, H(\Omega, P).$$

Let us use the above derived facts to solve some problems.

PROBLEM 1

It is known that the inhabitants of the city of Abadan always speak the truth, while inhabitants of the city of Belial always lie. A weary globetrotter knows that he is in one of these cities, but he does not know in which one. What is the least number of questions eliciting "yes" or "no" that will give him the information about his location? (It is known that the inhabitants of each city freely visit the other city.)

SOLUTION. The traveler may be in city A or B. Because he does not know where he is at all, the probability of being in either city is $\frac{1}{2}$. The uncertainty related to his location is hence

$$H = \frac{1}{2}\log 2 + \frac{1}{2}\log 2 = 1 \text{ bit}.$$

He has to ask one or more questions bringing him a total of one bit of information to nullify this uncertainty. However, a single question has two possible answers, "yes" or "no," and potentially can contain one bit of information. We demonstrated above that the question will contain a maximum of information if *both answers will be equally likely.* In this case, the answer will transmit

$$\frac{1}{2}\log 2 + \frac{1}{2}\log 2 = 1 \text{ bit}.$$

of information. These hits should be sufficient for an experienced globetrotter to find an appropriate question. If he asks someone, "Do you live in this city?", that will do the trick. In city A everybody answers "yes," in city B everybody answers "no." Through reasoning, we assumed that the traveler can equally likely meet an inhabitant of either city in either city.

PROBLEM 2

It is known that the inhabitants of the city of Cithaeron, which is located a short distance from Abadan and Belial, are equally likely to tell the truth or to lie. A weary globetrotter again finds himself in one of the cities A, B or C, but he does not know in which one. What is the least number of questions yielding "yes" or "no" for an answer that will give him information about his location and also where the person he is asking resides? (Inhabitants of all three cities travel widely among A, B and C.)

SOLUTION. There are nine different possibilities in the way the traveler can encounter someone: the traveler can be in one of the cities A, B or C and the other person can be an inhabitant of or visitor to one of these three cities. If the traveler is completely ignorant about his situation, then there is an uncertainty equal to log 9 bits. This uncertainty must be nullified by a certain number of questions, say m. In the solution of the first problem, we demonstrated that the maximum information in an answer to each question can be $\log 2 = 1$ bit, and this happens if and only if the answers "yes" and "no" are equally likely. Furthermore, the formula concerning extended sources (as derived above) shows that if each of m questions contains log 2 bits of information, then all m questions taken together will contain $m \log 2$ bits of information, provided they are independent. How many questions does one have to ask? The information contained in them ($m \log 2$) must be equal to or greater than the uncertainty of the traveler (log 9):

$$m \log 2 \geq \log 9$$

or

$$\log 2^m \geq \log 9$$

which is equivalent to $2^m \geq 9$. Obviously $m = 3$ is not enough and the traveler will have to ask *at least* four questions. As a matter of fact, exactly four questions will do the trick. He can ask
(a) Am I in one of the cities A or B?
(b) Am I in C?
(c) Do you live in C?
(d) Am I in A?

PROBLEM 3

Of 75 coins, 74 are of identical weight. One of them—a counterfeit coin—is a little lighter. How many weighings on the apothecary scales are necessary to discover the counterfeit coin?

SOLUTION. Each of the 75 coins is equally likely to be counterfeit, so the uncertainty to be removed by the weighings is equal to log 75. The experiment consisting of one weighing can have three possible outcomes—either the pans are in balance, or the left-hand pan goes down or the right-hand one does. Therefore, the information in one weighing is always log 3 at most, and reaches its maximum when all three outcomes are equally likely. If we repeat m weighings independently, then the total information is m log 3 at most and this information must be larger than the uncertainty of the selection of the counterfeit coin, which is log 75. Hence we have to repeat the weighings *at least* m times, where

$$m \log 3 \geq \log 75$$

or

$$\log 3^m \geq \log 75$$

or equivalently, $3^m \geq 75$. The solution of this inequality gives $m \geq 4$, so we have to perform at least four weighings. It is easy to see that it is enough to weigh the coins four times if one painstakingly follows the probabilistic requirement for the maximal information in each experiment. For each weighing, we have to put an equal number of coins on each pan. If for the first experiment we put 25 coins on each pan, then the light coin is equally likely to be in one of the three groups of 25 (either on the left-hand pan, the right-hand pan or not on the scales). Therefore all three outcomes of the experiment are equally likely (the left pan down, the right pan down, or balance), and the experiment gives the maximum information of log 3. We can easily determine after one weighing in which of the three groups is the counterfeit coin. Then we have 25 coins left to investigate. Imagine that for the second step we put k coins on each pan. Then the probability of the counterfeit coin being on each pan is $\frac{k}{25}$ and the probability that the counterfeit coin is not on the scales is $\frac{25 - 2k}{25}$. The maximization of the amount of information requires that the three numbers $\frac{k}{25}, \frac{k}{25}$ and $\frac{25 - 2k}{25}$ be as close as possible. This is obviously reached for $k = 8$. If we put eight coins on each pan and leave nine coins out, then one weighing will specify a group of eight or nine coins containing the counterfeit coin (if the scales are not in balance then the counterfeit coin is on the rising pan, otherwise it is not on the scales). Now we repeat the whole experiment with eight or nine coins, putting three coins on each pan. This will specify a group of two or three coins containing the counterfeit one, and one more repetition will give the final answer. Note that the entire solution has been achieved using ideas of information and entropy exclusively.

There are many other interesting problems that can be solved using the information theory. However, the most important applications of this theory are in communications (telephone, telegraph, communications satellites, etc.). Information theory helps to solve problems of the transmission of information sources through communications channels with noise, using appropriate methods of coding and so on. However, these methods require a more profound knowledge of probability theory and we cannot delve into them in this short survey.

Let us return to the definition of an information source. Can we consider the English language as a source of information from this point of view? Certainly we can—but it is a definite simplification, as we shall see later. The English alphabet consists of 26 symbols A,B,C, . . . ,Y,Z and of a special symbol "space." We can investigate the probabilities of the occurrence of each letter-symbol in an English sentence. These probabilities are typical for the English language, and differ considerably from French, Spanish, or other languages.

Probabilities of Symbols in English*

Symbol	Probability	Symbol	Probability
Space	0.1859	N	0.0574
A	0.0642	O	0.0632
B	0.0127	P	0.0152
C	0.0218	Q	0.0008
D	0.0317	R	0.0484
E	0.1031	S	0.0514
F	0.0208	T	0.0796
G	0.0152	U	0.0228
H	0.0467	V	0.0083
I	0.0575	W	0.0175
J	0.0008	X	0.0013
K	0.0049	Y	0.0164
L	0.0321	Z	0.0005
M	0.0198		

*From Reza, F. M. (1961). *An Introduction to Information Theory,* McGraw-Hill.

The entropy of this source is equal to $H = 4.03$ bits per symbol. The extended source with this alphabet and probabilities will then send messages of the following type.

T–NSREM–DIY–EESE– –F–O–SRIS–R– –UNNASHOR

Even though such a message is completely incomprehensible, it does capture a structure reminiscent of English. The main shortcoming is in the assumption that successive letters are independent. Obviously, this is not true. The probability of occurrence of every letter in English is strongly influenced by the previous letter. To demonstrate this, we can investigate another type of extended source which has the same alphabet but which behaves like a Markov chain (see Chapter XIV). The states of this Markov chain are the letters A,B,C, . . . ,Y,Z plus a "space" and the information source is determined by the transition probabilities $P_{\alpha\beta}$ that the letter α will be followed by the letter β where α and β are any two symbols of the English language. These probabilities can be estimated from actual English, and used again to generate an artificial English based on a Markov chain approximation. A typical sentence would thus appear:

*ACT–D–STE–MINTSAN–OLINS–TWID–OULY**

Obviously, the English structure is expressed much better and the method outlined clearly shows promise. Shannon suggested yet another method of analysis of English. He constructed a Markov chain with the states being whole English *words* rather than letters. The corresponding source of information would emit messages of the following type:

*REPRESENTING AND SPEEDILY IS AN GOOD APT OR COME CAN DIFFERENT NATURAL HERE**

The above mentioned methods belong to so-called *statistical linguistics.* Any language, real or artificial, is considered as a source of information in the sense given above (or perhaps in a somewhat more sophisticated sense). The source is then investigated from the point of view of probability theory, and analyzed as a probability space. What is it good for? It is possible to "teach" a computer to create its own English sentences and speak in a comprehensible manner. If all the necessary words and probabilities are stored in the computer's memory, then the

*Abramson, N. (1963). *Information Theory and Coding,* McGraw-Hill, New York.

machine can simulate the real language. However, this application is rather unimportant compared to problems of *deciding cases of disputed authorship.* There are many important literary works in English with anonymous authors. One important example is *The Federalist,* published in the years 1787–1788, urging the citizens of the state of New York to ratify the constitution of the United States. All 85 papers were anonymous but historians reliably established the authorship of many, leaving 12 undecided. It was clear that some of these papers were written by Alexander Hamilton, some by James Madison and perhaps some by both men jointly, as well as others; however, historians could not agree who wrote which. Mosteller and Wallace undertook an exhausting statistical analysis of the 12 disputed papers and come forth with an original and convincing solution of the problem.* A classic among disputed authorship cases is the question of whether Francis Bacon, or perhaps someone else, wrote some of Shakespeare's works. There are many similar problems that are analyzed using statistical linguistics.

STOCHASTIC MUSIC COMPOSITION

In the previous section, we mentioned Markov chains as tools for the analysis of English. The same methods can be used to analyze music. However, music has a rather complicated structure and we have to explain several fundamental notions of music theory.

Every song consists of three basic components: melody, harmony and rhythm. Melody is a sequence of single pitch tones giving the song an air. Harmony is simultaneous combinations of tones blended into chords pleasing to the ear, and rhythm is a pattern of regular or irregular pulses caused in music by the occurrence of beats. In order to create a song, a composer has to combine melody, harmony and rhythm in a sophisticated way so that the three components fit together and support each other. Its final effect is to give listeners a unique musical experience. However, in a simple approximation we can take these three components apart and investigate them separately. We shall try to describe each of them as a Markov chain. The resulting song will be obtained as a simultaneous realization of all three chains. We shall give here a very simple and elementary model that will capture only the most basic features of musical structures.

Let us start with harmony. In the first approximation, we can claim that the harmony of many songs consists of five basic chords. Assuming that we have a song in the key of C-major, the fundamental chords are as follows:

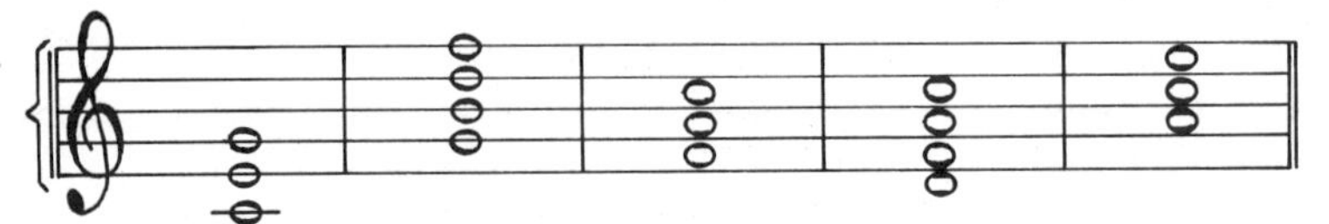

T–tonic   *D^7–dominant*   *S–subdominant*   *II^7–supertonic*   *VI–submediant*

The harmony of a song can often be described as a sequence of these chords, each of them being valid for one bar or time unit. Chords in the sequence are certainly not independent, but we can assume that they follow the pattern of a Markov chain. (In real music the harmony component is usually more complicated, but even this simple probabilistic approach yields interesting results.) The Markov chain for harmony therefore has five states: $\omega_1 = T$, $\omega_2 = D^7$, $\omega_3 = S$, $\omega_4 = II^7$ and $\omega_5 = VI$. The initial probability vector $\boldsymbol{p}$ and the transition probability matrix $\mathbf{P}$ could be as simple as

$$\boldsymbol{p} = \left(\frac{5}{6}, 0{,}0{,}0, \frac{1}{6}\right).$$

*Mosteller, F. R. and Wallace, D. (1964). *Inference and Disputed Authorship: The Federalist,* Addison-Wesley Publishing Co., Reading, Maine.

$$\mathbf{P} = \begin{array}{c} \\ T \\ D^7 \\ S \\ II^7 \\ VI \end{array} \begin{array}{c} \begin{array}{ccccc} T & D^7 & S & II^7 & VI \end{array} \\ \begin{pmatrix} 0 & \frac{2}{6} & \frac{2}{6} & \frac{1}{6} & \frac{1}{6} \\ \frac{4}{6} & 0 & 0 & 0 & \frac{2}{6} \\ \frac{2}{6} & \frac{3}{6} & 0 & 0 & \frac{1}{6} \\ 0 & \frac{5}{6} & \frac{1}{6} & 0 & 0 \\ 0 & 0 & \frac{2}{6} & \frac{4}{6} & 0 \end{pmatrix} \end{array}$$

Rhythm can also be considered as a Markov chain. Let us choose the time signature of 3/4 (that is, three beats per bar). There are many possible rhythm combinations available to the composer. However, let us pick seven elementary rhythms in 3/4 time signature, each with the validity of three quarter beats.

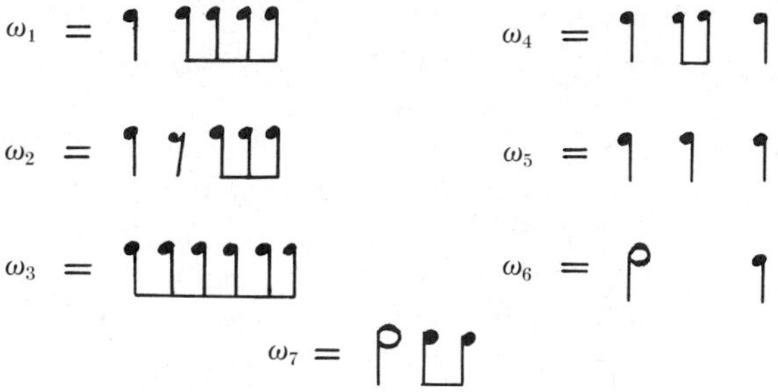

We can consider the rhythm of certain kinds of songs as a Markov chain with states ω_1 to ω_7, with the initial probability vector q and with the probability transition matrix $\mathbf{Q}$ given as

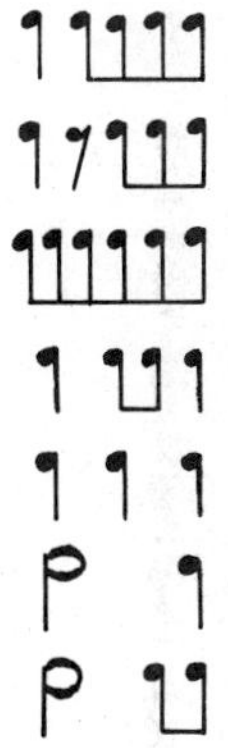

$$\mathbf{Q} = \begin{pmatrix} 0 & 0 & 0 & \frac{1}{6} & \frac{3}{6} & \frac{1}{6} & \frac{1}{6} \\ 0 & 0 & 0 & \frac{2}{6} & \frac{2}{6} & \frac{1}{6} & \frac{1}{6} \\ 0 & 0 & 0 & \frac{1}{6} & \frac{1}{6} & \frac{2}{6} & \frac{2}{6} \\ \frac{1}{6} & 0 & 0 & 0 & \frac{1}{6} & \frac{2}{6} & \frac{2}{6} \\ \frac{2}{6} & \frac{1}{6} & \frac{3}{6} & 0 & 0 & 0 & 0 \\ \frac{2}{6} & \frac{1}{6} & \frac{3}{6} & 0 & 0 & 0 & 0 \\ \frac{2}{6} & \frac{3}{6} & \frac{1}{6} & 0 & 0 & 0 & 0 \end{pmatrix}$$

$$q = \left(\frac{2}{6}, \frac{1}{6}, \frac{2}{6}, \frac{1}{6}, 0, 0, 0\right).$$

The melody of a song can also be described using a Markov chain, however, the necessary number of states, hence the whole matrix, is considerably larger and more complicated.

Imagine now that we shall undertake the peculiar farce of imitating the creative work of a composer writing a song. We have all the probabilities describing the harmony Markov chain, and we can attempt to create a new composition simply by tossing a die. This method is known as *stochastic simulation.* Imagine that we would like to write an eight bar song (the majority of simple songs consist of several eight bar sequences). We shall therefore use eight chords, which we can get from the probabilities for the harmony chain as follows: The first chord should be chosen according to the initial probabilities so that $P(T) = \frac{5}{6}$, $P(\text{VI}) = \frac{1}{6}$, with the probability of any other chord being 0. Let us accept these rules: if 1,2,3,4,5 occurs on the die, then we begin the song with the T chord. If 6 appears, we open the song with the VI chord. We then obtain the chords with the required probabilities. Assume that the roll of a die results in the number 3. Then the first bar will contain the T chord. We shall choose the second chord according to the transition probability matrix. We have probabilities $P(T \longrightarrow T) = 0$, $P(T \longrightarrow D^7) = \frac{2}{6}$, $P(T \longrightarrow S) = \frac{2}{6}$, $P(T \longrightarrow \text{II}) = \frac{1}{6}$, $P(T \longrightarrow \text{VI}) = \frac{1}{6}$. We shall make the following agreement: If a single roll of a die results in 1 or 2, we use the D^7 chord for the second bar; if 3 or 4 occurs, we use the S chord; if 5, II^7; and if 6, then the VI chord is used. So the selection of the chords for the second bar corresponds to the first row of the transition probability matrix. Imagine that the actual toss results in 6. Then the second bar will contain the VI chord. It is easy to see that if the six remaining throws result in 5,3,4,2,5,1, the whole sequence for the eight chords will be as follows: T, VI, II^7, D^7, T, II^7, D^7, T.

In the same way, we can "create" the rhythm and melody of a song, thus "composing" it by tossing a die. Actual experiments with the above matrices yielded interesting results. Among several dozen songs, some turned out memorably pleasant. The following song was obtained in one of these sessions. We can see that it is based on the harmony derived above while the rhythm part was obtained using the probabilities for rhythm if the sequence of tosses of a die yields 1,2,6,4,3,4,3,5. The upper line is the actual outcome of a sequence of tosses (compare the harmony with the above simulation of the eight chords). The lower line was added later so that the whole "song" can be played on a piano.

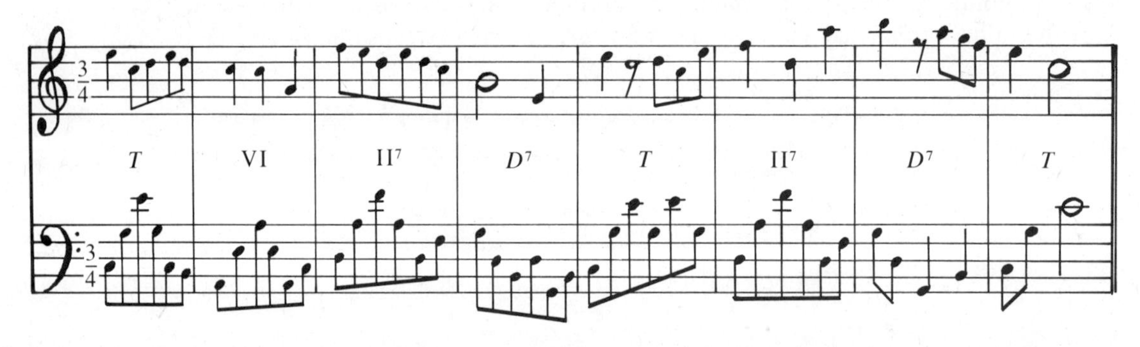

Song of a die

The composition technique of tossing dice and following certain numbers might look too mechanical and uninspired. However, people have been excited for centuries by mechanical methods of composing music. *Athanasius Kircher,* a German Jesuit, described in 1650 a device for composing music which was actually built later for *Samuel Pepys* (see Chapter XI, Example 7) and called "musirithmica mirifica." In 1757, Bach's pupil *Johann Philipp Kirnberger* published *The Ever-ready Composer of Polonaises and Minuets,* which used a die for randomness and unexpected variety. There have been many other attempts of this kind; the most famous system is attributed to *Wolfgang Amadeus Mozart.* In 1792, one year after Mozart's death, a Berlin publisher offered a pamphlet, *Musikalisches Würfelspiel,* explaining how a pair of dice can be used "to compose without the least knowledge of music" any number of German waltzes. Several years later, another pamphlet was offered for the dice composition of contredanses. These methods appear quite interesting and entertaining, but unfortunately some find the idea rather trifling and superficial. Do we have to compose music by tossing a die (or let us say, by generating it similarly on a computer) when so many composers can do a much better job and at the same time convey some emotional charge to the listeners? Certainly the creation and composition of *new* music using a simulation of Markov chains is not really worth the effort. However, probabilistic methods of the theoretical analysis of music are very useful and revealing. They can be employed for deciding cases of disputed authorship exactly as in the case of literary works (see the previous section). At the same time, transition probability matrices can characterize the *style* or *genre* of a certain kind of music—notions which are hard to describe quantitatively otherwise.

COUNTABLE SPACES

This chapter investigates an extension of the axiomatic model of probability, which was given in the previous chapters. We shall assume that the experiment that is studied has countably infinitely many distinguishable outcomes and, therefore, that a corresponding sample space is countably infinite. Unlike the finite case, an appropriate model of probability for countable spaces requires some knowledge of calculus, namely, a background familiarity with infinite series. In order to refresh the reader's memory, we shall usually quote the necessary results from calculus as needed. First of all, we have to make clear what we mean by a *countably infinite set* or a *countable set* (we shall use both expressions interchangeably).

> **DEFINITION 1:** A set S is said to be countable if there is one-to-one mapping of S onto the whole set of natural numbers $\{1,2,3, \ldots\}$.

It is obvious that the set of natural numbers itself is countable, and the same is true for each of its infinite subsets. Furthermore, it is known (and obvious) that the set of all integers and its every infinite subset is countable, and that the set of all rational numbers and its every infinite subset is countable (however, the latter is not so obvious). Every countable set can be written as a sequence $S = \{s_1,s_2,s_3, \ldots\}$.

We have already mentioned many examples of countable sample spaces. Let us recall several: A coin is tossed till a head appears. It can happen on the first toss, the second or the third toss, and so on, therefore an appropriate sample space is the set of all natural numbers itself (hence, it is countable). A die is rolled till two 6's in a row appear. A young man approaches girls one after another till he gets a date; we assume that he has a strong enough drive not to give up after being rejected several times, however, it can take a very long time before he succeeds and it is impossible to put an upper limit on the number of necessary trials. The number of cars passing through the main gate of Yellowstone National Park within 24 hours. The number of particles scintillating on the screen of a Geiger-Müller counter. The number of passengers arriving at JF Kennedy International Airport between 6 A.M. and 10 P.M. on a certain day. The number of dollars a gambler is ahead at a blackjack table (we assume that he is very rich and his bankroll is unlimited; he can be ahead or behind any number of dollars and a sample space can be the set of all integers). There are many other examples in everyday life that require an infinite sample space for their description, but those given above are sufficient to convey the idea of the main types of spaces which will be treated in this chapter. Given a (countable) sample space Ω, we shall define again an *event* as *any subset of* Ω (see Chapter III). We have to mention immediately that this approach is logically consistent in the case of finite as well as countable sample spaces. However, in the case of larger spaces (so-called uncountable spaces; for example, if Ω is the set of all real numbers between 0 and 1), this approach is completely wrong, even though frequently used in introductory textbooks on probability. A logically and mathematically consistent treatment of uncountable spaces requires an incomparably higher degree of mathematical sophistication and cannot even be discussed at this level.

Imagine that $A_1,A_2,A_3, \ldots$ is an infinite sequence of subsets of Ω. In agreement with Chapter II we shall define a *countable union (respectively, intersection) of the sets A_i's as the set of all elements of Ω which are contained in at least one of (respectively, in all) the sets A_i and denote it by $\bigcup_{i=1}^{\infty} A_i$ (respectively, $\bigcap_{i=1}^{\infty} A_i$).* All the basic properties of unions and intersections investigated in Chapter I remain unchanged. We shall say that *a sequence of sets* $A_1, A_2, A_3,$ $\ldots$ *is pairwise disjoint if for any* $i \neq j$, $A_i \cap A_j = \emptyset$. The following example shows the importance of countable unions in probability.

EXAMPLE 1

Let a fair coin be tossed till a head occurs. As an appropriate sample space take the set of all integers, $\Omega = \{1,2,3, \ldots\}$, where the outcome i means that $(i - 1)$ tails are followed by a head. Let further the event A_i mean that the first head occurred on the i-th trial. Then

$$A_i = \{i\} \qquad i = 1,2,3, \ldots$$

and the sequence of events $A_1, A_2, A_3, \ldots$ is pairwise disjoint. We can investigate the event F that the first head appears in less than five trials. Obviously,

$$F = A_1 \cup A_2 \cup A_3 \cup A_4 = \bigcup_{i=1}^{4} A_i.$$

Furthermore, we can investigate the event E that the first head appears on an even trial (second, fourth, sixth, etc.). Obviously,

$$E = \{2,4,6,8, \ldots\} = A_2 \cup A_4 \cup A_6 \cup A_8 \ldots = \bigcup_{i=1}^{\infty} A_{2i}$$

and therefore E is expressed as a countable union of certain A_i's. Similarly, if O means that the first head appears on an odd trial, then

$$O = \{1,3,5, \ldots\} = A_1 \cup A_3 \cup A_5 \ldots = \bigcup_{i=1}^{\infty} A_{(2i-1)}.$$

Let us assume now that a probability measure P is to be defined for all subsets of Ω, and let us investigate what properties it should have. We should obviously try to preserve the axioms A_1 to A_3 of Chapter V and their consequences. Therefore, we should have

$$P(F) = P(A_1 \cup A_2 \cup A_3 \cup A_4) = P\left(\bigcup_{i=1}^{4} A_i\right) = P(A_1) + P(A_2) + P(A_3) + P(A_4)$$
$$= \sum_{i=1}^{4} P(A_i).$$

This fact follows from the (finite) additivity of P and the disjointness of A_i's. However, we expect that something similar should hold also for the set E, which is a countable union of disjoint sets.

$$P(E) = P(A_2 \cup A_4 \cup A_6 \cup \ldots) = P\left(\bigcup_{i=1}^{\infty} A_{2i}\right) = P(A_2) + P(A_4) + P(A_6) + \ldots$$
$$= \sum_{i=1}^{\infty} P(A_{2i}).$$

Similarly, we expect that

$$P(O) = P(A_1 \cup A_3 \cup A_5 \cup \ldots) = P\left(\bigcup_{i=1}^{\infty} A_{2i-1}\right) = P(A_1) + P(A_3) + P(A_5) + \ldots$$

$$= \sum_{i=1}^{\infty} P(A_{2i-1}).$$

However, *this fact does not follow from axioms A_1 to A_3 even though it is very desirable.* We shall therefore replace A_3 by a stronger axiom A_3', which will guarantee the desired property.

DEFINITION 2: Let Ω be a countable sample space. A real valued function P defined for all subsets (events) of Ω is called a probability measure if the following axioms are satisfied:

A_1: For every $A \subset \Omega$, $P(A) \geq 0$.
A_2: $P(\Omega) = 1$.
A_3': For any sequence of pairwise disjoint sets $A_i \subset \Omega$,

$$P\left(\bigcup_{i=1}^{\infty} A_i\right) = \sum_{i=1}^{\infty} P(A_i).$$

The couple (Ω, P) is called a *probability measure space.*

COMMENT. The original axiom A_3 required that if A_1 and A_2 are disjoint sets, then the probability of their (finite) union is equal to the (finite) sum of their probabilities (the union and the sum being "finite" because they consist of two elements only). The new axiom A_3' requires that if $A_1, A_2, A_3, \ldots$ are pairwise disjoint sets, then the probability of their (countable) union is equal to the (countable) sum of their probabilities. For these reasons, the original A_3 is sometimes called the *axiom of finite additivity* while the new A_3' is called the *axiom of countable additivity* (we are adding probabilities of countably many sets).

THEOREM 1: Let Ω be a countable sample space and P be a real valued function defined for all subsets of Ω. If A_1, A_2 and A_3' are satisfied, then A_3 is also satisfied. ←

Proof: Let us show first that $P(\phi) = 0$. Define a sequence A_i, $i = 1,2,3, \ldots$ as follows:

$$A_1 = \Omega, \qquad A_2 = A_3 = A_4 = \ldots = \phi.$$

Then the sets A_i are pairwise disjoint and

$$\bigcup_{i=1}^{\infty} A_i = \Omega \cup \phi \cup \phi \cup \phi \cup \ldots = \Omega.$$

Using A_3', we get

$$1 = P(\Omega) = P\left(\bigcup_{i=1}^{\infty} A_i\right) = \sum_{i=1}^{\infty} P(A_i) = P(\Omega) + P(\phi) + P(\phi) + P(\phi) + \ldots = 1 + P(\phi) + P(\phi) + P(\phi) + \ldots.$$

Therefore $$0 = P(\phi) + P(\phi) + P(\phi) + \ldots$$

hence $$P(\phi) = 0.$$

Let us verify A_3. Assume that A_1 and A_2 are disjoint sets. Define

$$A_3 = A_4 = A_5 = \ldots = \phi.$$

Then the sets A_i, $i = 1,2,3, \ldots$ are pairwise disjoint and

$$P(A_1 \cup A_2) = P\left(\bigcup_{i=1}^{\infty} A_i\right) = \sum_{i=1}^{\infty} P(A_i) = P(A_1) + P(A_2) + 0 + 0 + \ldots = P(A_1) + P(A_2).$$

Therefore A_3 is satisfied. q.e.d.

COMMENT. Theorem 1 says that in the presence of A_1 and A_2, the axiom A_3' is *stronger* than A_3, that countable additivity implies finite additivity. Furthermore, we can now use all the theorems of Chapter V for countable probability measure spaces; they remain unchanged because they follow from A_1 to A_3.

EXAMPLE 2

Find the probability that in a sequence of tosses of a fair coin, the first head will appear on an odd trial.

SOLUTION. An appropriate sample space for this experiment was discussed in Example 1. Let A_i be the event that the first head will occur on the i-th trial, and O be the event that the first head will occur on an odd trial. Obviously A_i's are pairwise disjoint and

$$O = \bigcup_{i=1}^{\infty} A_{2i-1} = A_1 \cup A_3 \cup A_5 \cup \ldots .$$

Furthermore, A_i occurs if and only if $(i - 1)$ tails precede the first head. All the trials being independent, we have

$$P(A_i) = \frac{1}{2^{i-1}} \cdot \frac{1}{2} = \frac{1}{2^i}$$

and therefore the required probability is

$$P(O) = P\left(\bigcup_{i=1}^{\infty} A_{2i-1}\right) = \sum_{i=1}^{\infty} P(A_{2i-1}) = \sum_{i=1}^{\infty} \frac{1}{2^{2i-1}} = \frac{1}{2} + \frac{1}{2^3} + \frac{1}{2^5} + \cdots$$

$$= 2\sum_{i=1}^{\infty}\left(\frac{1}{4}\right)^i = 2\,\frac{\frac{1}{4}}{1 - \frac{1}{4}} = \frac{2}{3}.$$

$\left(\text{We used the fact that if } 0 < q < 1 \text{ then } \sum_{n=1}^{\infty} q^n = \frac{q}{1-q}.\right)$ Similarly, if E is the event that the first head will appear on an even trial, then $E = O'$ and

$$P(E) = 1 - P(O) = 1 - \frac{2}{3} = \frac{1}{3}.$$

EXAMPLE 3

Guided public tours of the spacecraft launching site leave every hour on the hour from the waiting room just behind the main gate. Picture taking is prohibited, therefore some (but not all) sightseers are screened by a metal detector for hidden cameras while going through the gate. Fritz would like to take some secret photos, and he decided to arrive on the hour, being thus the last person of the group and hoping that his camera will not be found. What is the probability that he will be stopped, given the fact (unknown to him) that every second visitor is checked, that the detector counter is set to zero after the departure of each group and the probability that k visitors will gather for the tour within an hour is

$$p_k = \frac{1}{ek!} \qquad k = 0,1,2,3, \ldots$$

where $e = 2.718282 \ldots$ is the Euler constant?

SOLUTION. His camera will not be discovered if 0,2,4,6, etc. people enter the waiting room before him. The required probability is therefore

$$P = \frac{1}{e0!} + \frac{1}{e2!} + \frac{1}{e4!} + \frac{1}{e6!} + \cdots = \frac{1}{e}\sum_{k=0}^{\infty}\frac{1}{(2k)!}.$$

Using the expansion for e^x,

$$e^x = 1 + \frac{x}{1!} + \frac{x^2}{2!} + \frac{x^3}{3!} + \cdots = \sum_{k=0}^{\infty}\frac{x^k}{k!}$$

we get

$$P = \frac{1}{e}\sum_{k=0}^{\infty}\frac{1}{(2k)!} = \frac{1}{e}\cdot\frac{1}{2}(e + e^{-1}) = .05677.$$

In the previous examples, we gave the probability measure P by specifying the probability p_k of each sample point ω_k. It is obvious that the situation in the countable case is the same as in the finite case (see Chapter V) and that the probability measure P is *uniquely determined* by any sequence of numbers p_k, $k = 1,2,3, \ldots$ such that

(a) $p_k \geq 0$. $\qquad k = 1,2,3, \ldots \qquad$ (b) $\sum_{k=1}^{\infty} p_k = 1.$

We shall give many examples of this type later on. The first condition is obviously satisfied in both Examples 2 and 3. Let us verify the second condition.

In Example 2, we have $p_k = \frac{1}{2^k}$ for $k = 1,2,3, \ldots$

$$\sum_{k=1}^{\infty} p_k = \sum_{k=1}^{\infty}\frac{1}{2^k} = \frac{\frac{1}{2}}{1-\frac{1}{2}} = 1.$$

In Example 3, $p_k = \dfrac{1}{ek!}$ for $k = 0,1,2,3, \ldots$, hence

$$\sum_{k=0}^{\infty} p_k = \frac{1}{e}\sum_{k=0}^{\infty}\frac{1}{k!} = \frac{1}{e}\cdot e = 1.$$

The conditional probability is defined in the same way as in the finite case. Also Bayes' theorem and the formula of total probability remain valid, the only change being the replacement of finite sums by infinite sums.

➡ **THEOREM 2:** Let $A_i \subset \Omega$, $i = 1,2,3, \ldots$ be a sequence of pairwise disjoint sets such that $\Omega = \bigcup_{i=1}^{\infty} A_i$, and let $B \subset \Omega$.
(a) Formula of total probability:

$$P(B) = \sum_{j=1}^{\infty} P(A_j)P(B/A_j).$$

(b) Bayes' theorem:

$$P(A_i/B) = \frac{P(A_i)P(B/A_i)}{\sum_{j=1}^{\infty} P(A_j)P(B/A_j)}. \qquad i = 1,2,3, \ldots$$

Proof: The proof is identical to that of Theorems 2 and 3 of Chapter VIII, only n is replaced by ∞.
q.e.d.

In the sequel of this chapter we shall make frequent use of the *binomial series.* For any positive integer k and for any real number y, we define

$$\binom{y}{k} = \frac{y(y-1)\ldots(y-k+1)}{k!}$$

where $0! = 1$. Further, we define

$$\binom{y}{0} = 1.$$

It is proved in calculus that for any real x and y such that $|x| < 1$, we can use the following binomial series:

$$(1+x)^y = \sum_{k=0}^{\infty}\binom{y}{k}x^k.$$

This series, together with Theorem 2, will be used in the following example.

EXAMPLE 4

Lara flips a fair coin till she gets a tail. If she had to flip k times, then she puts 3^k balls into a bag out of which $2k$ balls are blue and the rest are red. Then Ivan has to draw a ball from the bag.

(a) What is the probability that his ball is blue?
(b) If it is blue, what is the probability that Lara got a tail on the third trial?

SOLUTION. (a) Let A_k be the event that Lara gets the first tail on the k-th flip. Then

$$P(A_k) = 2^{-k}. \qquad k = 1,2,3, \ldots$$

Furthermore, let B be the event that Ivan gets a blue ball. We have to find $P(B)$. Using the formula of total probability, we get

$$P(B) = \sum_{j=1}^{\infty} P(A_j) \cdot P(B/A_j).$$

Obviously, if Lara gets a tail on the j-th trial, then there are $2j$ blue balls out of 3^j in the bag,

hence
$$P(B/A_j) = \frac{2j}{3^j}.$$

Hence
$$P(B) = \sum_{j=1}^{\infty} 2^{-j}\frac{2j}{3^j} = \frac{2}{6}\sum_{j=1}^{\infty}\frac{j}{6^{j-1}} = \frac{2}{6}\left(1 - \frac{1}{6}\right)^{-2} = \frac{12}{25}$$

as follows from the binomial series for $x = -\frac{1}{6}$ and $y = -2$.

(b) We have to find $P(A_3/B)$.

$$P(A_3) = 2^{-3}, \quad P(B/A_3) = \frac{2 \cdot 3}{3^3} = \frac{2}{3^2}.$$

$$P(A_3/B) = \frac{P(A_3)P(B/A_3)}{P(B)} = \frac{2^{-3} \cdot \frac{2}{3^2}}{\frac{12}{25}} = \frac{25}{432}.$$

The notions of a random variable and its probability distribution are defined in exactly the same way as in the finite case. (We shall give examples later.) However, we have to be more careful with the definition of the expectation of a random variable.

DEFINITION 3: Let X be a r.v. on a countable probability measure space, let x_i, $i = 1,2,3, \ldots$ be the set of all possible distinct values that X can attain and let $f(x_i) = P(\{X = x_i\})$, $i = 1,2,3, \ldots$ be the probability distribution of X. If the series

$$\sum_{i=1}^{\infty} |x_i| \, f(x_i) < +\infty$$

converges, then the expected value of X is defined as

$$E(X) = \sum_{i=1}^{\infty} x_i \, f(x_i).$$

COMMENT. We define $E(X)$ only if the series $\sum_{i=1}^{\infty} x_i f(x_i)$ converges absolutely, otherwise we say that $E(X)$ does not exist. It is known that absolute convergence implies ordinary convergence, and therefore, if $E(X)$ exists then it is finite,

$$-\infty < E(X) = \sum_{i=1}^{\infty} x_i f(x_i) < +\infty.$$

All the theorems on random variables from Chapters XI and XII remain true. We shall give now several examples of r.v.'s on countable spaces, their distributions and expectations.

GEOMETRIC DISTRIBUTION

Suppose we have an experiment with two outcomes: a success and a failure. Further, let the probability of a success be p, $0 < p < 1$, and the probability of a failure be $q = 1 - p$. Let us investigate independent repetitions of this experiment, and let the r.v. X denote the number of trials necessary to produce the first success (see Example 2). Obviously, X can attain values $1,2,3, \ldots, k, \ldots$. In order to get the first success on the k-th trial, the first $(k - 1)$ trials must result in a failure (with the probability q^{k-1}) and the k-th trial must result in a success (with the probability p). Hence

$$p_k = P(\{X = k\}) = pq^{k-1}. \qquad k = 1,2,3, \ldots$$

We can easily see (using the formula for the sum of geometric series) that

$$\sum_{k=1}^{\infty} p_k = p \sum_{k=1}^{\infty} q^{k-1} = p \cdot \frac{1}{1-q} = 1.$$

Let us calculate the expected value of the geometric distribution.

$$E(X) = \sum_{k=1}^{\infty} kp_k = p \sum_{k=1}^{\infty} kq^{k-1} = p(1 + 2q + 3q^2 + \ldots).$$

However, using the binomial series (see page 318), we get

$$(1-q)^{-2} = \sum_{k=0}^{\infty} \binom{-2}{k} (-q)^k = 1 + 2q + 3q^2 + \ldots .$$

Therefore
$$E(X) = p(1-q)^{-2} = pp^{-2} = p^{-1}.$$

Let us calculate the variance of the geometric distribution. We shall use the formula (see page 198)

$$V(X) = E(X^2) - (E(X))^2.$$

$$E(X^2) = \sum_{k=1}^{\infty} k^2 pq^{k-1} = p \sum_{k=1}^{\infty} [k(k+1) - k]q^{k-1} = p \sum_{k=1}^{\infty} k(k+1)q^{k-1} - p \sum_{k=1}^{\infty} kq^{k-1}.$$

The second expression is equal to p^{-1}. As for the first expression, we get

$$\sum_{k=1}^{\infty} k(k+1)q^{k-1} = 2 + 2 \cdot 3q + 3 \cdot 4q^2 + 4 \cdot 5q^3 + \ldots .$$

Using the binomial series, we get

$$2(1-q)^{-3} = 2\sum_{k=0}^{\infty}\binom{-3}{k}(-q)^k = 2\left(1 + 3q + \frac{3 \cdot 4}{2}q^2 + \frac{3 \cdot 4 \cdot 5}{2 \cdot 3}q^3 + \ldots\right)$$

$$= 2 + 2 \cdot 3q + 3 \cdot 4q^2 + 4 \cdot 5q^3 + \ldots .$$

Therefore $\quad E(X^2) = 2p(1-q)^{-3} - p^{-1} = 2pp^{-3} - p^{-1} = 2p^{-2} - p^{-1}.$

$V(X) = E(X^2) - (E(X))^2 = 2p^{-2} - p^{-1} - p^{-2} = qp^{-2}.$

THEOREM 3: For the geometric distribution, $E(X) = p^{-1}$ and $V(X) = qp^{-2}$. ←

EXAMPLE 5

When tossing a fair coin, $p = q = \frac{1}{2}$. Therefore, the expected number of tosses to produce a head is equal to $p^{-1} = 2$. The variance of the number of tosses till the first head appears is equal to $qp^{-2} = 2$.

EXAMPLE 6

Bank Craps. Bank craps, or, as it is commonly called, craps, is perhaps the most popular casino game in the United States and its fame has spread to the Caribbean, South America and Europe. Craps is a game of pure chance. The "stickman" will push two dice to a gambler and announce: "New 'shooter' coming out." Then the shooter will start rolling the dice. Anybody having the "pass line bet" will win his bet in the following ways:

(a) On the first roll of the dice, if 7 or 11 comes up.
(b) If the shooter rolls 4, 5, 6, 8, 9 or 10 on the first roll, this number becomes his "point." He then keeps on rolling till he repeats his point before he "sevens out," i.e., before he gets a total of 7 on the dice.

If the shooter sevens out before repeating his point, the pass line bet is lost. The bet is also lost when the first roll yields the sum of 2, 3 or 12. Find the probability of winning the pass line bet.

SOLUTION. In each toss, only the total sum on both dice counts. Following is the probability distribution of the sum on the dice (see Example 15, page 139).

Sum k	2	3	4	5	6	7	8	9	10	11	12
$P(k)$	$\frac{1}{36}$	$\frac{2}{36}$	$\frac{3}{36}$	$\frac{4}{36}$	$\frac{5}{36}$	$\frac{6}{36}$	$\frac{5}{36}$	$\frac{4}{36}$	$\frac{3}{36}$	$\frac{2}{36}$	$\frac{1}{36}$

The probability of winning is calculated as follows:

$$P(\text{win}) = P(\text{win on the first roll}) + P(\text{win on a successive roll}).$$

Now $\quad P(\text{win on the first roll}) = P(7) + P(11) = \frac{6}{36} + \frac{2}{36} = \frac{8}{36} = .22222.$

Using the formula of total probability, we get

$$P(\text{win on a successive roll}) = \sum_{k=4,5,6,8,9,10} P(k \text{ on the first roll})P(\text{repeat } k \text{ before } 7/k \text{ on the first roll}).$$

The probabilities $P(k \text{ on the first roll}) = P(k)$ for $k = 4, 5, 6, 8, 9, 10$ are given in the above table. However the conditional probabilities are a little more difficult to find. As a matter of fact, we can use the same approach as when calculating the geometric distribution, even though there is a slight difference. Let us assume that the point k has been thrown on the first roll. Then the probability of *not throwing* k or 7 in one of the following rolls is $Q(k) = 1 - P(k) - P(7) = 1 - P(k) - \frac{1}{6}$. The probability that the shooter will repeat his point k in the $(r + 1)^{\text{st}}$ throw and will not seven out in the meantime is $Q^r(k)P(k)$ (as in the geometric distribution), $r = 0,1,2,3, \ldots$. Hence

$$R(k) = P(\text{repeat } k \text{ before } 7/k \text{ on the first roll}) = \sum_{r=0}^{\infty} Q^r(k)P(k)$$

$$= P(k)\sum_{r=0}^{\infty} Q^r(k) = P(k)/(1 - Q(k)).$$

Following is the table of conditional probabilities.

k	$P(k)$	$R(k)$
4	$\frac{3}{36}$	$\frac{3}{9}$
5	$\frac{4}{36}$	$\frac{4}{10}$
6	$\frac{5}{36}$	$\frac{5}{11}$
8	$\frac{5}{36}$	$\frac{5}{11}$
9	$\frac{4}{36}$	$\frac{4}{10}$
10	$\frac{3}{36}$	$\frac{3}{9}$

Hence $P(\text{win on a successive roll}) = \frac{3}{36}\cdot\frac{3}{9} + \frac{4}{36}\cdot\frac{4}{10} + \frac{5}{36}\cdot\frac{5}{11} + \frac{5}{36}\cdot\frac{5}{11} + \frac{4}{36}\cdot\frac{4}{10} + \frac{3}{36}\cdot\frac{3}{9} = .27071.$

Therefore, the probability of winning the pass line bet in craps is

$$P(\text{win}) = .22222 + .27071 = .49293.$$

The probability of losing the pass line bet is

$$P(\text{lose}) = 1 - P(\text{win}) = 1 - .49293 = .50707.$$

Therefore, the house advantage is $.50707 - .49293 = .01414$ or about 1.414%, as one can find in any decent gambling handbook.

We have calculated the probabilities $R(k)$ using geometric series. It is interesting that we can also get them with the following argument. Let us say that $k = 4$ appeared on the first roll of the dice. From this moment, only sums of 4 or 7 matter and other outcomes are disregarded. The sum of 4 points can be thrown in three different ways (1–3, 2–2, 3–1) and if it happens before a 7 appears, the shooter wins. The sum of 7 points can be thrown in six different ways (1–6, 2–5, 3–4, 4–3, 5–2, 6–1) and if it happens before a 4 appears, the shooter loses. Therefore, the probability that a 4 appears before a 7 if a 4 appeared on the first roll is equal to $R(4) = \dfrac{3}{3+6} = \dfrac{3}{9}$. Similarly, one can calculate $R(k)$ for other values of $k = 5, 6, 8, 9, 10$.

Let us investigate now the "don't pass line bet." This bet is won if the first "come out" is not a double 6 and if the pass line bet is lost. If the first come out is a double 6, then the don't pass line bet is neither won nor lost, and the wager has to stay for the next game. Otherwise this bet is lost.

In order to find the probability of winning the don't pass line bet, we can consider that there are 35 possible outcomes with a pair of dice (the outcome 6–6 must be disregarded). If $P(k)$ denotes getting the sum of k points on the first roll and $R(k)$ denotes repeating 7 before k if k came out on the first toss, then using the same methods as above we can find the following values.

k	$P(k)$	$R(k)$
2	$\frac{1}{35}$	
3	$\frac{2}{35}$	
4	$\frac{3}{35}$	$\frac{6}{9}$
5	$\frac{4}{35}$	$\frac{6}{10}$
6	$\frac{5}{35}$	$\frac{6}{11}$
8	$\frac{5}{35}$	$\frac{6}{11}$
9	$\frac{4}{35}$	$\frac{6}{10}$
10	$\frac{3}{35}$	$\frac{6}{9}$

Using these figures, we can get

$$P(\text{win don't pass line bet}) = P(2) + P(3) + \sum_{k=4,5,6,8,9,10} P(k)R(k)$$

$$= \frac{1}{35} + \frac{2}{35} + 2\left(\frac{3}{35} \cdot \frac{6}{9} + \frac{4}{35} \cdot \frac{6}{10} + \frac{5}{35} \cdot \frac{6}{11}\right) = .49299.$$

$$P(\text{lose don't pass line bet}) = 1 - .49299 = .50701.$$

And finally, the house advantage is $.50701 - .49299 = .01402$, or about 1.402%. We can therefore see that the don't pass line bet is slightly better than the pass line bet, even though the difference is very small and could appear only in extremely long runs of games. However, this fact is well known to devoted gamblers from practical experience.

NEGATIVE BINOMIAL DISTRIBUTION

Let us consider a situation similar to the case of geometric distribution, however, this time let the r.v. X denote the number of trials till the m-th success appears ($m \geq 1$ is a given integer). Obviously, if $m = 1$, then we get the geometric distribution. The r.v. X represents the *waiting time* for the m-th success and is called the *negative binomial* r.v. Let us find its probability distribution. In order to get the m-th success on the k-th trial, we have to get $(m - 1)$ successes among the first $(k - 1)$ trials $\left(\text{with the probability } \binom{k-1}{m-1} p^{m-1}q^{k-m}\right)$, and a success on the k-th trial (with the probability p). Hence

$$p_k = P(\{X = k\}) = \binom{k-1}{m-1} p^m q^{k-m} \qquad \text{for } k \geq m.$$

Obviously, $\quad p_k = 0 \qquad$ for $1 \leq k \leq m$.

Using the binomial series, we get

$$\sum_{k=1}^{\infty} p_k = \sum_{k=m}^{\infty} \binom{p-1}{m-1} p^m q^{k-m}$$

$$= p^m\left(1 + mq + \frac{(m+1)m}{2} q^2 + \frac{(m+2)(m+1)m}{2 \cdot 3} q^3 + \ldots\right) = p^m(1-q)^{-m} = 1.$$

Let us find the expected value of the waiting time for the m-th success.

$$E(X) = \sum_{k=m}^{\infty} k\binom{k-1}{m-1} p^m q^{k-m} = p^m\left(m + (m+1)mq + (m+2)\frac{(m+1)m}{2} q^2 + \ldots\right)$$

$$= mp^m\left(1 + (m+1)q + \frac{(m+1)(m+2)}{2} q^2 + \ldots\right).$$

Using the binomial series, we have

$$(1-q)^{-(m+1)} = \sum_{k=0}^{\infty} \binom{-m-1}{k} (-q)^k = \left(1 + (m+1)\,q + \frac{(m+1)(m+2)}{2} q^2 + \ldots\right).$$

Hence $$E(X) = mp^m(1-q)^{-(m+1)} = mp^{-1}.$$

Obviously, for $m = 1$ we get the expected value of the geometric distribution, which is equal to p^{-1}. Let us find the variance of the waiting time for the m-th success.

$$E(X^2) = \sum_{k=m}^{\infty} k^2 \binom{k-1}{m-1} p^m q^{m-k} = p^m \sum_{k=m}^{\infty} [k(k+1) - k] \binom{k-1}{m-1} q^{m-k}$$

$$= p^m \sum_{k=m}^{\infty} k(k+1) \binom{k-1}{m-1} q^{m-k} - p^m \sum_{k=m}^{\infty} k \binom{k-1}{m-1} q^{m-k}.$$

The second expression is equal to mp^{-1} as calculated above.

$$\sum_{k=m}^{\infty} k(k+1)\binom{k-1}{m-1} q^{m-k} = m(m+1) + (m+1)(m+2)mq + (m+2)(m+3)\frac{(m+1)m}{2} q^2 + \dots$$

$$= m(m+1)\left(1 + (m+2)q + \frac{(m+2)(m+3)}{2} q^2 + \dots\right).$$

From the binomial series, we get

$$(1-q)^{-m-2} = \sum_{k=0}^{\infty} \binom{-m-2}{k} (-q)^k = 1 + (m+2)q + \frac{(m+2)(m+3)}{2} q^2 + \dots .$$

Hence $$E(X^2) = m(m+1)p^m(1-q)^{-m-2} - mp^{-1} = m(m+1)p^{-2} - mp^{-1}.$$

Finally, we get

$$V(X) = E(X^2) - (E(X))^2 = m(m+1)p^{-2} - mp^{-1} - m^2p^{-2} = mp^{-2} - mp^{-1} = mqp^{-2}.$$

THEOREM 4: For the negative binomial distribution,

$$E(X) = mp^{-1} \quad \text{and} \quad V(X) = mqp^{-2}.$$

Sometimes it is convenient to investigate the r.v. $Y = X - m$, which means the *total number of failures before the occurrence of the m-th success.* Obviously, the distribution of Y is given as

$$\tilde{p}_k = P(\{Y = k\}) = P(\{X = m + k\}) = \binom{m+k-1}{k} p^m q^k = \binom{-m}{k} p^m (-q)^k$$

for $k \geq 0$. Furthermore, we have

$$E(Y) = E(X) - m = mp^{-1} - m = mqp^{-1}. \qquad V(Y) = V(X) = mqp^{-2}.$$

EXAMPLE 7

When tossing a fair coin, find the probability of getting at most two tails before getting three heads. Find also the expected number of tails before getting three heads.

SOLUTION. If we call getting a head a success, then $m = 3, p = q = \frac{1}{2}$ and we have to consider the distribution of the r.v. Y defined above. The required probability is equal to

$$p = \tilde{p}_0 + \tilde{p}_1 + \tilde{p}_2 = \binom{2}{0}\left(\frac{1}{2}\right)^3 + \binom{3}{1}\left(\frac{1}{2}\right)^4 + \binom{4}{2}\left(\frac{1}{2}\right)^5 = \frac{1}{2}.$$

The expected number of tails before getting three heads is

$$E(Y) = 3 \cdot \frac{1}{2} \cdot \left(\frac{1}{2}\right)^{-1} = 3.$$

EXAMPLE 8

How many times would you expect to roll a fair die in order to get five 6's? Find also the variance of the number of trials.

SOLUTION. Using the formula for the expected value of the negative binomial distribution, we have

$$E(X) = mp^{-1} = 5 \cdot \left(\frac{1}{6}\right)^{-1} = 5 \cdot 6 = 30$$

and this conclusion is in agreement with common sense intuition. Furthermore,

$$V(X) = mqp^{-2} = 5 \cdot \frac{5}{6} 6^2 = 150.$$

POISSON DISTRIBUTION

Frequently we have to investigate so-called *rare events:* the number of mistakes a typist makes on one page; the number of people suffering from a certain rare disease in a community of a given size; the number of stars of a given intensity in an area of given dimensions, and so on. We can often approximate the distribution of such events by the so-called *Poisson distribution* (the reason is given in Theorem 10, page 336). However, the Poisson distribution also describes well such events as the number of cars passing through an intersection within a period of time; the number of customers requiring service during a period of time; the number of customers whose service is completed during a period of time; the number of machines in a factory which break down during a period of time; the number of radioactive particles scintillating on the screen of a Geiger-Müller counter, etc. This distribution is defined as follows: for a given constant $\lambda > 0$, let the probability that the r.v. X is equal to $k (k = 0,1,2,3, \ldots)$ be

$$p_k = P(\{X = k\}) = e^{-\lambda}\frac{\lambda^k}{k!}. \qquad k = 0,1,2,3, \ldots$$

Then we say that the r.v. X has the *Poisson distribution with parameter* λ. Using the expansion for e^x (see page 317), we get

$$\sum_{k=0}^{\infty} p_k = e^{-\lambda} \sum_{k=0}^{\infty} \frac{\lambda^k}{k!} = e^{-\lambda} e^{\lambda} = 1.$$

Let us find the expected value of the Poisson distribution.

$$E(X) = \sum_{k=0}^{\infty} k\, e^{-\lambda} \frac{\lambda^k}{k!} = e^{-\lambda} \lambda \sum_{k=1}^{\infty} \frac{\lambda^{k-1}}{(k-1)!} = \lambda\, e^{-\lambda} \sum_{k=0}^{\infty} \frac{\lambda^k}{k!} = \lambda\, e^{-\lambda}\, e^{\lambda} = \lambda.$$

The expected value of the Poisson distribution is therefore equal to λ. Let us find the variance.

$$E(X^2) = \sum_{k=0}^{\infty} k^2\, e^{-\lambda} \frac{\lambda^k}{k!} = e^{-\lambda} \sum_{k=0}^{\infty} [k(k-1) + k] \frac{\lambda^k}{k!}$$

$$= e^{-\lambda} \sum_{k=2}^{\infty} k(k-1) \frac{\lambda^k}{k!} + e^{-\lambda} \sum_{k=0}^{\infty} k \frac{\lambda^k}{k!}.$$

The second expression is equal to $E(X) = \lambda$. Furthermore,

$$\sum_{k=2}^{\infty} k(k-1) \frac{\lambda^k}{k!} = \lambda^2 \sum_{k=2}^{\infty} \frac{\lambda^{k-2}}{(k-2)!} = \lambda^2 \sum_{k=0}^{\infty} \frac{\lambda^k}{k!} = \lambda^2 e^{\lambda}.$$

$$E(X^2) = e^{-\lambda} \lambda^2 e^{\lambda} + \lambda = \lambda^2 + \lambda.$$

Finally,

$$V(X) = E(X^2) - (E(X))^2 = \lambda^2 + \lambda - \lambda^2 = \lambda.$$

Therefore, the variance of the Poisson distribution is also equal to λ.

THEOREM 5: For the Poisson distribution, ←

$$E(X) = \lambda \quad \text{and} \quad V(X) = \lambda.$$

EXAMPLE 9

A Geiger-Müller counter exposed to a stable radiation shows on its screen an average of four scintillating radioactive particles. Assuming that the number of particles on the screen is a random variable with the Poisson distribution, find the probability that there will be
(a) exactly three scintillating particles.
(b) at least four scintillating particles.
(c) an even number of scintillating particles.

SOLUTION. We assume that the number of particles on the screen has the Poisson distribution with parameter λ, that is, the probability that there will be exactly k particles is equal to

$$p_k = e^{-\lambda} \frac{\lambda^k}{k!}. \qquad k = 0,1,2,3, \ldots.$$

We know that the expected number of particles on the screen is λ (see Theorem 5), and because the average number of observed particles is four, we have $\lambda = 4$.

(a) The probability of observing exactly 3 particles is

$$p_3 = e^{-4}\frac{4^3}{3!} = .1954.$$

(b) The probability of observing at least 4 particles is

$$p = 1 - P(\text{less than } 4) = 1 - p_0 - p_1 - p_2 - p_3 = 1 - e^{-4}\frac{4^0}{0!} - e^{-4}\frac{4^1}{1!} - e^{-4}\frac{4^2}{2!} - e^{-4}\frac{4^3}{3!}$$

$$= 1 - .0183 - .0733 - .1465 - .1954 = .5665.$$

(c) The probability of observing an even number of particles (considering 0 as an even number) is

$$p = p_0 + p_2 + p_4 + p_6 + \ldots = e^{-\lambda}\left(1 + \frac{\lambda^2}{2!} + \frac{\lambda^4}{4!} + \frac{\lambda^6}{6!} + \ldots\right).$$

However,
$$e^{\lambda} = 1 + \frac{\lambda}{1!} + \frac{\lambda^2}{2!} + \frac{\lambda^3}{3!} + \frac{\lambda^4}{4!} + \frac{\lambda^5}{5!} + \frac{\lambda^6}{6!} + \ldots.$$

$$e^{-\lambda} = 1 - \frac{\lambda}{1!} + \frac{\lambda^2}{2!} - \frac{\lambda^3}{3!} + \frac{\lambda^4}{4!} - \frac{\lambda^5}{5!} + \frac{\lambda^6}{6!} + \ldots.$$

Adding these two series and dividing by two, we get

$$\frac{1}{2}(e^{\lambda} + e^{-\lambda}) = 1 + \frac{\lambda^2}{2!} + \frac{\lambda^4}{4!} + \frac{\lambda^6}{6!} + \ldots.$$

Therefore, the required probability is

$$p = e^{-\lambda}\frac{1}{2}(e^{\lambda} + e^{-\lambda}) = \frac{1}{2}(1 + e^{-2\lambda}) = \frac{1}{2}(1 + e^{-8}) = .50017.$$

EXAMPLE 10

Sunshine Village ski resort in Banff National Park in the Canadian Rockies is accessible only by a mountain bus from the dead end of the public highway. It is known that the first morning bus brings on the average 10 early skiers. Assume that all of them go immediately to the Strawberry chairlift built for threesomes. Find the probability that if they follow the rules (exactly three skiers per chair), they will fully occupy all the chairs used (no chair will go with only one or two of these skiers).

SOLUTION. We may assume that the number of early bird skiers is a Poisson random variable with the expected value 10, hence $\lambda = 10$. All the chairs will be fully occupied if there are 3, 6, 9, 12, etc., skiers. Hence the required probability is

$$p = p_3 + p_6 + p_9 + p_{12} \ldots = e^{-\lambda} \sum_{k=3,6,9,12\ldots} \frac{\lambda^k}{k!}.$$

We can evaluate this probability by adding numerically all its terms up to a required degree of accuracy. However, using series for e^x and $\cos x$ and multiplying them, we can easily see that

$$\sum_{k=3,6,9,12\ldots} \frac{\lambda^k}{k!} = -1 + \frac{e^{\lambda}}{3} + \frac{2}{3}e^{-\lambda/2}\cos\left(\frac{\sqrt{3}}{2}\lambda\right).$$

Hence the required probability is

$$p = e^{-\lambda}\left(-1 + \frac{e^{\lambda}}{3} + \frac{2}{3}e^{-\lambda/2}\cos\left(\frac{\sqrt{3}}{2}\lambda\right)\right).$$

After substituting $\lambda = 10$, we get

$$p = .33329.$$

PROBABILITY GENERATING FUNCTIONS

We shall introduce now a very powerful and useful tool for investigation of probability distributions. This method was discovered by the Swiss mathematician Leonhard Euler (1707–1783) and makes wide use of calculus, namely of the power series. (Students not familiar with this topic are advised to consult calculus texts before reading this section.) Let X be a random variable which can attain only nonnegative integer values with the following probabilities

$$P(\{X = i\}) = p_i. \qquad i = 0,1,2, \ldots .$$

For every value z, we can investigate the power series

$$P(z) = p_0 + p_1 z + p_2 z^2 + \ldots = \sum_{i=0}^{\infty} p_i z^i.$$

The variable z can be equal to any real number (but it is sometimes advantageous to consider z as a complex variable). It can be shown that if the power series $P(z)$ converges for some number z, then there exists a real number $R > 0$ such that $P(z)$ converges for all $|z| < R$, diverges for all $|z| > R$ and may or may not converge for $|z| = R$. The number R is usually called the *radius of convergence* of $P(z)$. The function $P(z)$ thus defined for all z from an appropriate domain is called the *probability generating function.*

THEOREM 6: The probability generating function of a nonnegative integer valued random variable converges for all values $|z| \leq 1$. ←

Proof:

$$|P(z)| = \left|\sum_{k=0}^{\infty} p_k z^k\right| \leq \sum_{k=0}^{\infty} p_k |z|^k \leq \sum_{k=0}^{\infty} p_k = 1.$$

q.e.d.

EXAMPLE 11

Let us find the probability generating function of the binomial random variable.

$$P(z) = \sum_{k=0}^{\infty} p_k z^k = \sum_{k=0}^{n} \binom{n}{k} p^k q^{n-k} z^k = (pz + q)^n.$$

EXAMPLE 12

Let us find the probability generating function of the negative binomial distribution on the numbers $k = m, m + 1, \ldots$ (waiting time for the m-th success).

$$P(z) = \sum_{k=0}^{\infty} p_k z^k = \sum_{k=m}^{\infty} \binom{k-1}{m-1} p^m q^{k-m} z^k = \sum_{j=0}^{\infty} \binom{m+j-1}{m-1} p^m q^j z^{m+j}$$

$$= \sum_{j=0}^{\infty} \frac{m(m+1)\ldots(m+j-1)}{j!} p^m q^j z^{m+j}$$

$$= p^m z^m \sum_{j=0}^{\infty} \binom{-m}{j} (-qz)^j = p^m z^m (1-qz)^{-m} = \left(\frac{pz}{1-qz}\right)^m.$$

The power series is convergent for $|qz| < 1$ or $|z| < \frac{1}{q}$, but $\frac{1}{q} > 1$ and therefore $P(z)$ is defined for some values $|z| > 1$.

It is very easy to find the probability generating function of the negative binomial distribution on the numbers $k = 0,1,2, \ldots$ (the total number of failures before the occurrence of the m-th success). The probabilities are identical as before, but the random variable attains the values $k = 0,1,2, \ldots$ instead of $k = m, m+1, m+2, \ldots$ and, therefore, this time

$$P(z) = \sum_{k=m}^{\infty} \binom{k-1}{m-1} p^m q^{k-m} z^{k-m} = z^{-m} \sum_{k=m}^{\infty} \binom{k-1}{m-1} p^m q^{k-m} z^k = z^{-m} \left(\frac{pz}{1-qz}\right)^m$$

$$= \left(\frac{p}{1-qz}\right)^m.$$

EXAMPLE 13

For the Poisson distribution, we have

$$P(z) = \sum_{k=0}^{\infty} p_k z^k = \sum_{k=0}^{\infty} e^{-\lambda} \frac{\lambda^k}{k!} z^k = e^{-\lambda} \sum_{k=0}^{\infty} \frac{(\lambda z)^k}{k!} = e^{-\lambda} e^{\lambda z} = e^{\lambda(z-1)}.$$

Even though the definition of the probability generating function is based on a pure mathematical formalism, the function itself has a quite interesting practical probabilistic meaning, as suggested by *David von Dantzig*. Let X be a nonnegative integer valued random variable with probability distribution $P(\{X = k\}) = p_k$, and let z be a real number, $0 \leq z \leq 1$. Let us perform (observe) the random variable X, and if $X = k$ then let us perform k alternative trials with the probability of a success z (and the probability of a failure $1 - z$). Then, using the formula of total probability, we can find that the probability that *no failure will occur* is equal to

$$\sum_{k=0}^{\infty} P(\{X = k\}) P(\{\text{no failure}/X = k\}) = \sum_{k=1}^{\infty} p_k z^k = P(z).$$

This interpretation yields a very interesting result. Let $X_1, X_2, \ldots, X_n$ be independent random variables with probability generating functions (abbreviated p.g.f.) $P_1(z)$, $P_2(z)$, $\ldots, P_n(z)$ and let the p.g.f. of the sum $\sum_{i=1}^{n} X_i$ be $P(z)$. For each random variable X_i, the probability that the combined experiment described above will yield no failures is $P_i(z)$, and therefore the probability that there will be no failures in all (independent) combined experiments is $P_1(z) \cdot P_2(z) \ldots P_n(z)$. However, this probability is equal to the probability that the random variable $\sum_{i=1}^{n} X_i$ will yield no failures in its combined experiment, hence it is equal to $P(z)$. Thus we have proved the following:

THEOREM 7: The probability generating function of a sum of independent random variables is equal to the product of the corresponding probability generating functions,

$$P(z) = \prod_{i=1}^{n} P_i(z).$$

However, we can give another, more formal proof of this statement.

Proof: We can see that the p.g.f. $P(z)$ of a random variable X can be expressed as

$$P(z) = \sum_{k=0}^{\infty} p_k z^k = E(z^X).$$

Therefore, if $X_1, X_2, \ldots, X_n$ are independent random variables, then the p.g.f. $P(z)$ of $\sum_{i=1}^{n} X_i$ is equal to

$$P(z) = E(z^{\sum_{i=1}^{n} X_i}) = E(z^{X_1} \cdot z^{X_2} \ldots z^{X_n}) = E(z^{X_1}) \cdot E(z^{X_2}) \ldots E(z^{X_n}) = P_1(z) \cdot P_2(z) \ldots P_n(z)$$

where $P_i(z)$ is the p.g.f. of X_i. The last observation follows from the fact that the expected value of a product of independent random variables is equal to the product of expected values (see Theorem 5, page 191). q.e.d.

EXAMPLE 14

For the alternative distribution (binomial distribution with $n = 1$), the p.g.f. is $pz + q$. The binomial random variable can be thought of as a sum of n independent alternative random variables (see page 209), hence the p.g.f. of the binomial distribution is $(pz + q)^n$ (see Example 11). Similarly, the geometric random variable has the p.g.f. $\frac{pz}{1 - qz}$ (the negative binomial variable for $m = 1$) and the negative binomial random variable is the sum of m independent geometric random variables, hence its p.g.f is $\left(\frac{pz}{1 - qz}\right)^m$ (see Example 12).

We shall now make some more profound observations of the properties of the p.g.f. We know that every p.g.f. $P(z)$ is convergent for all $|z| \leq 1$, and using the theory of power series we can differentiate it any number of times, term by term. Therefore, we get

$$\frac{d}{dz} P(z) = \sum_{k=1}^{\infty} k p_k z^{k-1}$$

and further

$$\frac{d^n}{dz^n} P(z) = \sum_{k=n}^{\infty} \frac{k!}{(k-n)!} p_k z^{k-n} \qquad \text{for } |z| < 1.$$

If we now substitute $z = 0$, we get

$$\frac{d^n}{dz^n} P(0) = n!\, p_n.$$

Therefore, if we know the p.g.f. and can differentiate it n times, we can easily find the original probability distribution as follows:

$$p_n = \frac{1}{n!} \frac{d^n}{dz^n} P(0).$$

Therefore, the probability distribution uniquely determines its probability generating function and vice versa: the probability distribution can be uniquely recovered from its probability generating function. This is the reason we call $P(z)$ the p.g.f. We have thus proved:

→ **THEOREM 8:** There is a unique one-to-one relation between the probability distribution of a nonnegative integer valued random variable and its probability generating function.

EXAMPLE 15

For the binomial distribution, we have $P(z) = (pz + q)^n$ (see Example 11). Therefore,

$$\frac{d^k}{dz^k} P(z) = n(n-1) \ldots (n-k+1) p^k (pz+q)^{n-k}$$

$$\frac{d^k}{dz^k} P(0) = \frac{n!}{(n-k)!} p^k q^{n-k}.$$

Therefore,
$$p_k = \frac{1}{k!} \frac{d^k}{dz^k} P(0) = \binom{n}{k} p^k q^{n-k}.$$

EXAMPLE 16

For the negative binomial distribution on the numbers 0,1,2, . . . , we have $P(z) = \left(\frac{p}{1-qz}\right)^m$ (see Example 12). Hence,

$$\frac{d^k}{dz^k} P(0) = (-m)(-m-1) \ldots (-m-k+1) p^m (-q)^k$$

and
$$p_k = \frac{1}{k!} \frac{d^k}{dz^k} P(0) = \binom{-m}{k} p^m (-q)^k.$$

EXAMPLE 17

For the Poisson distribution, we have $P(z) = e^{\lambda(z-1)}$ (see Example 13). Hence,

$$\frac{d^k}{dz^k} P(z) = e^{-\lambda} \lambda^k e^{\lambda z}.$$

Furthermore,
$$p_k = \frac{1}{k!} \frac{d^k}{dz^k} P(0) = e^{-\lambda} \frac{\lambda^k}{k!}.$$

EXAMPLE 18

Find the probability that the sum on four fair dice will be 12, using counting techniques and p.g.f.

SOLUTION 1. The following table gives the number of points on four dice, and the number of ways to get them.

No. of points	No. of ways
1 1 4 6	$\frac{4!}{2!} = 12$
1 1 5 5	$\frac{4!}{2!2!} = 6$
1 2 3 6	$4! = 24$
1 2 4 5	$4! = 24$
1 3 3 5	$\frac{4!}{2!} = 12$
1 3 4 4	$\frac{4!}{2!} = 12$
2 2 2 6	$\frac{4!}{3!} = 4$
2 2 3 5	$\frac{4!}{2!} = 12$
2 2 4 4	$\frac{4!}{2!2!} = 6$
2 3 3 4	$\frac{4!}{2!} = 12$
3 3 3 3	$\frac{4!}{4!} = 1$
	125

When tossing four fair dice, we can get 6^4 equally likely outcomes, of which 125 will give the sum 12. Therefore, the required probability is

$$p = \frac{125}{6^4}.$$

SOLUTION 2. Let X be the number of points on one fair die. Then the corresponding p.g.f. is

$$P(z) = \frac{1}{6}(z + z^2 + z^3 + z^4 + z^5 + z^6) = \frac{z(1 - z^6)}{6(1 - z)}.$$

The p.g.f. $Q(z)$ for the sum on four dice will be (Theorem 7):

$$Q(z) = (P(z))^4 = \frac{z^4(1 - z^6)^4}{6^4(1 - z)^4} = \frac{z^4}{6^4}(1 - 4z^6 + 6z^{12} - 4z^{18} + z^{24}) \sum_{k=0}^{\infty} \binom{-4}{k} (-z)^k.$$

The probability that the sum will equal 12 we get from the coefficient at z^{12}, which is equal to

$$\frac{1}{6^4}\left(1 \cdot \binom{-4}{8} - 4 \cdot \binom{-4}{2}\right) = \frac{125}{6^4}.$$

Therefore,
$$\frac{d^{12}}{dz^{12}} Q(0) = 12! \frac{125}{6^4}$$

and the required probability is

$$p = \frac{1}{12!} \frac{d^{12}}{dz^{12}} Q(0) = \frac{125}{6^4}.$$

We can see that this solution is faster and less tedious than the solution involving combinatorial methods.

The previous example suggests a fast and easy way to calculate the distribution of the sum of n identically distributed independent random variables $Y = X_1 + X_2 + \ldots + X_n$, where every X_2 has the distribution as a r.v. X.

(a) Find the p.g.f. $P(z)$ of X.
(b) Raise it to the n-th power $(P(z))^n$.
(c) Expand $(P(z))^n$ into a power series, and find the coefficient at z^k. This coefficient is equal to the required probability $P(\{Y = k\})$.

We have discussed in some detail how to use the p.g.f. of a random variable in order to get its probability distribution. However, using a similar method, we can find the expectation and the variance of a random variable X directly from its p.g.f. Assume now that $\frac{d}{dz} P(z) = P'(z)$ and $\frac{d^2}{dz^2} P(z) = P''(z)$ exist for $z = 1$ (it may or may not happen because the existence of $P(z)$ for $|z| \geq 1$ is not guaranteed). Then

$$P'(1) = \sum_{k=0}^{\infty} k p_k = E(X).$$

$$P''(1) = \sum_{k=0}^{\infty} k(k-1) p_k = \sum_{k=0}^{\infty} k^2 p_k - \sum_{k=0}^{\infty} k p_k = E(X^2) - E(X).$$

Hence, $V(X) = E(X^2) - (E(X))^2 = P''(1) + P'(1) - (P'(1))^2 = P''(1) + P'(1)[1 - P'(1)]$.

We have thus proved the following:

➡ **THEOREM 9:** If $P'(1)$ and $P''(1)$ exist, then

$$E(X) = P'(1).$$
$$V(X) = P''(1) + P'(1)[1 - P'(1)].$$

EXAMPLE 19

For the binomial distribution, we have

$$\frac{d^k}{dz^k}P(z) = \frac{n!}{(n-k)!}p^k(pz+q)^{n-k}$$

(see Example 15). Hence for $k = 1$, we get

$$P'(1) = np(p+q)^{n-1} = np = E(X).$$

For $k = 2$,

$$P''(1) = n(n-1)p^2(p+q)^{n-2} = n(n-1)p^2.$$

$$V(X) = P''(1) + P'(1)[1 - P'(1)] = n(n-1)p^2 + np(1-np) = npq.$$

EXAMPLE 20

For the Poisson distribution, we have

$$\frac{d^k}{dz^k}P(z) = e^{-\lambda}\lambda^k e^{\lambda z}$$

(see Example 17). For $k = 1$, we get

$$P'(1) = e^{-\lambda}\lambda e^{\lambda} = \lambda = E(X).$$

For $k = 2$, we get

$$P''(1) = e^{-\lambda}\lambda^2 e^{\lambda} = \lambda^2.$$

$$V(X) = P''(1) + P'(1)[1 - P'(1)] = \lambda^2 + \lambda(1-\lambda) = \lambda.$$

Compare this result with Theorem 5.

THE POISSON APPROXIMATION TO THE BINOMIAL DISTRIBUTION

Let us assume that we have a sequence of binomial experiments with the probability of a success p, and with the number of repetitions n such that the number of repetitions n keeps increasing above all limits. At the same time, let us assume that p keeps decreasing for each experiment in such a way that the product np is constant, say $np = \lambda > 0$. Let us investigate the behavior of the binomial probabilities $\binom{n}{k}p^k q^{n-k}$ with increasing n,

$$\lim_{n\to\infty}\binom{n}{k}p^k q^{n-k} = \lim_{n\to\infty}\frac{n(n-1)\ldots(n-k+1)}{k!}\left(\frac{\lambda}{n}\right)^k\left(1-\frac{\lambda}{n}\right)^{n-k}$$

$$= \lim_{n\to\infty}\frac{\lambda^k}{k!}\left(\frac{n-1}{n}\right)\left(\frac{n-2}{n}\right)\ldots\left(\frac{n-k+1}{n}\right)\left(1-\frac{\lambda}{n}\right)^n\left(1-\frac{\lambda}{n}\right)^{-k} = \frac{\lambda^k}{k!}e^{-\lambda}.$$

Thus we can see that with increasing n, the binomial distribution converges to the Poisson distribution.

→ **THEOREM 10:** If in a sequence of binomial experiments the number of repetitions n increases above all limits, so that $np = \lambda > 0$ is constant, then

$$\lim_{n\to\infty} \binom{n}{k} p^k q^{n-k} = e^{-\lambda} \frac{\lambda^k}{k!}.$$

COMMENT. Even though we proved Theorem 10 above, using the binomial and Poisson distributions, we shall give an alternative proof using the probability generating functions.

Proof: The p.g.f. of the binomial distribution is $(pz + q)^n$. Assuming that n tends to infinity so that $np = \lambda > 0$ is constant, we get

$$\lim_{n\to\infty} (pz + q)^n = \lim_{n\to\infty} \left(\frac{\lambda}{n} z + 1 - \frac{\lambda}{n}\right)^n = \lim_{n\to\infty} \left(1 + \frac{\lambda(z-1)}{n}\right)^n = e^{\lambda(z-1)}.$$

Thus we can see that the p.g.f. of the binomial distribution converges to the p.g.f. of the Poisson distribution. Using Theorem 8, we conclude that the binomial distribution also must converge to the Poisson distribution. q.e.d.

EXAMPLE 21

A company surveying for oil estimates the probability of an experimental drilling yielding an oil discovery is .01. Find the probability that if the company performs 300 (independent and mutually remote) drilling experiments it will get at most 4 new finds.

SOLUTION. The probability P of at most $k = 4$ successes in $n = 300$ independent repetitions, if the probability of each success is $p = .01$, should be calculated using the binomial distribution:

$$P = \binom{300}{0}(.01)^0(.99)^{300} + \binom{300}{1}(.01)^1(.99)^{299} + \binom{300}{2}(.01)^2(.99)^{298}$$

$$+ \binom{300}{3}(.01)^3(.99)^{297} + \binom{300}{4}(.04)^4(.99)^{296}.$$

To evaluate this expression represents a really formidable numerical obstacle. Nevertheless, armed with enduring patience, we can find that

$$P = .8161.$$

However, using the Poisson approximation to the binomial distribution with $\lambda = np = 3$, we can easily find that P is approximately

$$P \approx e^{-3}\left(\frac{3^0}{0!} + \frac{3^1}{1!} + \frac{3^2}{2!} + \frac{3^3}{3!} + \frac{3^4}{4!}\right) = .8153.$$

The Poisson approximation, therefore, gives fairly accurate answers and at the same time it is incomparably easier to calculate numerically.

EXAMPLE 22

Following is the table of binomial probabilities

$$B(k) = \sum_{i=0}^{k} \binom{n}{i} p^i q^{n-i}$$

and Poisson probabilities

$$P(k) = \sum_{i=0}^{k} e^{-\lambda} \frac{\lambda^i}{i!}$$

for $p = .1$, $n = 100$ and $\lambda = np = 10$.

k	Binomial $B(k)$	Poisson $P(k)$
0	.000027	.000045
1	.000322	.000499
2	.001945	.002769
3	.0078	.0103
4	.0237	.0292
5	.0576	.0671
6	.1172	.1301
7	.2061	.2202
8	.3209	.3328
9	.4513	.4579
10	.5832	.5830
11	.7030	.6968
12	.8018	.7916
13	.8761	.8645

It can be seen that the approximation of the binomial distribution by the Poisson distribution is already fairly good if $n = 100$ and $p = .1$. If $n \geq 100$ and $p \leq .1$, then the approximation may be expected to be even better.

EXERCISES

(Answers in Appendix to exercises marked)*

*1. Roll a fair die till you get a 6. Let the r.v. X denote the number of the toss yielding the 6. Find the probability distribution of X, $E(X)$ and $V(X)$.

2. Roll a fair die till you get three 6's (not necessarily in a row). Let the r.v. Y denote the number of the toss yielding the third 6. Find the probability distribution of Y, $E(Y)$ and $V(Y)$.

*3. Consider the Petersburg Paradox (Example 9, Chapter XII, page 185) in the case where the game has to continue till heads appear (without the stopping rule). Let X be the payoff in this game. Investigate $E(X)$.

4. Consider the game of bank craps (Example 6, page 321). Assume that the new shooter rolled i points on the first shot, where $i = 4, 5, 6, 8, 9, 10$. In this case he had to keep on rolling. Given the fact that he has won the bet (he repeated his point before getting a 7), find the expected number of tosses he needed to do it.

*5. For a negative binomial r.v. X, $E(X) = 20$, $V(X) = 180$. Find m, p and q.

*6. The probability that the heavyweight boxing champ Joe Altobelo will knock out his challenger Sid Cattlet in one round is .3, while Sid will knock out Joe in a single round with probability .2. According to the rules, both will fight till one of them is knocked out. Assuming independent successive rounds, find the probability that Sid will defeat Joe.

7. Assume that the number of cars arriving at a service station within one minute has a Poisson distribution with the parameter $\lambda = 2$. Find

the probability that within one minute there will be
(a) at most one
(b) exactly two
(c) at least three
cars arriving at the station.

*8. Assume that the number of virus particles in 1 milliliter of a patient's blood has a Poisson distribution with $\lambda = 4$. Find the probability that 1 milliliter of sample blood will contain the following number of virus particles:
(a) at most two.
(b) exactly three.
(c) an even number.

9. A radioactive substance emits on the average three α-particles per minute. Assuming that the number of particles X emitted during the time t is given by the Poisson distribution

$$P(\{X = k\}) = e^{-\lambda t}\frac{(\lambda t)^k}{k!}$$

find the probability that within three minutes, the following particles will be emitted:
(a) at least two
(b) exactly nine
(c) an odd number.

*10. Find the probability that the sum on three fair dice will be 9 using
(a) counting techniques.
(b) probability generating functions.

11. Discard face cards from a standard deck of cards and assign each card its point value (an ace counts 1). Then draw four cards with replacement from a well shuffled deck. Find the probability that the total will be equal to 20.

12. Let $S_n = X_1 + X_2 + \ldots + X_n$ be the sum of n mutually independent random variables uniformly distributed and attaining the values $1,2, \ldots ,N$. Show that the p.g.f. of S_n is

$$P(z) = \left[\frac{z(1 - z^N)}{N(1 - z)}\right]^n$$

and further

$$P(\{S_n = j\}) = N^{-n}\sum_{i=0}^{\infty}(-1)^i\binom{n}{i}\binom{j - Ni - 1}{n - 1}.$$

*13. Let X be a r.v. with the p.g.f. $P(z)$. Express $E(X^3)$ and $E(X^4)$ in terms of $P(z)$.

14. Using the p.g.f., derive the formulas for the mean and variance of the geometric distribution.

15. Using the p.g.f., derive the formulas for the mean and variance of the negative binomial distribution.

16. Using the p.g.f., derive the formulas for the mean and variance of the uniform distribution on the numbers $1,2, \ldots ,N$. (See Exercise 12.)

17. Let X_i be independent binomial r.v.'s representing n_i repetitions of the same experiment with the probability of success p; $i = 1,2$. Using the p.g.f., show that the r.v. $(X_1 + X_2)$ has the binomial distribution with parameters $(n_1 + n_2)$ and p.

18. Let X_i be independent Poisson r.v.'s with the parameter λ_i; $i = 1,2$. Using the p.g.f., show that the r.v. $(X_1 + X_2)$ has the Poisson distribution with the parameter $(\lambda_1 + \lambda_2)$.

19. Let X_i be independent negative binomial r.v.'s with the parameters p and m_i; $i = 1,2$. Using the p.g.f., show that the r.v. $(X_1 + X_2)$ has the negative binomial distribution with the parameters p and $(m_1 + m_2)$.

20. Using the binomial series, show that for any two real numbers m, n and for any nonnegative integer

$$\binom{m + n}{k} = \sum_{i=0}^{k}\binom{m}{i}\binom{n}{k - i}.$$

21. The probability that a typist will type a page of a mathematical textbook without a mistake is .01. If the textbook contains 500 typed pages, find the probability of the following number of perfect pages:
(a) less than 3.
(b) at most 5.

*22. It has been observed that on the average 1 out of every 10 hi-fi amplifiers produced by the SAKE company is defective and has to be repaired under the warranty. If Sam's Stereo Mart sells 150 of these units, find the
(a) expected number of units that will be returned for guaranteed repairs.
(b) probability that at most five units will be returned for repairs.

23. The probability that a letter will not be delivered because of the chaos in the postal system is .05. If you send 100 holiday cards to your friends, find the probability that at least 2 will not be delivered.

*24. The "ace-deuce" bet in craps (see Example 6) is identical to the "don't pass line bet" except that "2-1" or "1-2" on the first "come out" instead of double 6 is a stand-off, and the wager has to be left for the next game. Find the probability of winning the ace-deuce bet, and the house advantage.

25. The "big-six" bet in craps (see Example 6) is won if a 6 appears before 7; the "big-eight" bet is won if an 8 appears before 7. Find the corresponding probability of winning for each bet, and the house advantage.

*26. Toss a coin. If a head appears, take a symmetrical tetrahedron die with the numbers 1, 2, 3, 4 on it, otherwise take an ordinary cube die. Keep rolling the die till a 1 appears. What is the probability that you have to roll the die an even number of times?

*27. What is the probability that in a sequence of tosses of a pair of ordinary dice, either the sum 5 or 8 appears before 7 appears?

NORMAL DISTRIBUTION AND CENTRAL LIMIT THEOREM

This chapter deals with rather involved ideas, and for this reason we will start with some simple illustrative examples. If a fair coin is tossed four times, what is the probability P of getting exactly two heads? The answer is given by Theorem 1 of Chapter XI: If X is a binomial random variable with parameters p and n, then its distribution is

$$P(\{X = k\}) = \binom{n}{k} p^k q^{n-k}. \qquad 0 \leq k \leq n.$$

In our case, $n = 4$, $k = 2$, $p = \frac{1}{2}$, hence the required probability is

$$P = P(\{X = 2\}) = \binom{4}{2}\left(\frac{1}{2}\right)^2\left(\frac{1}{2}\right)^2 = .375.$$

The solution is very simple and even the numerical calculation can be performed very easily. But if we ask for the probability P that in 200 tosses one gets exactly 100 heads, the situation is quite different. Conceptually the problem is the same as before, and can be solved using again the binomial distribution for $n = 200$, $k = 100$ and $p = \frac{1}{2}$.

$$P = \binom{200}{100}\left(\frac{1}{2}\right)^{100}\left(\frac{1}{2}\right)^{100} = \binom{200}{100}\left(\frac{1}{2}\right)^{200}.$$

A *numerical evaluation* of the formula in this case is hopelessly difficult. One can use tables of logarithms to find $\left(\frac{1}{2}\right)^{200}$ but there is no way to find $\binom{200}{100}$ numerically with a simple method. One could perhaps use a small or medium sized computer (indeed, some of them have subroutines for binomial distribution) to get $P = .05635$ exactly to five decimal places. However, if we have to evaluate the binomial distribution for very large values of n, then even computers get stuck owing to the nature of their memory. Fortunately it has already been shown by *De Moivre* (1667–1754) and *Laplace* (1749–1827) that for large values of n, the binomial distribution $\binom{n}{k} p^k q^{(n-k)}$ can be very closely approximated by a certain kind of curve, which is usually referred to as the *normal distribution curve.*

In order to get a real feeling for this type of approximation, we shall investigate the (binomial) distribution of the number of heads X_n in n independent tosses of a fair coin for some values of n. In this case the binomial distribution has the following form:

$$P(\{X_n = k\}) = \binom{n}{k}\frac{1}{2^n}. \qquad 0 \leq k \leq n.$$

The expected value of X_n and its standard deviation are as follows:

$$E(X_n) = \frac{n}{2} \qquad \sigma(X_n) = \frac{\sqrt{n}}{2}.$$

We would like to compare the distributions of X_n for different values of n, but these distributions are incomparable because they have different expected values and standard deviations. For this reason (as well as for other reasons which will become clear later), it is expedient to investigate the distribution of the *standardized* r.v. S_n given as follows:

$$S_n = \frac{X_n - E(X_n)}{\sigma(X_n)} = \frac{X_n - \frac{n}{2}}{\frac{\sqrt{n}}{2}} = \frac{2}{\sqrt{n}} X_n - \sqrt{n}.$$

It follows from Theorem 2 and Theorem 8 of Chapter XII that

$$E(S_n) = \frac{E(X_n) - \frac{n}{2}}{\frac{\sqrt{n}}{2}} = 0, \qquad \sigma(S_n) = \frac{\sqrt{V(X_n)}}{\frac{\sqrt{n}}{2}} = 1.$$

Therefore all random variables S_n, $n = 1,2,3 \ldots$ have the same mean equal to 0 and the same standard deviation equal to 1. Hence, instead of comparing the distributions of X_n, it is more reasonable to compare the distributions of S_n. Furthermore, we know that for each particular value of X_n or S_n, its probability is decreasing with increasing n; for this reason (and for other reasons that will become clear later) we shall multiply these probabilities by a "normalizing factor," $\frac{\sqrt{n}}{2}$. See the accompanying tables of these normalized distributions for $n = 1,4,9,16$.

$n = 1$

Values of X_1	k	0	1
Values of S_1	$2k - 1$	-1	1
Probabilities	$\binom{1}{k}\frac{1}{2}$	$\frac{1}{2}$	$\frac{1}{2}$
Normalized probabilities	$\frac{1}{2}\binom{1}{k}\frac{1}{2}$	$\frac{1}{4}$	$\frac{1}{4}$

$n = 4$

Values of X_4	k	0	1	2	3	4
Values of S_4	$k - 2$	-2	-1	0	1	2
Probabilities	$\binom{4}{k}\frac{1}{2^4}$	.0625	.25	.375	.25	.0625
Normalized probabilities	$\binom{4}{k}\frac{1}{2^4}$	.0625	.25	.375	.25	.0625

$n = 9$

Values of X_4	k	0	1	2	3	4	5	6	7	8	9
Values of S_9	$\frac{2}{3}k - 3$	-3	$-2\frac{1}{3}$	$-1\frac{2}{3}$	-1	$-\frac{1}{3}$	$\frac{1}{3}$	1	$1\frac{2}{3}$	$2\frac{1}{3}$	3
Probabilities	$\binom{9}{k}\frac{1}{2^9}$	.0020	.0176	.0703	.1641	.2461	.2461	.1641	.0703	.0176	.0020
Normalized probabilities	$\frac{3}{2}\binom{9}{k}\frac{1}{2^9}$	.0029	.0264	.1055	.2461	.3691	.3691	.2461	.1055	.0264	.0029

$n = 16$

Values of X_{16}	k	0	1	2	3	4	5	6	7	8	9	10	11	12	13	14	15	16
Values of S_{16}	$\frac{1}{2}k - 4$	-4	$-3\frac{1}{2}$	-3	$-2\frac{1}{2}$	-2	$-1\frac{1}{2}$	-1	$-\frac{1}{2}$	0	$\frac{1}{2}$	1	$1\frac{1}{2}$	2	$2\frac{1}{2}$	3	$3\frac{1}{2}$	4
Probabilities	$\binom{16}{k}\frac{1}{2^{16}}$	.0000	.0002	.0018	.0085	.0278	.0667	.1222	.1746	.1964	.1746	.1222	.0667	.0278	.0085	.0018	.0002	.0000
Normalized probabilities	$2\binom{16}{k}\frac{1}{2^{16}}$	.0000	.0004	.0036	.0170	.0555	.1333	.2444	.3491	.3928	.3491	.2444	.1333	.0555	.0170	.0036	.0004	.0000

From the above graphs, we can see that the normalized probabilities are with increasing n approaching closely to an ideal bell-shaped curve. It can be shown that this so-called normal curve is defined by the equation

$$y = \frac{1}{\sqrt{2\pi}} e^{-x^2/2}$$

where $e = 2.71828183 \ldots$ is Euler's constant, and $\pi = 3.14159265 \ldots$ is Ludolf's number. Let us calculate some values of the normal curve:

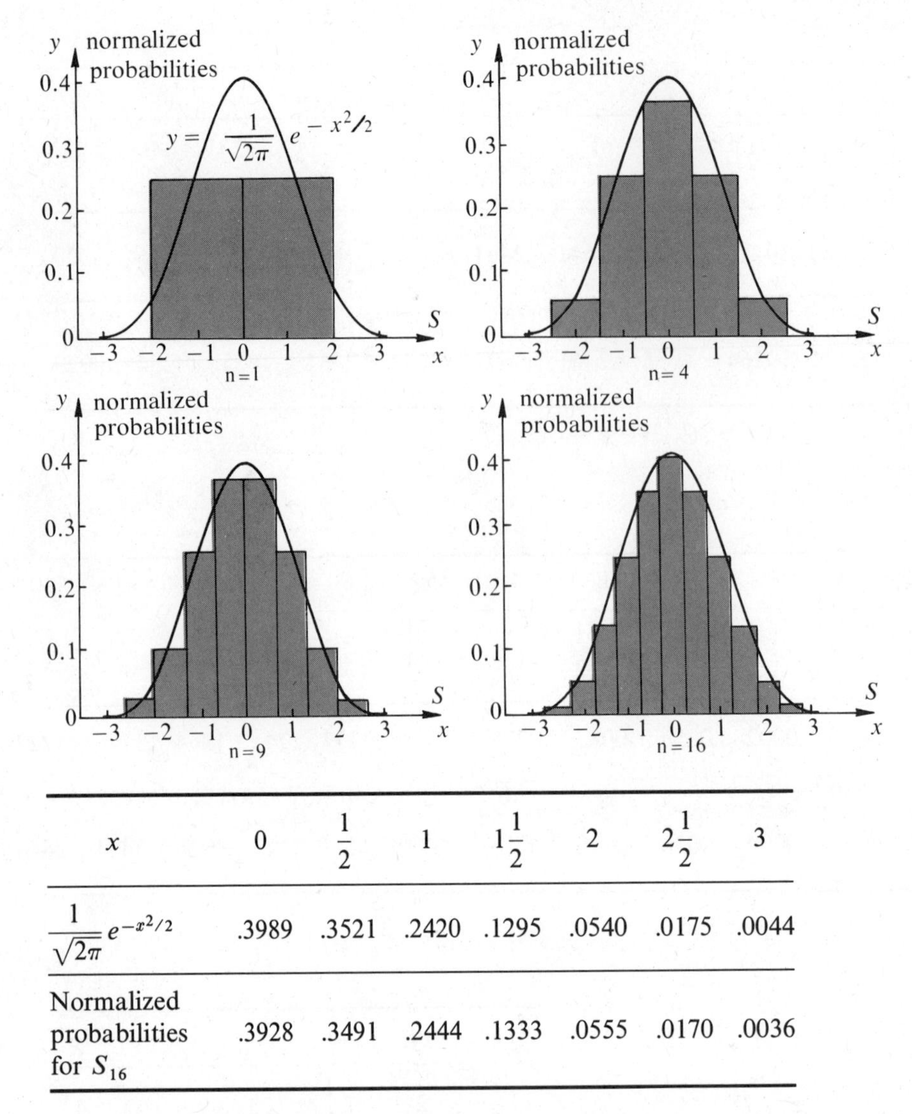

x	0	$\frac{1}{2}$	1	$1\frac{1}{2}$	2	$2\frac{1}{2}$	3
$\frac{1}{\sqrt{2\pi}} e^{-x^2/2}$	.3989	.3521	.2420	.1295	.0540	.0175	.0044
Normalized probabilities for S_{16}	.3928	.3491	.2444	.1333	.0555	.0170	.0036

Comparing the normal curve with the normalized probabilities for S_{16} (see the table of the distribution of X_{16} and S_{16}) we can see already that for $n = 16$ the approximation is quite good. For larger n, we may expect an even better approximation.

EXAMPLE 1

Using the normal distribution curve, find the probability $P(k)$ that 100 tosses of a fair coin will result in exactly k heads for $k = 50, 51, 52, 53, 54$. Compare with exact probabilities.

SOLUTION. The probability $P(k)$ of getting exactly k heads is

$$P(k) = \binom{100}{k} \frac{1}{2^{100}}.$$

We know that the normalized probabilities $\frac{\sqrt{n}}{2} P(k)$ can be approximated by a normal curve.

Furthermore, value k is standardized as $\left(\frac{2}{\sqrt{n}} k - \sqrt{n}\right) = \left(\frac{2}{\sqrt{100}} k - \sqrt{100}\right)$ $= (.2k - 10)$. (See the formula for S_n.)

Therefore,
$$\frac{\sqrt{100}}{2}P(k) \approx \frac{1}{\sqrt{2\pi}}e^{-(.2k-10)^2/2}$$

or equivalently,
$$P(k) \approx \frac{1}{5\sqrt{2\pi}}e^{-(.2k-10)^2/2}.$$

k	50	51	52	53	54
Exact value of $P(k)$	.0796	.0780	.0735	.0666	.0580
Normal approximation of $P(k)$	.0798	.0782	.0737	.0666	.0579

We can see that the agreement between exact and approximate values is extremely fine. At the same time, calculations of approximate values take only a few seconds (using tables of e^x or a pocket calculator) while calculations of exact values take a long time, even on a small computer.

We have numerically shown that a binomial distribution with $p = \frac{1}{2}$ can be for large n closely approximated by the normal distribution curve. This is true for any value of p, as follows from the famous theorem cited next. However, before we state the theorem, we shall introduce the following notation: Let a_n, b_n, $n = 1,2,3, \ldots$ be two sequences of real numbers. Then

$$a_n \sim b_n \quad \text{means that} \quad \lim_{n\to\infty}\frac{a_n}{b_n} = 1.$$

Practically it means that for very large n, the term a_n is "very close" to b_n, and sometimes we can replace a_n by b_n without making too large an error.

THEOREM 1: *Local DeMoivre-Laplace Limit Theorem.* Let $0 < p < 1$ and $q = 1 - p$. Furthermore, let

$$x = \frac{k - np}{\sqrt{npq}} \qquad 0 \le k \le n.$$

Then
$$\sqrt{npq}\binom{n}{k}p^k q^{n-k} \sim \frac{1}{\sqrt{2\pi}}e^{-x^2/2}.$$

(The proof of this theorem is on page 357.)

Theorem 1 says that for very large values of n, the expression $\sqrt{npq}\binom{n}{k}p^k q^{n-k}$ can be approximated by $\frac{1}{\sqrt{2\pi}}e^{-x^2/2}$ without making too large an error or equivalently, that $\binom{n}{k}p^k q^{n-k}$ can be approximated by $\frac{1}{\sqrt{npq}\sqrt{2\pi}}e^{-x^2/2}$. Let us investigate how good this approximation is.

EXAMPLE 2

Toss a perfect die 300 times. Find the probability $P(k)$ of getting exactly k 6's for $k = 50$, 51, 52 using exact calculations, and also using normal approximation (Theorem 1).

SOLUTION. In this case, $p = \frac{1}{6}, q = \frac{5}{6}, n = 300.$

$$P(k) = \binom{300}{k}\left(\frac{1}{6}\right)^k\left(\frac{5}{6}\right)^{300-k}.$$

Using Theorem 1, we get

$$x = \frac{k - np}{\sqrt{npq}} = \frac{k - 50}{\sqrt{\frac{250}{6}}} \qquad k = 50, 51, 52$$

and we approximate $P(k)$ by the expression

$$\frac{1}{\sqrt{\frac{250}{6}}\sqrt{2\pi}}e^{-x^2/2}.$$

The actual numerical values are in the following table:

k	50	51	52
Exact value of $P(k)$	.0617	.0605	.0579
Approximation of $P(k)$	.0618	.0611	.0589

We can see that the agreement between exact binomial probabilities and their approximations using a normal curve is really very good. Theorem 1 thus appears to be very useful. However, we are using it only as a starting point on our way to another theorem that will be even simpler and more convenient.

Let us introduce now the following notation. For any real number x, we define the function $\varphi(x)$ as

$$\varphi(x) = \frac{1}{\sqrt{2\pi}}e^{-x^2/2}$$

and we define $\varphi(x)$ as the area under the normal curve to the left of x.

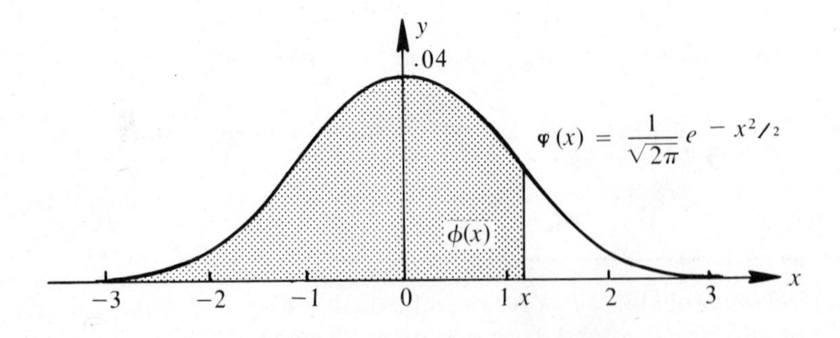

Furthermore, for any two real numbers $a < b$, we define the number $A(a,b)$ as the area under the normal curve between a and b.

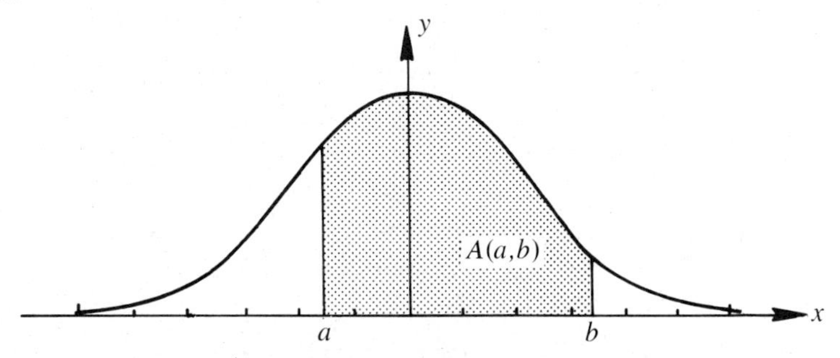

Values of $\varphi(x)$ can be easily calculated using tables of exponential function or a calculator with e^x function. However, we shall also need values of $\phi(x)$ and $A(a,b)$, which cannot be easily calculated. A numerical table of values $\phi(x)$ for some values of x can be found on page 360.

We can see from the table that the total area under the normal curve is equal to 1. Furthermore, the area to the left of x is equal to the area to the right of $(-x)$. Hence

$$\phi(x) + \phi(-x) = 1$$

or

$$\phi(-x) = 1 - \phi(x).$$

For this reason, it is enough to tabulate $\phi(x)$ only for $x > 0$.

Furthermore, we can see that for any two numbers $a < b$, we have

$$A(a,b) = \phi(b) - \phi(a).$$

EXAMPLE 3

Directly from the table on page 360, we can read

$$\phi(0) = .5. \qquad \phi(1) = .8413. \qquad \phi(1.5) = .9332.$$

Furthermore,

$$\phi(-1) = 1 - \phi(1) = 1 - .8413 = .1587.$$

$$\phi(-2.53) = 1 - \phi(2.53) = 1 - .9943 = .0057.$$

Let us calculate some areas $A(a,b)$.

$$A(1.33,\ 3.02) = \phi(3.02) - \phi(1.33) = .9993 - .9082 = .0911.$$

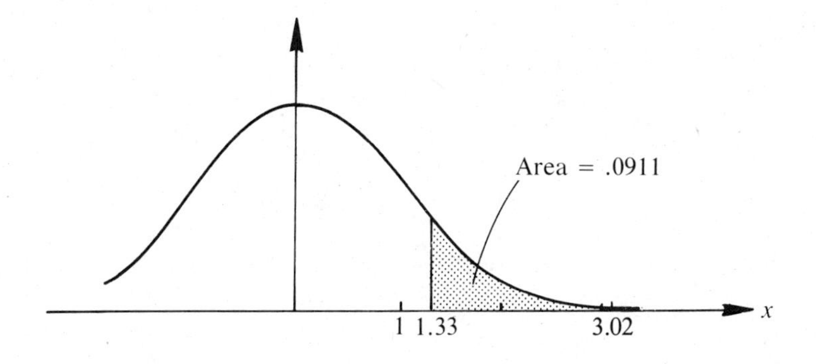

$A(-.91, 1.25) = \phi(1.25) - \phi(-.91) = \phi(1.25) - 1 + \phi(.91) = .8944 - 1 + .8186 = .7130.$

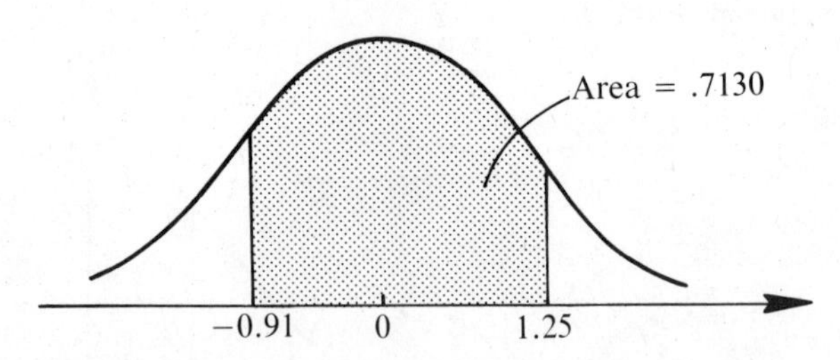

$$A(-2.1, -.11) = \phi(-.11) - \phi(-2.1) = 1 - \phi(.11) - 1 + \phi(2.1) = \phi(2.1) - \phi(.11)$$
$$= .9821 - .5438 = .4383.$$

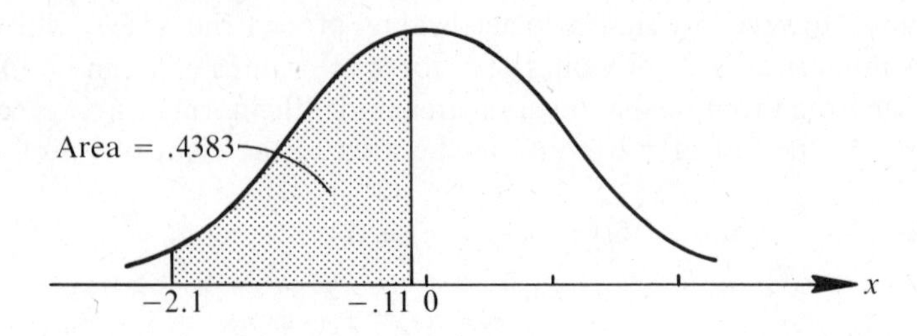

We have spent some time on the investigation of areas under the normal curve because we shall need it in the further discussion of approximation of the binomial distribution. Let X_n be a binomial random variable with parameters p and n. Furthermore, let $a < b$ be real numbers and let S_n be the standardized random variable, i.e.,

$$S_n = \frac{X_n - np}{\sqrt{npq}}.$$

The distribution of X_n is

$$P(\{X_n = k\}) = \binom{n}{k} p^k q^{n-k}. \qquad 0 \leq k \leq n.$$

For any given $0 \leq k \leq n$ define

$$x_k = \frac{k - pn}{\sqrt{npq}}.$$

We shall investigate the probability $P(\{a < S_n < b\})$, which is equal to the sum of those $\binom{n}{k} p^k q^{n-k}$ for which $a < x_k = \dfrac{k - np}{\sqrt{npq}} < b$, i.e.,

$$P(\{a < S_n < b\}) = \sum_{a<x_k<b} \binom{n}{k} p^k q^{n-k} = \sum_{a<x_k<b} \frac{1}{\sqrt{npq}} \left(\sqrt{npq} \binom{n}{k} p^k q^{n-k}\right).$$

However, from Theorem 1 we know that $\sqrt{npq}\binom{n}{k} p^k q^{n-k}$ can be for large n replaced by

$\varphi(x_k) = \dfrac{1}{\sqrt{2\pi}} e^{-x_k^2/2}.$

Hence
$$P(\{a < S_n < b\}) \sim \sum_{a<x_k<b} \frac{1}{\sqrt{npq}}\varphi(x_k).$$

Let y_k be the midpoint between x_{k-1} and x_k; furthermore, let z_k be the midpoint between x_k and x_{k+1}. Then the distance between y_k and z_k is

$$|z_k - y_k| = \frac{x_{k+1} + x_k}{2} - \frac{x_k + x_{k-1}}{2}$$
$$= \frac{(k+1) - np + (k - np)}{2\sqrt{npq}} - \frac{k - np + ((k-1) - np)}{2\sqrt{npq}} = \frac{1}{\sqrt{npq}}.$$

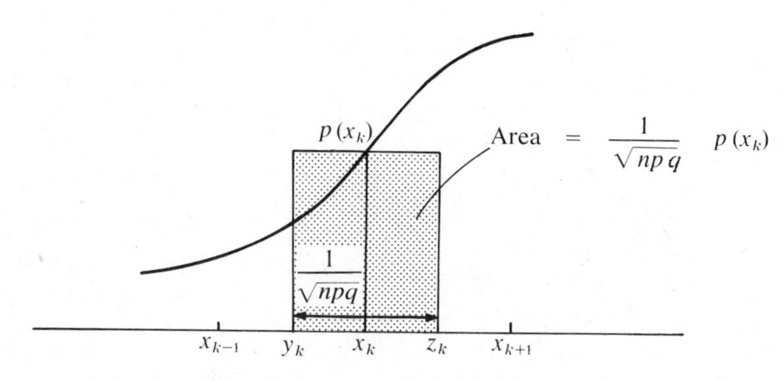

Therefore the expression $\frac{1}{\sqrt{npq}}\varphi(x_k)$ is equal to the area of the rectangle with the base (y_k, z_k) and with the height $\varphi(x_k)$. The expression

$$\sum_{a<x_k<b} \frac{1}{\sqrt{npq}}\varphi(x_k)$$

is thus equal to the area under the step function defined by division points x_k and by values of the normal curve $\varphi(x_k)$ as follows:

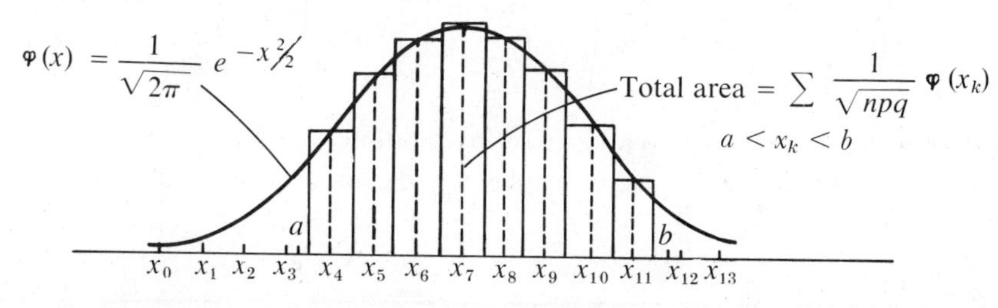

However, with increasing n we may expect that this area will be very close to the area $A(a,b)$ under the normal curve between a and b. The above discussion is an intuitive illustration for the validity of the next theorem.

THEOREM 2: *Integral DeMoivre-Laplace Limit Theorem.* For any two real numbers $a < b$ and for the binomial r.v. X_n with parameters n and p, we have

$$P\left(\left\{a < \frac{X_n - np}{\sqrt{npq}} < b\right\}\right) \sim A(a,b).$$

(The proof of this theorem is on page 358.)

Theorem 2 shows how we can approximate the distribution of the standardized r.v. $S_n = \frac{X_n - np}{\sqrt{npq}}$ by the normal curve. However, in practical situations we usually have to find the distribution of the binomial r.v. X_n. It can be done easily, also using Theorem 2. Let $0 \leq \alpha < \beta \leq n$ be two integers and let us investigate the probability $P(\{\alpha \leq X_n \leq \beta\})$. We can see that the two following events are identical:

$$\{\alpha \leq X_n \leq \beta\} = \left\{\alpha - \frac{1}{2} < X_n < \beta + \frac{1}{2}\right\}.$$

Hence,

$$\begin{aligned}
P(\{\alpha \leq X_n \leq \beta\}) &= P\left(\left\{\alpha - \frac{1}{2} < X_n < \beta + \frac{1}{2}\right\}\right) \\
&= P\left(\left\{\frac{\alpha - \frac{1}{2} - np}{\sqrt{npq}} < \frac{X_n - np}{\sqrt{npq}} < \frac{\beta + \frac{1}{2} - np}{\sqrt{npq}}\right\}\right) \\
&\sim A\left(\frac{\alpha - \frac{1}{2} - np}{\sqrt{npq}}, \frac{\beta + \frac{1}{2} - np}{\sqrt{npq}}\right) \\
&= \phi\left(\frac{\beta + \frac{1}{2} - np}{\sqrt{npq}}\right) - \phi\left(\frac{\alpha - \frac{1}{2} - np}{\sqrt{npq}}\right)
\end{aligned}$$

Thus, using Theorem 2, we have proved the following theorem.

→ **THEOREM 3:** Let X_n be a binomial r.v. with parameters n and p, and further let α, β be two integers, $0 \leq \alpha < \beta \leq n$. Then

$$P(\{\alpha \leq X_n \leq \beta\}) \sim \phi\left(\frac{\beta + \frac{1}{2} - np}{\sqrt{npq}}\right) - \phi\left(\frac{\alpha - \frac{1}{2} - np}{\sqrt{npq}}\right).$$

Theorem 3 shows a very simple way to approximate binomial probabilities by a normal curve. Let us calculate some numerical examples in order to see how good this approximation is.

EXAMPLE 4

Let us toss a coin n times, and let X_n be the number of heads. Calculate the probabilities

$$P(\{E(X_n) - \sigma(X_n) \leq X_n \leq E(X_n) + \sigma(X_n)\})$$

using the exact formula and normal approximation for $n = 4, 16, 36, 64, 100, 144, 196, 256, 324$. Compare both methods. (Refer also to Example 31, Chapter XII.)

SOLUTION. In this case $p = q = \frac{1}{2}$,

$$E(X_n) = \frac{n}{2}, \qquad \sigma(X_n) = \frac{\sqrt{n}}{2},$$

and we can put $\alpha = E(X_n) - \sigma(X_n) = \dfrac{n - \sqrt{n}}{2}$.

$$\beta = E(X_n) + \sigma(X_n) = \frac{n + \sqrt{n}}{2}.$$

The exact binomial probabilities are

$$P(\{\alpha \le X_n \le \beta\}) = \sum_{k=\alpha}^{\beta} \binom{n}{k} \frac{1}{2^n}.$$

Normal approximation is given as

$$\phi\left(\frac{\beta + \frac{1}{2} - np}{\sqrt{npq}}\right) - \phi\left(\frac{\alpha - \frac{1}{2} - np}{\sqrt{npq}}\right) = \phi\left(\frac{2\beta + 1 - n}{\sqrt{n}}\right) - \phi\left(\frac{2\alpha - 1 - n}{\sqrt{n}}\right)$$
$$= \phi\left(\frac{\sqrt{n} + 1}{\sqrt{n}}\right) - \phi\left(\frac{-\sqrt{n} - 1}{\sqrt{n}}\right).$$

			$P(\{E(X_n) - \sigma(X_n) \le X_n \le E(X_n) + \sigma(X_n)\})$	
n	$\alpha = E(X_n) - \sigma(X_n)$	$\beta = E(X_n) + \sigma(X_n)$	Exact value	Normal approximation
4	1	3	.8750	.8664
16	6	10	.7899	.7888
36	15	21	.7570	.7568
64	28	36	.7396	.7394
100	45	55	.7287	.7286
144	66	78	.7214	.7212
196	91	105	.7161	.7160
256	120	136	.7120	.7120
324	153	171	.7088	.7088

From the accompanying table, we can see that the normal approximation is already quite good for $16 \le n < 64$. For $64 \le n < 196$ it is very good, and for $n > 196$ it is excellent. However, the quality of approximation does not depend only on n, it depends on p as well.

EXAMPLE 5

Toss a fair die n times and find the probability that the number of 6's will be between $\left(\frac{n}{6} - 1\right)$ and $\left(\frac{n}{6} + 10\right)$ included for $n = 12, 36, 102, 144, 198, 258, 324$. Use the exact formula for the binomial distribution and the normal approximation, and compare them. Compare the results with Example 4 also.

SOLUTION. Let X_n be the number of 6's in n tosses. Let us put

$$\alpha = \frac{n}{6} - 1 \qquad \beta = \frac{n}{6} + 10.$$

The exact binomial probabilities are $\left(p = \frac{1}{6}, q = \frac{5}{6}\right)$

$$P(\{\alpha \leq X_n \leq \beta\}) = \sum_{k=\alpha}^{\beta} \binom{n}{k} \left(\frac{1}{6}\right)^k \left(\frac{5}{6}\right)^{n-k}$$

Normal approximation is given as

$$\phi\left(\frac{\beta + \frac{1}{2} - np}{\sqrt{npq}}\right) - \phi\left(\frac{\alpha - \frac{1}{2} - np}{\sqrt{npq}}\right) = \phi\left(\frac{\frac{n}{6} + 10 + \frac{1}{2} - \frac{n}{6}}{\sqrt{n \cdot \frac{1}{6} \cdot \frac{5}{6}}}\right) - \phi\left(\frac{\frac{n}{6} - 1 - \frac{1}{2} - \frac{n}{6}}{\sqrt{n \cdot \frac{1}{6} \cdot \frac{5}{6}}}\right)$$

$$= \phi\left(\frac{63}{\sqrt{5n}}\right) - \phi\left(\frac{-9}{\sqrt{5n}}\right).$$

			$P(\{\alpha \leq X_n \leq \beta\})$	
n	$\alpha = \frac{n}{6} - 1$	$\beta = \frac{n}{6} + 10$	Exact value	Normal approximation
12	1	12	.8878	.8774
36	5	16	.7393	.7488
60	9	20	.6875	.6985
102	16	27	.6412	.6521
144	23	34	.6108	.6218
198	32	43	.5793	.5897
258	42	53	.5494	.5594
324	53	64	.5210	.5299

We can see that in this case, the approximation is fairly good but considerably worse than that in Example 4. It has been empirically found that the quality of approximation depends on the product np. For $5 < np \leq 10$ the approximation is fair, for $np > 10$ it is usually good.

EXAMPLE 6

Determination of Sample Size. Let X_n be a binomial r.v. with parameters n and p. In Chapter XIII on the Law of Large Numbers, we proved that if n is increasing then the fraction $\frac{X_n}{n}$ approaches in a sense the probability p. Imagine now that p is unknown. How many times should we perform the experiment so that we have a large probability, say at least .95, that $\frac{X_n}{n}$ will differ from the true (but unknown) value of p by less than some given small number ϵ, where for example $\epsilon = .01$ or $.001$, etc.?

We require that
$$P\left(\left\{\left|\frac{X_n}{n} - p\right| < \epsilon\right\}\right) \geq .95.$$

However,

$$P\left(\left\{\left|\frac{X_n}{n} - p\right| < \epsilon\right\}\right) = P\left(\left\{-\epsilon < \frac{X_n}{n} - p < \epsilon\right\}\right) = P\left(\left\{-\epsilon < \sqrt{\frac{pq}{n}}\left(\frac{X_n - np}{\sqrt{npq}}\right) < \epsilon\right\}\right)$$

$$= P\left(\left\{-\epsilon\sqrt{\frac{n}{pq}} < \frac{X_n - np}{\sqrt{npq}} < \epsilon\sqrt{\frac{n}{pq}}\right\}\right) \sim \phi\left(\epsilon\sqrt{\frac{n}{pq}}\right) - \phi\left(-\epsilon\sqrt{\frac{n}{pq}}\right)$$

as follows from Theorem 2. Therefore, our requirement can be written as

$$P\left(\left\{\left|\frac{X_n}{n}-p\right|<\epsilon\right\}\right)\sim\phi\left(\epsilon\sqrt{\frac{n}{pq}}\right)-\phi\left(-\epsilon\sqrt{\frac{n}{pq}}\right)\geq .95.$$

However, from the table on page 360, we can see that $\phi(1.96)-\phi(-1.96)=.95$. Therefore, our requirement will be satisfied if

$$\epsilon\sqrt{\frac{n}{pq}}\geq 1.96$$

or

$$n\geq\left(\frac{1.96}{\epsilon}\right)^2 pq.$$

If we have some approximate value of p, we can use it here to get a determination of the necessary sample size. However, if p (and hence q) is completely unknown, then we can use the fact that

$$pq\leq\frac{1}{4}$$

as shown on page 224. Then our requirement will be satisfied if

$$n\geq\frac{1}{4}\left(\frac{1.96}{\epsilon}\right)^2.$$

It should be clear that the "high probability" .95 led to the choice of 1.96. Should one stipulate another high probability, say .99, then we can satisfy the requirement if we replace 1.96 by a suitably chosen number.

EXAMPLE 7

How many times should one toss a fair coin so that with probability at least .95 the relative frequency of heads will differ from $\frac{1}{2}$ by less than ϵ, where $\epsilon = .2, .1, .05, .01, .005, .001$?

SOLUTION. According to Example 6, the necessary number of tosses is

$$n\geq\left(\frac{1.96}{\epsilon}\right)^2 pq=\frac{1}{4}\left(\frac{1.96}{\epsilon}\right)^2.$$

Following is the table of the number of tosses:

ϵ	.2	.1	.05	.01	.005	.001
$n\geq$	25	97	385	9604	38416	960400

We can see that even a small improvement in the accuracy of the estimation ϵ will result in a very high increase in the number of necessary tosses.

EXAMPLE 8

Skunks in a certain area suffer from heavy infestations of lungworms. During the summer months, usually up to one third of the male population is diseased. How many animals should the regional wildlife biologist check this summer if he wants the sample proportion of tested diseased animals to be within $\epsilon = .05$ of the actual unknown population proportion p with at least a 90% probability? (That is, the unknown proportion should differ from the sample proportion by less than .05.)

SOLUTION. If X_n is the number of diseased animals in a sample of n male skunks, and if the unknown actual proportion of infested males in the population is p, then we require (as in Example 6) that

$$P\left(\left\{\left|\frac{X_n}{n} - p\right| < .05\right\}\right) \sim \phi\left(.05\sqrt{\frac{n}{pq}}\right) - \phi\left(-.05\sqrt{\frac{n}{pq}}\right) \geq .9.$$

However from the table on page 360, we get

$$\phi(1.645) - \phi(1.645) = .9.$$

Hence the biologist's requirement will be satisfied if

$$0.5\sqrt{\frac{n}{pq}} \geq 1.645$$

or

$$n \geq \left(\frac{1.645}{.05}\right)^2 pq.$$

We do not know the actual value of p, but from past experience it is known that $p \leq \frac{1}{3}$, hence $pq \leq \frac{1}{3} \cdot \frac{2}{3} = \frac{2}{9}$.

Thus we get

$$n \geq \left(\frac{1.645}{.05}\right)^2 \cdot \frac{2}{9} = 240.5.$$

The wildlife biologist should test at least 241 (independently chosen) male skunks in order to be at least 90% sure that the unknown proportion will be within .05 of the sample proportion of infested animals.

EXAMPLE 9

Test of Loaded Dice. A gambler suspects that a certain casino die has been loaded. He complains to the casino owner and they argue. The casino owner claims that the die is perfect, hence the probability of a 6 for each toss is $p = \frac{1}{6}$. However, the gambler has an alternative claim that the die is loaded, hence $p \neq \frac{1}{6}$. They decide to toss the die $n = 120$ times and count the number of 6's, which they denote by X. Should X be unreasonably high or unreasonably low, they agree that it is evidence against the owner's claim $p = \frac{1}{6}$. However, they know that even if the die is perfect and $p = \frac{1}{6}$, they can still get a very high or very low number of 6's, which could lead to rejection of the owner's claim and thus to a wrong decision. In order to insure themselves against this error, they would like to have a very small probability of such a false decision, say at most $\alpha = .05$. They would like to find two numbers, say $x_1 < x_2$, so that if $X < x_1$ or $X > x_2$, they will reject the owner's claim $p = \frac{1}{6}$. At the same time, they want:

$$P\left(\left\{\text{reject owner's claim } p = \frac{1}{6} \text{ if actually } p = \frac{1}{6}\right\}\right)$$

$$= P(\{\text{error}\}) = P\left(\left\{X < x_1 \text{ or } x > x_2 \text{ if } p = \frac{1}{6}\right\}\right) \leq .05.$$

Using Theorem 3, they can get

$$.05 \geq P\left(\left\{X < x_1 \text{ or } X > x_2 \text{ if } p = \frac{1}{6}\right\}\right) = 1 - P\left(\left\{x_1 \leq X \leq x_2 \text{ if } p = \frac{1}{6}\right\}\right)$$

$$\sim 1 - \left[\phi\left(\frac{x_2 + \frac{1}{2} - np}{\sqrt{npq}}\right) - \phi\left(\frac{x_1 - \frac{1}{2} - np}{\sqrt{npq}}\right)\right] \text{ where } p = \frac{1}{6}.$$

Hence the requirement is

$$\phi\left(\frac{x_2 + \frac{1}{2} - \frac{n}{6}}{\sqrt{\frac{5n}{36}}}\right) - \phi\left(\frac{x_1 - \frac{1}{2} - \frac{n}{6}}{\sqrt{\frac{5n}{36}}}\right) \geq .95.$$

From the table on page 360, we get

$$\phi(1.96) - \phi(-1.96) = .95$$

hence we can put

$$\frac{6\left(x_2 + \frac{1}{2} - \frac{n}{6}\right)}{\sqrt{5n}} = 1.96. \qquad \frac{6\left(x_1 - \frac{1}{2} - \frac{n}{6}\right)}{\sqrt{5n}} = -1.96.$$

We know that $n = 120$ and we can solve the equations for x_1 and x_2.

$$x_2 = 19.5 + \frac{1.96\sqrt{600}}{6} = 27.5.$$

$$x_1 = 20.5 - \frac{1.96\sqrt{600}}{6} = 12.5.$$

Therefore, if 120 tosses of the die result in 12 or less 6's or in 28 or more 6's, then the owner's claim should be rejected in favor of the gambler's claim. The probability of a false decision in this case is at most .05.

EXAMPLE 10

Hypothesis Testing. In the previous example, we derived a method to test a certain type of claim concerning the parameter p of the binomial distribution. Let us rephrase the problem in general terms. An experiment can result in a success with an unknown probability p or in a failure with an unknown probability $q = 1 - p$. Two people argue about the actual value of p. One claims that p is equal to some particular value p_0 $\left(\text{say } p = \frac{1}{6} \text{ or } p = \frac{1}{2}, \text{ etc.}\right)$. This claim is called the *null hypothesis* and is usually denoted as

$$H_0 : p = p_0.$$

The other claims that p is not equal to p_0 $\left(\text{say } p \neq \frac{1}{6} \text{ or } p \neq \frac{1}{2} \text{ etc.}\right)$. This claim is called the *alternative hypothesis* and is usually denoted as

$$H_A : p \neq p_0.$$

In order to settle this argument, both sides agree to perform n independent repeated trials, and count the number of successes X. A significantly low or significantly high value of X is considered to be evidence against H_0 and in such a case H_0 will be rejected in favor of H_A. However, even if H_0 is true, one can get a very low or a very high value of X, and thus reject H_0. This type of false decision is called a *Type* I *error.* Both sides agree that this should happen only with a very low probability α (say $\alpha = .05$ or $\alpha = .01$, etc.). The number α is called the *probability of Type* I *error* or the *significance level.*

We would like to find two numbers $x_1 < x_2$ so that we can reject H_0 if $X < x_1$ or $X > x_2$, and at the same time

$$\begin{aligned} P(\{\text{reject } H_0 \text{ if } H_0 \text{ is true}\}) &= P(\{\text{Type I error}\}) \\ &= P(\{X < x_1 \quad \text{or} \quad X > x_2 \quad \text{if } H_0 \text{ is true}) \\ &= 1 - P(\{x_1 \leq X \leq x_2 \quad \text{if } H_0 \text{ is true}\}) \leq \alpha \end{aligned}$$

or equivalently $$P(\{x_1 \leq X \leq x_2 \quad \text{if } H_0 \text{ is true}\}) \geq 1 - \alpha.$$

However, if H_0 is true then $p = p_0$. Hence by Theorem 3,

$$P(\{x_1 \leq X \leq x_2 \quad \text{if } H_0 \text{ is true}\}) \sim \phi\left(\frac{x_2 + \frac{1}{2} - np_0}{\sqrt{np_0q_0}}\right) - \phi\left(\frac{x_1 - \frac{1}{2} - np_0}{\sqrt{np_0q_0}}\right) \geq 1 - \alpha$$

where $q_0 = 1 - p_0$.

For any given number $0 < \alpha < 1$, we can find a number z_α from the table on page 360, so that

$$\phi(z_\alpha) = 1 - \alpha$$

or equivalently $$\phi(z_{\alpha/2}) = 1 - \frac{\alpha}{2}.$$

In this case

$$\begin{aligned} A(-z_{\alpha/2}, z_{\alpha/2}) = \phi(z_{\alpha/2}) - \phi(-z_{\alpha/2}) &= 1 - \frac{\alpha}{2} - (1 - \phi(z_{\alpha/2})) \\ &= 1 - \frac{\alpha}{2} - \left(1 - \left(1 - \frac{\alpha}{2}\right)\right) = 1 - \alpha. \end{aligned}$$

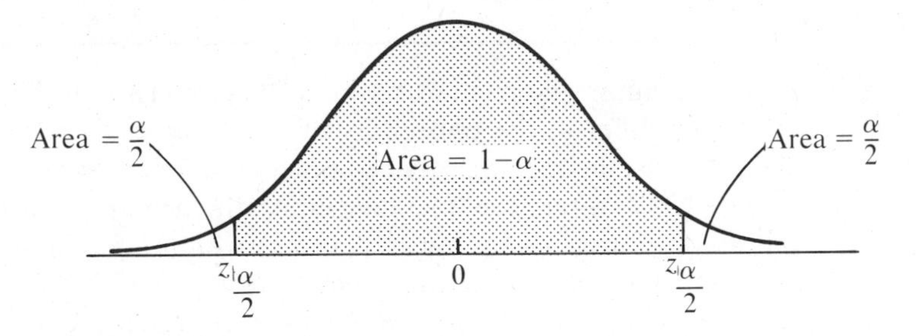

For example, we can find the following values

α	.1	.05	.02	.01
$z_{\alpha/2}$	1.645	1.96	2.33	2.575

It is clear that if we put

$$\frac{x_2 + \frac{1}{2} - np_0q_0}{\sqrt{np_0q_0}} = z_{\alpha/2} \qquad \frac{x_1 - \frac{1}{2} - np_0}{\sqrt{np_0q_0}} = -z_{\alpha/2}$$

then our requirements will be satisfied. Solving for x_1 and x_2, we get

$$x_1 = np_0 + \frac{1}{2} - z_{\alpha/2}\sqrt{np_0q_0}, \qquad x_2 = np_0 - \frac{1}{2} + z_{\alpha/2}\sqrt{np_0q_0}.$$

We can summarize the whole procedure as follows:

$$H_0: p = p_0$$
$$H_A: p \neq p_0$$

Do not reject H_0 if $np_0 + \frac{1}{2} - z_{\alpha/2}\sqrt{np_0q_0} \leq X \leq np_0 - \frac{1}{2} + z_{\alpha/2}\sqrt{np_0q_0}$, otherwise reject H_0.

Then $P(\{\text{Type I error}\}) = \alpha$.

We can see that Type I error is an intrinsic component of our testing procedure. However, we have to realize that there could also be another type of error. It could easily happen that $p \neq p_0$, hence H_0 is false. But we could still have the number of successes satisfying the inequality $x_1 \leq X \leq x_2$, and thus we would have to accept (not reject) H_0. This error is called *Type* II *error.*

	H_0 true	H_0 false
Accept H_0	correct	Type II error
Reject H_0	Type I error	correct

Type II error and its probability depend on the actual value of p.

EXAMPLE 11

An air rifle pellet manufacturer markets three types of pellets. They all have the same physical dimensions, but they differ in the uniformity of these dimensions, hence in their quality. It is required that 80% of the medium quality pellets be within the tolerance limits. If a higher proportion of pellets is within the tolerance limits, the batch is sold as expensive precise target ammunition. If a lower proportion of pellets is within the limits, the batch is sold as cheap ammunition for plinking. The testing laboratory measured 200 pellets selected at random from a very large shipment, and 142 were within the tolerance limits. Test the claim that the shipment consists of medium quality pellets at the significance level $\alpha = .05$.

SOLUTION. According to the procedure derived in Example 10, we have

$$H_0: p = .8$$
$$H_A: p \neq .8$$

where p denotes the actual (unknown) proportion of pellets within the tolerance limits. For $\alpha = .05$, we have $z_{\alpha/2} = 1.96$.

$$x_1 = np_0 + \frac{1}{2} - z_{\alpha/2}\sqrt{np_0q_0} = 200 \cdot (.8) + \frac{1}{2} - 1.96\sqrt{200 \cdot (.8) \cdot (.2)}$$
$$= 160.5 - 11.0874 = 149.4126.$$
$$x_2 = np_0 - \frac{1}{2} + z_{\alpha/2}\sqrt{np_0q_0} = 200 \cdot (.8) - \frac{1}{2} + 1.96\sqrt{200 \cdot (.8) \cdot (.2)}$$
$$= 159.5 + 11.0874 = 170.5874.$$

Therefore the claim of medium quality should be rejected if in a sample of 200 pellets there are more than 170 or less than 150 pellets within the tolerance limits. Because there are 142 pellets within the limits, we have to reject the hypothesis of medium quality.

EXAMPLE 12

Imagine that in Example 11 the actual proportion of pellets within the tolerance limits is 90%. Find the probability of Type II error.

SOLUTION. $P(\{\text{Type II error}\}) = P(\{\text{Accept } H_0: p = .8 \text{ if actually } p = .9\}) = P(\{150 \leq X \leq 170 \text{ if } p = .9\})$ where X is the number of pellets within the limits. Using Theorem 3, we have

$$P(\{\text{Type II error}\}) \sim \phi\left(\frac{170 + \frac{1}{2} - 200 \cdot (.9)}{\sqrt{200 \cdot (.9) \cdot (.1)}}\right) - \phi\left(\frac{150 - \frac{1}{2} - 200 \cdot (.9)}{\sqrt{200 \cdot (.9) \cdot (.1)}}\right)$$
$$= \phi(-2.24) - \phi(7.19) = .0126 - 0 = .0126.$$

Therefore, the probability of Type II error is only about 1.26%.

EXAMPLE 13

Sign Test. A buyer for a retail store chain can get supplies from two different wholesalers W_1 and W_2. In order to get the best deal, he has to sign an exclusive contract with one of the wholesalers. They both offer the same selection of goods, but their prices are usually different.

Each wholesaler claims that his prices are best and that he cannot be "undersold." The buyer doubts these claims because both cannot be best. He thinks that actually both W_1 and W_2 offer equally good deals because the differences in their prices are usually very small. To verify his own "claim," he selects 50 items at random and compares their prices. He finds that on 19 of these 50 items, W_1 offers a better price while on the rest he can get a better deal from W_2. Does this mean that there is an actual difference between the wholesalers?

SOLUTION. Let us assume that the proportion of items in the whole catalog for which the buyer can get a better deal from W_1 is p (its value could perhaps be found exactly, but it is nearly impossible to get quotations on all possible items from both wholesalers). The buyer's claim can be expressed as $p = \frac{1}{2}$ (he thinks that he can get a better price on about 50% of items from W_1 and on the rest from W_2). The joint claim of the wholesalers is $p \neq \frac{1}{2}$. Let us assume that the buyer would be willing to reject his own claim if it is actually true with a probability .05 at most.

Using the method of Example 10, we get

$$H_0: p = .5$$

$$H_A: p \neq .5 \qquad \alpha = .05$$

$$x_1 = 50 \cdot \frac{1}{2} + \frac{1}{2} - 1.96\sqrt{50 \cdot \frac{1}{2} \cdot \frac{1}{2}} = 18.6$$

$$x_2 = 50 \cdot \frac{1}{2} - \frac{1}{2} + 1.96\sqrt{50 \cdot \frac{1}{2} \cdot \frac{1}{2}} = 31.4.$$

The buyer should reject his claim if W_1 offers a better price on 18 or less items or 32 or more items out of 50. Since he got better prices on 19 items from W_1, he *should not reject* his claim that both wholesalers offer equally good deals. Nevertheless, he may want to do some more detailed investigation of the price structure.

COMMENT. This method is called a sign test because the buyer has taken into consideration only the signs of the differences in prices ("who offers a better deal"), and not the actual values of the differences. The investigation of the differences can be done in a similar way but the necessary probabilistic techniques would lead us beyond the scope of this text.

Proof of Theorem 1: We shall use *Stirling's formula* which is derived in calculus courses.

$$n! \sim \left(\frac{n}{e}\right)^n \sqrt{2\pi n}.$$

Then

$$\binom{n}{k} p^k q^{n-k} \sim \frac{\left(\frac{n}{e}\right)^n \sqrt{2\pi n}\, p^k q^{n-k}}{\left(\frac{k}{e}\right)^k \sqrt{2\pi k}\left(\frac{n-k}{e}\right)^{n-k} \sqrt{2\pi(n-k)}} = \sqrt{\frac{n}{2\pi k(n-k)}} \quad \left(\frac{np}{k}\right)^k \left(\frac{nq}{n-k}\right)^{n-k}$$

We have defined x as follows:

$$x = \frac{k - np}{\sqrt{npq}} \qquad 0 \leq k \leq n$$

hence
$$k = np + x\sqrt{npq} \qquad n - k = nq - x\sqrt{npq}.$$

Therefore we can see that

$$\lim_{n\to\infty} \frac{k}{np} = \lim_{n\to\infty}\left(1 + \frac{x\sqrt{npq}}{np}\right) = 1$$

thus $k \sim np$.

Similarly, we get $n - k \sim nq$.

Thus we have

$$\binom{n}{k} p^k q^{n-k} \sim \sqrt{\frac{n}{2\pi k(n-k)}}\left(\frac{np}{k}\right)^k\left(\frac{nq}{n-k}\right)^{n-k} \sim \frac{1}{\sqrt{2\pi npq}}\left(\frac{np}{k}\right)^k\left(\frac{nq}{n-k}\right)^{n-k}.$$

If we introduce the following notation:

$$f(n,k) = \left(\frac{np}{k}\right)^k\left(\frac{nq}{n-k}\right)^{n-k}$$

then we have
$$\binom{n}{k} p^k q^{n-k} \sim \frac{1}{\sqrt{2\pi npq}} f(n,k).$$

Let us now use the Taylor series:

$$\log(1 + t) = t - \frac{t^2}{2} + \frac{t^3}{3} + \ldots + (-1)^{n-1}\frac{t^n}{n} \ldots \quad |t| < 1$$

for the expansion of $\log\left(\frac{np}{k}\right)^k$:

$$\log\left(\frac{np}{k}\right)^k = k\log\left(1 - \frac{x\sqrt{npq}}{k}\right) = k\left(-\frac{x\sqrt{npq}}{k} - \frac{x^2npq}{2k^2} - \ldots\ldots\ldots\right).$$

Similarly,

$$\log\left(\frac{nq}{n-k}\right)^{n-k} = (n-k)\log\left(1 + \frac{x\sqrt{npq}}{n-k}\right) = (n-k)\left(\frac{x\sqrt{npq}}{n-k} - \frac{x^2npq}{2(n-k)^2} + \ldots\right).$$

This is true for $\left|\frac{x\sqrt{npq}}{k}\right| < 1$ and $\left|\frac{x\sqrt{npq}}{n-k}\right| < 1$ but these inequalities are satisfied for large enough n. Hence

$$\log f(n,k) = \log\left(\frac{np}{k}\right)^k + \log\left(\frac{nq}{n-k}\right)^{n-k} \sim -\frac{x^2npq}{2k} - \frac{x^2npq}{2(n-k)} = -\frac{x^2n^2pq}{2k(n-k)} \sim -\frac{x^2}{2}.$$

Therefore,
$$f(n,k) \sim e^{-x^2/2}$$

hence
$$\sqrt{npq}\binom{n}{k} p^k q^{n-k} \sim \frac{1}{\sqrt{2\pi}} e^{-x^2/2}.$$
q.e.d.

Proof of Theorem 2: We have $S_n = \dfrac{X_n - np}{\sqrt{npq}}$ and

$$x_k = \frac{k - np}{\sqrt{npq}} \quad \text{for} \quad 0 \leq k \leq n.$$

We have shown on page 397 that

$$P(\{a < S_n < b\}) = \sum_{a<x_k<b} \binom{n}{k} p^k q^{n-k} = \sum_{a<x_k<b} \frac{1}{\sqrt{npq}}\left(\sqrt{npq}\binom{n}{k} p^k q^{n-k}\right) \sim \sum_{a<x_k<b} \frac{1}{\sqrt{npq}} \varphi(x_k).$$

Furthermore
$$x_{k+1} - x_k = \frac{k+1-np}{\sqrt{npq}} - \frac{k-np}{\sqrt{npq}} = \frac{1}{\sqrt{npq}}$$

hence
$$P(\{a < S_n < b\}) \sim \sum_{a<x_k<b} \varphi(x_k)(x_{k+1} - x_k)$$

where
$$\varphi(x) = \frac{1}{\sqrt{2\pi}} e^{-x^2/2}.$$

But the right-hand sum is a Riemann sum for the definite integral $\int_a^b \varphi(x)\,dx$ (except, perhaps, some terms corresponding to the intervals around a and b which certainly converge to 0; see the picture on page 347). However, we have defined $A(a,b)$ as the area from a to b under the graph of $\varphi(x)$.

Hence
$$\int_a^b \varphi(x)\,dx = A(a,b).$$

Therefore
$$P(\{a < S_n < b\}) \sim A(a,b).$$
q.e.d.

EXERCISES

(Answers in Appendix to exercises marked)*

*1. Using the local limit theorem, find the probability that 225 tosses of a fair coin will yield 100 heads.

2. Using the local limit theorem, find the probability that 300 tosses of a fair coin will yield 140 heads.

*3. Using the local limit theorem, find the probability that 256 tosses of a fair die will yield 50 6's.

4. Using the local limit theorem, find the probability that 400 tosses of a fair die will yield 65 1's.

5. Find the following areas under the normal curve:
   *(a) $A(-\infty,1.76)$ (b) $A(-\infty,2.13)$
   *(c) $A(-\infty,-1.26)$ (d) $A(-\infty,-1.82)$
   *(e) $A(0.3,1.56)$ (f) $A(1,2)$
   *(g) $A(-2,1)$ (h) $A(-1,1)$

*6. Find the probability of getting at least 120 but less than 135 heads when tossing a fair coin 225 times.

7. Find the probability of getting between 40 and 60 6's (40 and 60 included) when tossing a fair die 300 times.

8. The probability that a student-driver will pass the road test on the first trial is estimated to be .6. Find the probability that, among 100 such drivers, at least 50 but no more than 65 will pass the road test.

*9. The probability that an item will be defective is .002. Find the probability that a shipment of 10,000 items will contain at most 25 defective items.

*10. How many times do you have to toss a fair coin so that with the probability at least 90% the relative frequency of tails will differ from $\frac{1}{2}$ by less than .01?

11. You do not know whether the coin is fair or not. How many times should you toss it so that with probability at least 95% the relative frequency of heads will not differ from the actual (unknown) probability of heads by more than .02?

12. How many times should you toss a fair die so that with probability at least 90% the relative frequency of 6's will differ from $\frac{1}{6}$ by at most .025?

13. A certain proportion of ball bearings produced by a machine is undersized. How many ball bearings should you test so that you can be 90% sure that the proportion of tested undersized balls will not differ more than .05 from the actual proportion?

*14. A bachelor has three dark brown and five black (single) socks in a drawer. Each morning he chooses a pair of socks at random, each evening he washes the socks and throws them back into the drawer. Find the probability that during one year (365 days) he will be wearing matching socks at least half of the time.

15. To verify whether or not a coin is fair, toss it 100 times and count the number of heads. Then test $P(H) = .5$ at the level of significance $\alpha = .05$. It might be interesting to use an old United States dime, consisting of several layers of different materials.

APPENDIX I

TABLE: AREAS UNDER THE NORMAL CURVE

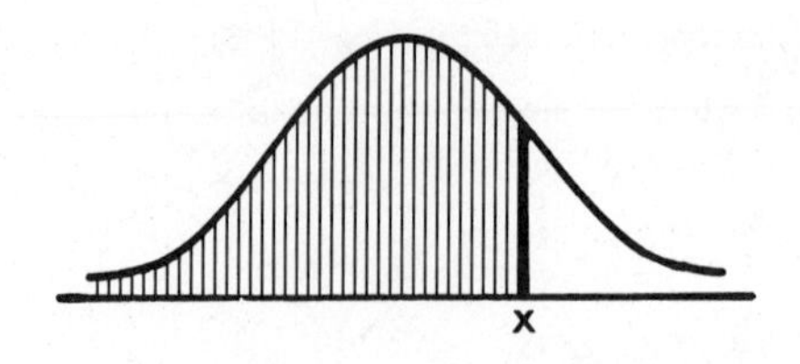

X	0	1	2	3	4	5	6	7	8	9
.0	.5000	.5040	.5080	. 5120	.5160	.5199	.5239	.5279	.5319	.5359
.1	.5398	.5438	.5478	.5517	.5557	.5596	.5636	.5675	.5714	.5753
.2	.5793	.5832	.5871	.5910	.5948	.5987	.6026	.6064	.6103	.6141
.3	.6179	.6217	.6255	.6293	.6331	.6368	.6406	.6443	.6480	.6517
.4	.6554	.6591	.6628	.6664	.6700	.6736	.6772	.6808	.6844	.6879
.5	.6915	.6950	.6985	.7019	.7054	.7088	.7123	.7157	.7190	.7224
.6	.7257	.7291	.7324	.7357	.7389	.7422	.7454	.7486	.7517	.7549
.7	.7580	.7611	.7642	.7673	.7703	.7734	.7764	.7794	.7823	.7852
.8	.7881	.7910	.7939	.7967	.7995	.8023	.8051	.8078	.8106	.8133
.9	.8159	.8186	.8212	.8238	.8264	.8289	.8315	.8340	.8365	.8389
1.0	.8413	.8438	.8461	.8485	.8508	.8531	.8554	.8577	.8599	.8621
1.1	.8643	.8665	.8686	.8708	.8729	.8749	.8770	.8790	.8810	.8830
1.2	.8849	.8869	.8888	.8907	.8925	.8944	.8962	.8980	.8997	.9015
1.3	.9032	.9049	.9066	.9082	.9099	.9115	.9131	.9147	.9162	.9177
1.4	.9192	.9207	.9222	.9236	.9251	. 9265	.9278	.9292	.9306	.9319
1.5	.9332	.9345	.9357	.9370	.9382	.9394	.9406	.9418	.9430	.9441
1.6	.9452	.9463	.9474	.9484	.9495	.9505	.9515	.9525	.9535	.9545
1.7	.9554	.9564	.9573	.9582	.9591	.9599	.9608	.9616	.9625	.9633
1.8	.9641	.9648	.9656	.9664	.9671	.9678	.9686	.9693	.9700	.9706
1.9	.9713	.9719	.9726	.9732	.9738	.9744	.9750	.9756	.9762	.9767
2.0	.9772	.9778	.9783	.9788	.9793	.9798	.9803	.9808	.9812	.9817
2.1	.9821	.9826	.9830	.9834	.9838	.9842	.9846	.9850	.9854	.9857
2.2	.9861	.9864	.9868	.9871	.9874	.9878	.9881	.9884	.9887	.9890
2.3	.9893	.9896	.9898	.9901	.9904	.9906	.9909	.9911	.9913	.9916
2.4	.9918	.9920	.9922	.9925	.9927	.9929	.9931	.9932	.9934	.9936
2.5	.9938	.9940	.9941	.9943	.9945	.9946	.9948	.9949	.9951	.9952
2.6	.9953	.9955	.9956	.9957	.9959	.9960	.9961	.9962	.9963	.9964
2.7	.9965	.9966	.9967	.9968	.9969	.9970	.9971	.9972	.9973	.9974
2.8	.9974	.9975	.9976	.9977	.9977	.9978	.9979	.9979	.9980	.9981
2.9	.9981	.9982	.9982	.9983	.9984	.9984	.9985	.9985	.9986	.9986
3.	.9987	.9990	.9993	.9995	.9997	.9998	.9998	.9999	.9999	1.0000

APPENDIX II

ANSWERS TO SELECTED EXERCISES

CHAPTER II

2. (a) $\{1\}$
   (b) $\{3,5,7,9,11\}$
   (c) $\{-2,3\}$
   (d) ϕ
   (e) $\{i\pi: i \text{ is an integer}\}$
12. (a) Yes
    (b) No
    (c) Yes
15. 2^{mn}

CHAPTER III

4. (a) $A \cap B$
   (b) $A \cup B$
   (c) $A - B$
   (d) $A \Delta B$
   (e) moose has not been hit
7. (a) $A \cap B \cap C$
   (b) $A' \cap B' \cap C'$
   (c) $A \cap (B \cup C)$
   (d) $A \Delta B \Delta C$
   A majority (300) answered A—Yes, B—No, C—No.

CHAPTER V

2. (a) $p_1 = .1,\ p_3 = .3.$
   (b) $p_1 = \frac{3}{7}, p_2 = \frac{2}{7}, p_3 = \frac{1}{7}, p_4 = \frac{1}{7}.$
3. (a) $P(A' \cap B) = \frac{1}{6}, P(A \cap B') = \frac{2}{6}.$
   $P(A \cup B) = \frac{4}{6}, P(A' \cap B') = \frac{2}{6}.$
4. (b) $P(A \cap C) = .2.$
7. $P(A) = \frac{14}{15}, P(B) = \frac{10}{15}, P(A \cap B) = \frac{9}{15}.$
9. (a) .25 (b) .05 (c) .2 (d) .08
11. $P(A) = .6, P(B) = .3, P(C) = .1.$

CHAPTER VII

2. (a) 224,640,000
   (b) 15,210,000
4. $\frac{6}{18}$

8. $\frac{5}{18}$

12. (a) $\frac{1}{14}$ (b) $\frac{1}{35}$

15. (a) $\frac{1}{6}$ (b) $\frac{7}{66}$ (c) $\frac{15}{22}$

18. $\frac{3}{11}$

21. (a) 286 (b) 165 (c) 80

23. $\frac{12}{27}$

25. $p_1 = .0556$, $p_2 = .0025$, $p_3 = .85 \times 10^{-4}$, $p_4 = 2 \times 10^{-6}$, $p_5 = 2 \times 10^{-8}$.

26. .9721

28. (a) $\binom{n}{2}$ (b) $\binom{n}{2} - n$

30. (a) $\frac{40}{143}$ (b) $\frac{9}{143}$

33. (a) $\frac{25}{323}$ (b) .3225

37. 0 if k is odd, $\binom{k}{\frac{k}{2}} 2^{-k}$ if k is even.

39. (a) .2 (b) .4

40. $\frac{3}{11}$

42. (a) $\frac{(n-2)^{r-1}}{(n-1)^{r-1}}$ (b) $\frac{(n-1)!}{(n-1)^r(n-1-r)!}$

CHAPTER IX

1. .2927

3. The tout's.

6. (a) $\frac{15}{52}$ (b) $\frac{55}{156}$ (c) $\frac{14}{39}$

10. .0826

14. (a) .33 (b) $.\overline{27}$ (c) Sea bass (d) .07236

16. .5

19. $\frac{1}{3}$

21. .1471

24. $\frac{3}{15}$

28. $n \geq 5$

31. $P(2 \text{ engine}) = \frac{600}{625}$; $P(4 \text{ engine}) = \frac{608}{625}$.

36. $P(\text{one hit}) = \frac{7}{18}$; $P(B/\text{one hit}) = \frac{2}{7}$.

41. .618976
46. .089265
48. .1333
59. .00329

CHAPTER X

3. (a) .5 (b) .25 (c) 0 (d) .5 (e) .5
5. (a) .02578 (b) .29923 (c) .00706

8.

X_1 \ X_2	0	1	2	3		
0	1	3	3	1	8	out
1	3	6	3	0	12	of
2	3	3	0	0	6	27
3	1	0	0	0	1	
	8	12	6	1		

10.

X \ Y	0	1	2		
0	6	4	0	10	out
1	12	11	2	25	of
2	3	6	1	10	45
	21	21	3		

12.

X \ Y	0	1	2	
1	0	.1	.1	.2
2	.1	.3	0	.4
3	0	.3	.1	.4
	.1	.7	.2	1

X, Y are not independent.

17. (a) .6626 (b) .0877 (c) .6626

CHAPTER XI

1. .8059
3. .3963
5. .3504
8. (a) .0074 (b) .0093
11. (a) .2306 (b) .1419
13. (a) .0019 (b) .0005 (c) .0024
16. He is fair, $P = .0853$.
19.

i	0	1	2	3
$f(i)$	.2970	.5030	.1869	.0136

21. .2211

25.

i	0	1	2	3	4
$f(i)$	$\frac{1}{35}$	$\frac{3}{35}$	$\frac{6}{35}$	$\frac{10}{35}$	$\frac{15}{35}$

CHAPTER XII

1. $P = \frac{32}{63}$. $E(X) = -2$.
2. $E(X) = 1.96$. $V(X) = .7584$.
6. $E(X) = 6\frac{1}{2}$¢ per board; it pays off.
8. $E(X) = 1.4444$. $V(X) = .4691$.
11. Jackpot \$29.14
14. \$.315
15. $p_1 + p_2 + p_3$
16. (a) $E = 4$, $V = 1$. (e) 7
20. (a) $V(T) = 16$. $V(U) = 4$. (c) 2
25. $E(X) = 1.5$. $V(X) = \frac{15}{28}$.
27. $E(X) = 35$. $V(X) = 29.167$.
30. $E(X) = .9$. $V(X) = .6850$.
35. $E = 7.2$.
38. $n = 25$, $p = .4$.
40. $E = 3$, $V = 2.1176$.

CHAPTER XIII

3. $P \geq \frac{24}{25}$
4. $LB = 98.36$, $UB = 201.64$.
6. $n_1 \geq 125$, $n_2 \geq 500$.

CHAPTER XIV

2. Denote ω_1 = late, ω_2 = on time.

(a) $$\boldsymbol{P} = \begin{pmatrix} \frac{1}{3} & \frac{2}{3} \\ \frac{1}{2} & \frac{1}{2} \end{pmatrix} \quad p = \left(\frac{3}{4}, \frac{1}{4}\right).$$

(b) $$\boldsymbol{P}^{(2)} = \begin{pmatrix} \frac{4}{9} & \frac{5}{9} \\ \frac{5}{12} & \frac{7}{12} \end{pmatrix}$$

$$\boldsymbol{P}^{(3)} = \begin{pmatrix} \frac{23}{54} & \frac{31}{54} \\ \frac{31}{72} & \frac{41}{72} \end{pmatrix} \quad \boldsymbol{P}^{(4)} = \begin{pmatrix} \frac{139}{324} & \frac{185}{324} \\ \frac{185}{432} & \frac{247}{432} \end{pmatrix}$$

(c) $\mathrm{p}^{(1)} = \left(\frac{3}{8}, \frac{5}{8}\right)$. $p^{(2)} = \left(\frac{7}{16}, \frac{9}{16}\right)$. $p^{(3)} = \left(\frac{41}{96}, \frac{55}{96}\right)$.

(d) Limit distribution is $\left(\frac{3}{7}, \frac{4}{7}\right)$. Man will be late $\frac{3}{7}$ of the time and on time $\frac{4}{7}$ of the time.

5. If Akklavik's sculptures are taken as states of the chain then

$$P = \begin{pmatrix} \frac{1}{4} & \frac{3}{4} & 0 & 0 \\ \frac{1}{6} & \frac{3}{6} & \frac{2}{6} & 0 \\ 0 & \frac{1}{2} & \frac{5}{12} & \frac{1}{12} \\ 0 & 0 & 1 & 0 \end{pmatrix}$$

Limit distribution is $\left(\frac{4}{35}, \frac{18}{35}, \frac{12}{35}, \frac{1}{35}\right)$.

10. $$\boldsymbol{P} = \begin{pmatrix} \frac{1}{3} & \frac{2}{3} & 0 & 0 & 0 \\ \frac{1}{10} & \frac{5}{10} & \frac{4}{10} & 0 & 0 \\ 0 & \frac{4}{15} & \frac{8}{15} & \frac{3}{15} & 0 \\ 0 & 0 & \frac{1}{2} & \frac{13}{30} & \frac{2}{30} \\ 0 & 0 & 0 & \frac{4}{5} & \frac{1}{5} \end{pmatrix} \qquad \boldsymbol{P}^{(2)} = \begin{pmatrix} \frac{8}{45} & \frac{25}{45} & \frac{12}{45} & 0 & 0 \\ \frac{25}{300} & \frac{127}{300} & \frac{124}{300} & \frac{24}{300} & 0 \\ \frac{12}{450} & \frac{124}{450} & \frac{221}{450} & \frac{87}{450} & \frac{6}{450} \\ 0 & \frac{120}{900} & \frac{435}{900} & \frac{307}{900} & \frac{38}{900} \\ 0 & 0 & \frac{60}{150} & \frac{76}{150} & \frac{14}{150} \end{pmatrix}$$

12. (a) $\boldsymbol{P}^{(2)} = \begin{pmatrix} .84 & .16 \\ .2 & .8 \end{pmatrix}$

Limit distribution is $\left(\frac{5}{9}, \frac{4}{9}\right)$.

(b) $\boldsymbol{P}^{(2)} = \begin{pmatrix} .04 & .96 \\ 0 & 1 \end{pmatrix}$

Matrix is not regular.

CHAPTER XV

1. (a) Minmax = maxmin = 2, strictly determined.
   (b) Maxmin = 2, minmax = 5, not strictly determined.
   (c) minmax = maxmin = 1, strictly determined.
   (d) minmax = maxmin = 3, strictly determined.
5. (a) $1 \leq x \leq 3$.
   (b) $3 \leq x \leq 5$.
6. (a) maxmin = minmax = 3.
   (b) $p_1^0 = \frac{8}{9}, p_2^0 = \frac{1}{9}, q_1^0 = \frac{4}{9}, q_2^0 = \frac{5}{9}, v = \frac{49}{9}$.
   (c) $p_1^0 = \frac{2}{3}, p_2^0 = \frac{1}{3}, q_1^0 = \frac{1}{3}, q_2^0 = \frac{2}{3}, v = \frac{5}{3}$.
7. (a) $p_1^0 = \frac{8}{12}, p_2^0 = \frac{4}{12}, q_1^0 = \frac{1}{2}, q_2^0 = q_3^0 = 0, q_4^0 = \frac{1}{2}, v = 3$.
   (c) $p_1^0 = \frac{2}{5}, p_2^0 = \frac{3}{5}, q_1^0 = \frac{1}{5}, q_2^0 = \frac{4}{5}, q_3^0 = 0, v = \frac{12}{5}$.

9.

	11	12	13	21	22	23	31	32	33	minima
11	0	2	2	−3	0	0	−4	0	0	−4
12	−2	0	0	0	3	3	−4	0	0	−4
13	−2	0	0	−3	0	0	0	4	4	−3
21	3	0	3	0	−4	0	0	−5	0	−5
22	0	−3	0	4	0	4	0	−5	0	−5
23	0	−3	0	0	−4	0	5	0	5	−4
31	4	4	0	0	0	−5	0	0	−6	−6
32	0	0	−4	5	5	0	0	0	−6	−6
33	0	0	−4	0	0	−5	6	6	0	−5
maxima	4	4	3	5	5	4	6	6	5	

minmax = 3, maxmin = −3, not strictly determined.

12. $p_1^0 = p_2^0 = \frac{1}{2}$, $p_3^0 = 0$, $v = 1$.

CHAPTER XVII

1. $P(\{X = k\}) = \frac{1}{6}\left(\frac{5}{6}\right)^{k-1}$, $E(X) = 6$, $V(X) = 30$.
3. $E(X) = +\infty$
5. $m = 2$, $p = \frac{1}{10}$, $q = \frac{9}{10}$.
6. .4
8. (a) .2381 (b) .1954 (c) .5002
10. .1157
13. $E(X^3) = P'''(1) + 3P''(1) + P(1)$.
22. (a) $\lambda = 15$.
    (b) .0028
24. $P(win) = .4781$, house advantage is 4.38%.
26. $\frac{34}{77}$
27. .6

CHAPTER XVIII

1. .013263
3. .0314066
5. (a) .9608 (c) .1038 (e) .3227 (g) .8185
6. .1737
9. .8834
10. $n = 6766$.
14. .5

INDEX